Advanced Structured Materials

AF180597

Volume 227

Series Editors

Andreas Öchsner, Faculty of Mechanical and Systems Engineering, Esslingen University of Applied Sciences, Esslingen, Germany

Lucas F. M. da Silva, Department of Mechanical Engineering, Faculty of Engineering, University of Porto, Porto, Portugal

Holm Altenbach, Faculty of Mechanical Engineering, Otto von Guericke University Magdeburg, Magdeburg, Germany

Common engineering materials are reaching their limits in many applications, and new developments are required to meet the increasing demands on engineering materials. The performance of materials can be improved by combining different materials to achieve better properties than with a single constituent, or by shaping the material or constituents into a specific structure. The interaction between material and structure can occur at different length scales, such as the micro, meso, or macro scale, and offers potential applications in very different fields.

This book series addresses the fundamental relationships between materials and their structure on overall properties (e.g., mechanical, thermal, chemical, electrical, or magnetic properties, etc.). Experimental data and procedures are presented, as well as methods for modeling structures and materials using numerical and analytical approaches. In addition, the series shows how these materials engineering and design processes are implemented and how new technologies can be used to optimize materials and processes.

Advanced Structured Materials is indexed in Google Scholar and Scopus.

P. M. Visakh
Editor

Rubber Based Bionanocomposites

Applications

Springer

Editor
P. M. Visakh
Department of Bioengineering
Ege University
Izmir, Turkey

ISSN 1869-8433 ISSN 1869-8441 (electronic)
Advanced Structured Materials
ISBN 978-3-031-78559-7 ISBN 978-3-031-78557-3 (eBook)
https://doi.org/10.1007/978-3-031-78557-3

© The Editor(s) (if applicable) and The Author(s), under exclusive license to Springer Nature Switzerland AG 2025

This work is subject to copyright. All rights are solely and exclusively licensed by the Publisher, whether the whole or part of the material is concerned, specifically the rights of translation, reprinting, reuse of illustrations, recitation, broadcasting, reproduction on microfilms or in any other physical way, and transmission or information storage and retrieval, electronic adaptation, computer software, or by similar or dissimilar methodology now known or hereafter developed.
The use of general descriptive names, registered names, trademarks, service marks, etc. in this publication does not imply, even in the absence of a specific statement, that such names are exempt from the relevant protective laws and regulations and therefore free for general use.
The publisher, the authors and the editors are safe to assume that the advice and information in this book are believed to be true and accurate at the date of publication. Neither the publisher nor the authors or the editors give a warranty, expressed or implied, with respect to the material contained herein or for any errors or omissions that may have been made. The publisher remains neutral with regard to jurisdictional claims in published maps and institutional affiliations.

This Springer imprint is published by the registered company Springer Nature Switzerland AG
The registered company address is: Gewerbestrasse 11, 6330 Cham, Switzerland

If disposing of this product, please recycle the paper.

Contents

Chapter 1
Rubber Based Bionanocomposites: Applications: State of Art and New Challenges

P. M. Visakh

Abstract In this chapter we provide some discussions on different topics such as cellulose-based rubber nanocomposites, cellulose based high performance rubber micro- or nano-composites, chitin based rubber bionanocomposites, chitin in rubber-based blends and micro composites, starch based bionanocomposites, polylactic acid based rubber composites and nanocomposite and bacterial cellulose/BC/rubber nanocomposites, cellulose in rubber-based blends and composites and applications of cellulose/rubber based bionanocomposites.

1.1 Cellulose-Based Rubber Nanocomposites

Many academicians and business people are interested in this field of study because these materials could be an option for the ever-dwindling non-renewable resources, environmental contamination, global warming, and energy crisis. To be specific, cellulose has emerged as a potential option due to its widespread availability from various sources, starch, alginate, chitin, chitosan, and gelatin (Trache 2018). The Natural rubbers (NRs), have excellent chemical and physical characteristics like suppleness, elasticity, resilience, abrasion and impact resistance etc. Additionally, NR is extremely compostable and mouldable. The NRs are frequently found in the tires of vehicles, cars, and aeroplanes. In reality, nitrocellulose (NC), which is cellulose in the shape of nanostructures, has emerged in recent years as one of the most promising sustainable materials. NC is highly preferred reinforcing material due to its excellent properties like renewable nature, higher surface area volume, biocompatible and non-toxic nature, high tensile and elastic modulus. BNC is almost identical in properties to CNC and CNF, but based on the bacterial growth conditions, it is frequently lightweight, non-toxic, and controls microfibril formation. BNC is well-known in biological fields like wound repair and regenerative medicine. Rubber was

P. M. Visakh (✉)
Natural Bioactive Materials Laboratory, Department of Bioengineering, Ege University, Bornova, Izmir, Turkey
e-mail: visagam143@gmail.com

© The Author(s), under exclusive license to Springer Nature Switzerland AG 2025
P. M. Visakh (ed.), *Rubber Based Bionanocomposites*, Advanced Structured Materials 227, https://doi.org/10.1007/978-3-031-78557-3_1

combined with CB and single-wall carbon nanotubes (SWNT), reducing dampening parameters and raising the glass transition temps. SWNT displayed more B than CB, indicating improved interaction between rubber and SWNT (Zhang et al. 2016). The nanocrystalline cellulose (NCC) was obtained from different natural sources such as kenaf, ramie, Napier grass, cotton cellulose etc. The thermal durability of NR/SWCNT nanocomposites created through latex stage mixing is largely unaffected at smaller SWCNT concentrations. MWCNT-reinforced NR nanocomposites were created by Jose et al. (2017) using the two-roll mixture, and their temperature behavior in nitrogen was investigated.

Strong acid decomposition can easily produce high aspect ratio nanoparticles called cellulose nanocrystals (CNCs) from cellulose microfibrils. They have special qualities brought about by the chemical of their surfaces, the crystallinity, and the three-dimensional structure. Aerogels, nanocomposite materials, optically active films, and other uses have all made use of CNCs. In a very comprehensive study Agrebi and coworkers investigated the chemical bonding taking place between natural rubber (NR) reinforced with cellulose nanowhiskers (CNW) and nanofibrillated cellulose (NFC) by means of FTIR in order to better understand potential chemical interactions that occur between the reinforcement and the matrix (Agrebi et al. 2019). Surface functionalization of nanocellulose is widely studied in order to improve its compatibility and reinforcing efficiency in rubber matrix. Kanoth et al. studied the reinforcing efficiency of functionalized cellulose nanocrystals (CNC) in NR (Parambath Kanoth et al. 2015). CNC functionalized by mercapto groups was crosslinked with NR by UV irradiation and this reaction was examined by real time FTIR analysis. Presently nanocellulose use as a reinforcement in rubber is largely been investigated from a commercial point of view. It has been established that nanocellulose forms a network kind of structure affecting the mobility of rubber chains and which ultimately influences its reinforcement effect in rubber matrix. With a view to understand the surface dynamics of the solid–solid interface in nanocomposites. In particular, in nanocellulose filled rubber nanocomposites, the transverse nuclear magnetic relaxation time, T2, is influenced by chemical and physical crosslinking on the particle surface which is used in calculating the fraction of polymer chains with limited mobility, on the adsorbed polymer layer that surrounds the filler particles.

The use of nanocellulose largely investigated for reinforcing rubbers has necessitated the development of methodologies that can precisely describe the degree of nanocellulose dispersion in the rubber matrix and invariably the structural morphology of the resulting composites. X-ray scattering investigations in the form of wide-angle X-ray diffraction (WAXD) and small angle X-ray scattering (SAXS) can contribute to understand these aspects and provide for a highly detailed picture of the structural morphology of cellulose-based rubber nanocomposites. In the nanocomposites the presence of nanocellulose promoted the strain-induced crystallization that was apparent from the rise in the intensity of peaks and suppression of amorphous halo with the optimum structural arrangement observed for 15 phr of nanocellulose in WAXD patterns. This structural property relationship was addressed in research by Wongvasana and colleagues that compared nanoclay (NC) and cellulose nanofibers

(CNF) reinforced natural rubber nanocomposites (Wongvasana et al. 2022). The study demonstrated that the reinforcement effect of CNF was distinct from that NC on the basis of WAXD observations and developed a model explaining the reinforcement effect of CNF. SANS method is also used for determining the dispersion of the nanocellulose into the matrix. Chazeau utilized this approach to analyze the dispersion behavior of tunicate whiskers in plasticized poly(vinyl chloride) and reported an isotropic dispersion without the formation of any aggregates (Chazeau et al. 2000). In a recent study Yasin et al. investigated the impact of an ionic liquid ([EMIM]OAc) on cellulose nanofiber (CNF) and CNF filled natural rubber (NR) composites and vulcanizates (Yasin et al. 2021). The study exhibited that the presence of an ionic liquid enhanced the dispersion of cellulose nanofillers in rubber matrix and boosted the density of crosslinking and thereby enhanced the mechanical strength of the vulcanized natural rubber.

1.2 Cellulose Based High Performance Rubber Micro- or Nano-composites

The use of cellulosic nanofillers for elastomers is a relatively new area of research, as NR have traditionally been the preferred choice for investigating their characteristics. NR is classified as polyisoprene, which is extracted from latex and in its uncured form has poor mechanical properties and limited applications, recent research suggests that cellulose fillers may offer a viable alternative to CB and Sil fillers, showing potential for improving mechanical properties in tire and thread manufacturing applications. The incorporation of cellulose fillers into rubber matrices holds promise for enhancing mechanical properties and performance, offering a greener option for tire and thread manufacturing while reducing reliance on synthetic fillers. Cellulose inherently contains both amorphous and crystalline regions, and the removal of most amorphous regions can result in a stronger and stiffer cellulose, increasing its potential as a reinforcing filler for various applications. The two main cellulose nanofillers, derived primarily from natural plant sources, yield CNC and CNF, depending on the extraction processes (Khui et al. 2023). The polysaccharide comprising of straight chains of D-glucose units is a vital primary part in numerous plants, for instance tree, hemp, wheat straw what's more, bamboo, where the elementary units are aligned to form bundle of microfibrils and ultimately a cellulosic fiber network, complexed with additional polymers hemicellulose and lignin bound to the cellulose surface and captured inside the microfibrils (Sun et al. 2023). The melt blending process combines rubber and cellulose into a masterbatch before heating and mechanical processing, also known as dry mixing. Dry mixing aids in dispersing cellulose in the matrix, influencing the melt blending process and promoting intercalation between cellulose and rubber (Narimissa et al. 2015). Inadequate heating and mechanical processing may result in filler agglomeration, poor matrix-filler intercalation, and consequently, adverse effects on the final product's properties (Rahman et al. 2022).

Cellulose-reinforced NR or silicon micro- and nanocomposites are prepared from cured latex matrices in a liquid state, with fillers typically incorporated and mechanically processed as a masterbatch (Yasin et al. 2021; Hussain et al. 2020). Masterbatch mixing involves various mechanical processes such as ultrasonication, shear blending, and homogenization, facilitating exfoliation and intercalation between the latex matrix and cellulose fillers. Polar cellulosic nanofillers face further challenges in rubber matrices, particularly in terms of agglomeration, restricting processing, vulcanization kinetics, and mechanical attributes. These bottlenecks hinder the full potential of cellulose in rubber nanocomposites. The unique characteristics of cellulose nanofillers extend their influence beyond mere mechanical reinforcement. Cellulose's interaction with rubber matrices also influences the nonlinear mechanical behavior, such described earlier as the Payne effect. The crosslinked networks impose restrictions on the mobility of rubber chains, while the interaction between filler and elastomer matrix acted as physical crosslink points, contributing to higher G' values for rubber nanocomposites. Cellulose-based nanofillers present a unique reinforcing opportunity due to their abundant hydroxyl groups, with inexhaustible bioresources availability. As mentioned before, cellulosic nanofillers have been found to enhance tearing, abrasion resistance, and tensile properties of natural and synthetic rubbers, such as nitrile butadiene rubber (NBR) and styrene butadiene rubber (SBR) nanocomposites (Mekonnen et al. 2019; Chen et al. 2014; Jardin et al. 2020; Neto et al. 2016).

1.3 Chitin Based Rubber Bionanocomposites

Chitin is a common polysaccharide similar to cellulose, with a repeating structure of (1,4)-β-N-acetyl glycosaminoglycan. It is the second most abundant biopolymer after cellulose. Chitin is mainly found in the exoskeletons of shellfish and insects, as well as in the cell walls of mushrooms (Ifuku and Saimoto 2012). Chitin nanofibers (ChNFs) with diameters less than 100 nm and aspect ratios exceeding 100 have a very high surface-to-volume ratio, creating highly porous meshes with unique properties compared to micro-sized fibers. Egbujuo et al. (2020) prepared natural rubber (NR) vulcanizates from natural rubber and chitin (crab shell) using a two-roll mill. Utilization of chitin powder as a filler in natural rubber vulcanizates: In comparison with carbon black filler. International Review of Applied Sciences and Engineering. The effects of the chitin at different contents (0–40 phr) on the mechanical properties of the NR/Chitin vulcanizates with carbon black as reference filler have been investigated. Petchnui et al. (2024) explored promising NRL/ChNF composite materials for use in energy-harvesting devices Triboelectric nanogenerators (TENGs) from biowaste of shrimp shell-extracted chitin nanofiber (ChNF) and natural rubber (NRL) at loading levels of 0.1 and 0.2 wt% (NRL/ChNF). NRL/ChNF composites demonstrated an increased dielectric constant of 3.3 lecat 0.1 MHz surface potential difference from 0.182 to 1.987 V, and a higher Young's modulus (3.4 GPa). Yongwang Liu et al. first prepared CNCs with aqueous dispersions by acid hydrolysis of raw chitin (Liu et al. 2018). CNC dispersions were compounded with natural rubber (NR) latex to prepare NR/CNC composites. The tensile strength of NR/CNCs composite containing 5 wt.% CNCs is 5.75 MPa which is 6.25 multiples of pure NR. Mathew et al. (2023) reported a nanocomposite from chitin nanowhiskers from shrimp shells and acrylonitrile-butadiene synthetic rubber. The results showed tensile strength and tear strength of nanocomposites by 116 and 54% and higher content chitin nanowhiskers significantly improve the thermal stability of studied nanocomposites. It is widely used in the insulation industries especially architecture. Silica nanoparticles were used as fillers to improve the physical and mechanical properties of 1,4-cis pBR. Nanosized reinforcements in novel chitin-based rubber biomaterial composites may introduce new features to the material such as increased strength, lighter and more porous physical properties, tunable optical emission, color variations, enhanced chemical reactivity, improved electrical conductivity, and superparamagnetic or other magnetic properties. Additionally, these composites may offer new functionalities that are not available at the micro- or macroscale (Yang et al. 2007; Hule and Pochan 2007). All-chitin nanocomposites are particularly intriguing due to the potential of creating biobased materials for medical use with enhanced mechanical and barrier properties while retaining characteristics such as nontoxicity, antibacterial activity, biocompatibility, and biodegradability of chitosan (Shelma et al. 2008; Mathew et al. 2009).

Chitosan, a bio-based renewable polymer with essential functional groups, serves as an effective dry strength agent due to its structural similarity to cellulose. Research has shown that adding chitosan at 0.25–0.5 o.d. wt% of pulp greatly enhances the mechanical properties of hand sheets, except for tearing strength (Song et al. 2018). Chitin-based rubber biocomposites are being considered as a potential replacement for current materials in the aerospace industry. Additionally, green tires are gaining popularity due to their durability, safety, fuel efficiency, and environmental benefits (Sattayanurak et al. 2019; Thomas et al. 2021).

1.4 Chitin in Rubber-Based Blends and Micro Composites

The challenge of ensuring that chitin/rubber composites can withstand harsh temperature environments is crucial (Markovic and Visakh 2017). Additionally, the large size of chitin in chitin/rubber-based blends and micro composites often fails to effectively improve the properties of these composites. Johns and Rao conducted a study on natural rubber latex/chitosan (NR/CS) synthesis by blending with different compositions using solution casting followed by compression. They found that blending improved the thermal stability of chitosan. The blend NR95 CS5 showed the maximum value for better interfacial adhesion between chitosan and natural rubber. Zhang and Cao developed an environmentally friendly antibacterial composite from natural rubber latex foam (NRLF) by synthesizing chitin using the Dunlop method (Zhang and Cao 2020). They tested the mechanical properties and antibacterial activity of the NRLF composites and compared them with those of pure NRLF. Additionally, Kodsangma et al. developed a new antimicrobial bio-thermoplastic elastomer (TPSC/ENR) by blending thermoplastic starch (TPS) with rubber (Kodsangma et al. 2020). Thermoplastic starch-chitosan (TPSC) was prepared by the solution mixing of cassava starch, chitosan (CTS), and glycerol in acidified water (lactic acid 1wt%) at 80 °C, followed by melt mixing at 130 °C. Abdul Wahab et al. (2022) prepared heat-healable devulcanized EPDM waste-NR (DEPDMw-NR) blends with nano-chitin (0–7 phr) loading, by dry milling chitin. They evaluated the effect of nano-chitin addition on tensile properties before and after self-healing. Products made from NRL, such as gloves, condoms, transfusion tubes, and urinary catheters, are widely used in the medical and health industry. However, they are prone to bacterial or fungal contamination when in direct contact with human skin and mucous membranes during use (Wahba et al. 2013).

The presence of functional groups on the surface of filler particles can significantly influence the filler-matrix interactions and, consequently, the strength of the reinforcing effect (Galimberti et al. 2014). Elasto-gel films comprising silk sericin (SRC), natural rubber (NR), and chitin whisker (CTW) for wound treatment. However, the pure SRC film was difficult to fabricated because of its weak structural feature and this overcome by using NR as a binder. In another study Haris and Raju synthesized comprising chitosan (CTS) grafted onto the backbone of epoxidized natural rubber (CTS-g-ENR) (Haris and Raju 2014). Kawano et al. reported a self-assembled chitin nanofiber (CNF) film via regeneration from an ion gel with an ionic liquid, followed by sonication and filtration (Kawano et al. 2017). Somseemee et al. (2022) prepared a new biocomposite made from epoxidized natural rubber (ENR) reinforced with chitosan (CHI) and cellulose nanocrystals (CNCs) was created without using conventional crosslinking agents. The biocomposite with 1–2 phr of A-CNCs had the best combination of strength, self-healing, and TENG performance, and can be used as a positive electrode. Chitin/rubber blend films offer a sustainable alternative to petroleum-based plastic packaging, as well as the optimization of the production process, to make it fast, efficient, and inexpensive (Zhang et al. 2024). Chitin/chitosan are cationic polysaccharides derived from waste in the marine food processing industry, engineers and chemists consider chitin-based rubber biocomposites as a very good remedy and replacement for the existing material (Jayalakshmy 2017). Akito kowano et al. investigated the properties of natural rubber (NR)-chitin nanofiber (CHNF) composite. The result showed that the mechanical properties of NR composite increase with an increase in the weight ratio of CHNFs. The studies of Ding et al. on chitin and NR concluded that CHNFs with a high aspect ratio and good stability enhanced the strength and toughness of the composite (Ding et al. 2019).

1.5 Starch Based Bionanocomposites

Polymers resulting from renewable resources such as starch, rubber, cellulose, or chitins provide an effective solution to accomplish these requirements. A variety of factors, such as the dwindling of fossil fuel reserves, rising costs, environmental concerns related to the impact of degradation or incineration on global warming, recyclable processes susceptible to contamination, and potential toxicity of end products, have led to increased interest in using biopolymers as replacements for those derived from petroleum (Nagalakshmaiah et al. 2019). Starch, a biodegradable polymer, is composed of two primary constituents, amylose and amylopectin, with their proportions varying based on the plant source, such as corn, cassava, potatoes, sago, rice, and others (Thomas et al. 2021). Among all types of starch, rice starch (RS) is a popular option for various food and non-food uses owing to its non-allergenic nature, easy digestibility, white tint, tiny grain size (3–10 mm), and an extensive option of amylose/amylopectin ratios (Thakur et al. 2018). Starch-based films display essential physical uniqueness, counting lucidity, absence of distinctive aroma, and partial permeability to humidity, oxygen, carbon dioxide, and savoring substances. These unique traits can yield outcomes similar to those achieved with a controlled atmosphere. This particular composition permits sago starch films to showcase a slower swelling charge in water improving their ability of sago starch (SS) as a biopolymer film for packaging. Riyajan et al. (2012) prepared gelatinized cassava starch and handled them with potassium persulfate (K2S2O8) to shape the tailored amylum with lesser molecular weight.

The thermal in addition to mechanical qualities of NR were stepped forward with grafting. The NR embedded starch was additionally employed to encapsulate urea which could then be released with the aid of engorgement with water, consequently providing for the controlled release of urea. Wu et al. (2004) suggested a relative observations at the mechanical characteristics of starch-bolstered NR composites, indicated the use of co-coagulation and direct mixing (-roll mill) strategies.

The starch carries a polysaccharide sequence of amylose and amylopectin that are taken up by microbes inside the soil to supply enzymes including amylases and glucosidases which can be used to interrupt the glucosidic linkages in starch. As the starch granules reside in among the rubber debris, the breakage of glucosidic linkages in starch can also disrupt the rubber particles, as a result boost up the degradation method. Biodegradable polymers, e.g. starch-based rubber bionanocomposites, are drastically employed as frameworks for tissue engineering, drug delivery systems, wound dressings, surgical sutures, and implants because of their mechanical strength, compatibility with living tissues, ability to break down naturally, and the production of harmless, degradable by-products (Gomes et al. 2002; Lu et al. 2009; Taherimehr et al. 2021).

Thermoplastic starch (TPS)/beta-tricalcium phosphate (β-TCP) NPs Thermoplastic starch (TPS)/beta-tricalcium phosphate (β-TCP) NPs were advanced. The findings revealed that addition of β-TCP at 10%, advanced the tensile electricity (from 1.67 to 4.8 MPa) and young's modulus (from 66.54 to 390.5 MPa), and reduced elongation at ruin (78.56–18.03%) of TPS. Starch is one of the most promising biopolymers for the production of food packaging trays for several reasons. It is bio derived and fully biodegradable; its cost is very low and its production is abundant worldwide. Nevertheless, up to now the introduction of starch in the food packaging market has been restricted because it presents several drawbacks: native starch after being transformed becomes brittle and breaks easily, it is hydrophilic and therefore moist products cannot be packed and its density is higher (1400 kg/m^3) than common synthetic polymers employed in the food packaging market such as polystyrene (1070 kg/m^3). The NR matrix and starch were made more compatible by using the PBA. The NR composites treated with PBA g-starch and their tensile characteristics, tear strength, and hardness. By using acid hydrolysis, Valodkar and Thakore (2010) were able to extract starch nanocrystals. To alter NR, starch nanocrystals, pure starch, and waxy corn starch were utilized.

Two roll-mill methods were used to prepare the composites. Potato starch nanocrystals have been separated by Bouthegourd et al. (2011) and Rajisha et al. (2014), and the NR Latex was reinforced with these nanocrystals. It was decided to vary the nanocrystal concentrations between 5 and 20 wt%. The existence of potato starch nanocrystals in the NR matrix was verified by the FTIR measurements. Due to their ability to improve a variety of tire characteristics, starch-based rubber nanocomposites have several promising uses in the tire industry. Tire formulations can be made more resilient to wear by adding starch-based nanocomposites. Tire life can be increased and abrasion resistance can be increased by using nanocomposites to reinforce the rubber matrix. By augmenting the mechanical characteristics of tires, like their elasticity and tensile strength.

1.6 Polylactic Acid Based Rubber Composites and Nanocomposites

Polylactic acid (PLA) is the most promising from a commercial standpoint. Production of lactic acid polymer (PLA) typically involves ring-opening polymerization of cyclic lactic dimer. PLA can be made by directly poly-condensing lactic acid. Through the process of bacterial fermentation, lactic acid may be extracted from sugar cane or maize starch (Reddy et al. 2008; John et al. 2009; Vignesh Kumar et al. 2022). Producing PLA via various techniques results in a wide range of characteristics (Farah et al. 2016). The glass transition of PLA is around 58–59 °C. The virtually constant melting point (Tm) of PLA produced by the direct technique is around 163 °C. However, owing to the catalyst and impurities, ring-opened PLA is less stable and has a lower Tm of 178 °C (Chin et al. 2021; Xu 2004; Grignard et al. 2019). Natural rubber composites are used in the production of medical devices, including seals, gaskets, and components, due to their biocompatibility, flexibility, and resistance to wear (Chang et al. 2021; Urtekin et al. 2023; Magaña et al. 2022). Due to their resilience, natural rubber composites are employed in seismic bearings, providing shock absorption and to provide stability in buildings and bridges, especially in earthquake-prone areas. Natural rubber composites are utilized to manufacture isolation pads and expansion joints in construction to accommodate movement, dampen vibrations, and provide flexibility while maintaining structural integrity (Fan et al. 2024; Stevenson 1992; Teh et al. 2022). The cellulose fibers reinforced polymer composites were found to have a density of about 0.9 g/cm^3, good sound reduction, an impact strength of 25 kJ/m^2, and a stiffness of about 3000 MPa and also these properties can be improved even more (Kamath et al. 2017; Johny et al. 2023; Mylsamy et al. 2024). Researchers have discovered that sugar palm fiber shows a lot of promise as a fiber reinforcement for polymer matrix composites (Isma'ila Mukhtar et al. 2016; Palanisamy et al. 2024; Almeshaal et al. 2022). Polylactic acid (PLA) natural rubber nano composites have a wide range of potential applications due to their unique properties (Bijarimi et al. 2013; Bitinis et al. 2013; Jamshidian et al. 2010).

These composites typically exhibit improved mechanical strength, durability, and thermal stability compared to traditional rubber materials. Here are some potential applications for PLA natural rubber nano composites. PLA natural rubber nano composites can be used in the production of consumer goods such as sports equipment, footwear, and electronic device components. The improved mechanical properties make them ideal for high-performance applications. Numerous food and non-food products have been packed using plastics due to the material's adaptability and widespread use (Singha et al. 2019; Ranakoti et al. 2022; Yuan et al. 2019). PLA composites can be used in aerospace and defence applications for manufacturing components such as seals, gaskets, and vibration isolation materials due to their ability to withstand challenging environments and mechanical stress. Manufacturing processes for PLA natural rubber nano composites typically involve melt blending or solution mixing of PLA and natural rubber, followed by the incorporation of nano-sized materials such as nanoparticles or nanofibers to enhance the mechanical properties of the composite (Bhat et al. 2021; Paek et al. 2022; Hawary et al. 2023). PLA/natural rubber composites can be utilized in structural elements such as beams, columns, and panels, offering good strength, impact resistance, and durability (Varma et al. 2024).

PLA/natural rubber composites can be used to develop water filtration membranes for various applications, including wastewater treatment, desalination, and purification of drinking water. The flexibility and biocompatibility of natural rubber, combined with the mechanical strength and biodegradability of PLA, make these composites suitable for producing durable and environmentally friendly water filtration membranes (Bharagava et al. 2023). PLA/natural rubber composites may find use in electrodialysis membranes for applications such as salt removal and ion exchange processes. The electrical insulating properties of natural rubber coupled with the formability and chemical resistance of PLA make these composites suitable for electrodialysis applications (Jasso-Salcedo et al. 2023).

1.7 Bacterial Cellulose/BC/Rubber Nanocomposites

Bacterial Cellulose (BC) is a biopolymer produced by certain bacteria, notable for its high purity, crystallinity, and mechanical strength. When combined with rubber, a versatile elastomeric material, BC/rubber composites exhibit improved properties suitable for a range of applications. This explores the interdisciplinary nature of BC/rubber composites, bridging materials science, biotechnology, and engineering. Currently, there is a significant amount of research and attention dedicated to isolating and producing nanoscale cellulose reinforced rubber and polymeric materials (Klemm et al. 2011; Siro and Plackett 2010).

BC is composed of only cellulosic material which is free from hemicellulose, pectic polysachhrides or lignin-polymers. The bacteria release cellulose into a liquid culture medium forming nanofibers with size varying from 25 to 100 nm. These nanofibers are responsible for forming a thin cellulosic skin the culture (Jonas and Farah 1998; Ross et al. 1991; Iguchi et al. 2000; Gindl and Keckes 2004). The predominant bacterial species utilized for cellulose synthesis are derived from the genus Gluconacetobacter (formerly known as Acetobacter), specifically Gluconacetobacter xylinus. Additional species are Gluconacetobacter hansenii, Gluconacetobacter europaeus, and Sarcina ventriculi (Walling et al. 2023). BC production may be accomplished with many growing methods: Depositing BC over natural fibers has been shown to have a promising effect when dealing with thermal processing of composites. BC coated natural fiber has the potential to bridge the CTE mismatch of the fibers and the matrix. According to the previous studies, through the cultivation of cellulose-releasing bacteria with natural fibers in a suitable culture, BC can be selectively deposited onto the fiber surfaces (Cichocki and Thomason 2002; Baley et al. 2006).

In another study, acrylated epoxidised soya oil (AESO) was used to infuse the natural fiber preforms, which were then polymerized to create BC/sisal fiber composites (Lee et al. 2012; Diniz et al. 2004). The fiber fraction of sisal/polyAESO and BC/sisal/polyAESO was fixed at 4%. The use of sisal fibers in polyAESO resulted in a substantial enhancement of the tensile mod from 400 MPa in pristine polyAESO to 3200 MPa in polyAESO composites reinforced with 4% sisal fibers. In comparison to all cellulose nanoscale composites to all-cellulose composites manufactured using membrane filters, the nanocomposites have greater tensile modulii at a shorter dissolving time than that of the composite matrix. The presence of rigid BC web in the nanocomposites is responsible for this effect. Gea et al. (2010), conducted a study where they cultivated BC in the presence of PVA and compared it to BC that had been infused with PVA.

Following the purification process, the BC loaded PVA had 96.3 wt%, whereas the BC infused PVA had 98.6 wt% loading of BC. The tensile mod of BC loaded PVA was lower as compared to that of BC infused PVA due to the reduced loading volume in the matrix made in situ. The tensile performance of BC loaded PVA was close to 210 MPa, which is closer to that of the pristine BC, compared to the 110 MPa of BC infused PVA. BC has several applications, including its usage as a thickening and stabilizing agent in the manufacturing of polymeric foam. Blomfeldt and his co-workers (2010, 2011) included 1 wt% of BC added to foam made from wheat gluten. The foam was created by dispersing gluten and water followed by cooling the mixture at −25 °C for a duration of 6 h, and finally undergoing freeze-drying. The addition of BC serves to scale up the solution viscosity and minimize fluctuations in foam density. Xu et al. (2016) conducted dynamic mechanical tests on composites of natural rubber with CB, as well as with CB and multi-walled CNTs at various CNTs loadings.

The thermoplastic rubber blend filled with acidified CNTs exhibited the maximum storage modulus at −90 °C, whereas the rubber containing untreated CNTs had the lowest storage modulus (Boonmahitthisud and Chuayjuljit 2012). The biocompatibility and mechanical features of these composites offer benefits for implanted drug delivery devices that need both flexibility and durability (Das et al. 2024). BC/rubber composites provide exceptional strength and flexibility, making them suitable for application in bone implants. Uniform dispersion is crucial for achieving excellent composite characteristics by ensuring that BC is evenly distributed inside the rubber matrix. Inadequate dispersion can cause particles to clump together, which can create areas of weakness in the composite material (Jabbari and Babaeipour 2023).

The BC/rubber composites' biodegradability renders them an eco-friendly substitute for disposable items and packaging materials, hence mitigating plastic waste (Shokatayeva et al. 2019). These composites may be produced using sustainable methods, so helping to minimize their environmental footprint. BC/rubber composites are suitable for enhancing the cushioning, flexibility, and durability of shoe soles. The material's capacity to adapt to the contour of the foot promotes comfort. BC's biodegradability renders these composites a sustainable choice for manufacturing footwear (Adak 2024).

1.8 Cellulose in Rubber-Based Blends and Composites

The nanoparticles extracted from cellulose, in the form of nanocrystals and fibrils, are characterized by distinguished larger surface area, higher aspect ratio, and the more capability of percolating networks within a polymer matrix (Moon et al. 2011). Incorporating CNPs into natural rubber provides several benefits. As biocompatible, ecofriendly, renewable, and biodegradable, the extracted cellulosic materials make natural rubber nanocomposites more environmentally friendly and sustainable compared to traditional petroleum-based rubber composites (Siró and Plackett 2010). By utilizing cellulose nanomaterials, these nanocomposites can benefit from the biodegradability and sustainability of cellulose, making them environmentally friendly alternatives to traditional rubber composites. The preparation of crystalline cellulose particles is feasible through adjustment of the reaction parameters to alter the dimensions and forms of the product. If the obtained particles have mixtures of micro- and nonoscale cellulose particles, it is recommended to change conditions to get the desired quality of nanoparticles; this highlights benefits for the environment and the economy (Abitbol et al. 2023).

The treatment of fibers aims to address these issues by enhancing interfacial compatibilities and facilitating dispersion. Common surface modification techniques (Trache et al. 2020) include chemical grafting (e.g. silylation, esterification, and polymer grafting), physical adsorption (e.g. surfactants, polymers), and oxidation (e.g. TEMPO-mediated oxidation). Chemical modification is frequently used for enhancing the combination between particles of natural rubber and nanocellulose on interfacial plane, and to overcome problems in case of compounding with polar systems (Gelir et al. 2017). Since OH groups are considered the only functional groups found in nanocellulose, the reactivity of these groups is important after modification. The research on natural rubber, fibers, and their composites has given a great attention because of the sustainability and environmental effect. Both the micro-/and nano-cellulose are usually taken as sustainable fillers for enhancing the characteristics of the used rubber (Low et al. 2021; Roy et al. 2021). The lignocellulosic fibrous material is an alternative renewable eco-friendly solution that can be applied due to their low-cost and attractive potential features with desirable physical and mechanical properties for more functional materials (Ates et al. 2020). Different agricultural waste, e.g. wood aggregate, bast and leaf, cereal straw, palm tree, residue, and aquatic chips, could be considered as environmental-friendly filling materials in construction mortars (Laborel-Préneron et al. 2016).

A green sponge composite were prepared using natural rubber and cellulose fiber via the Dunlop processing technique, and applied as composites with different improved properties; the water absorptivity which reached 400% is very vital for storage of water (Suppanucroa et al. 2023). As the main factor affecting the capability of natural rubber to keep water molecules, the physical properties, specially swelling, were studied depending on fillers. Composites made of cellulosic fibrils, elastic natural rubber, organoclay, natural rubber, and liquid natural rubber were hybrided by melt-blending with different loading percentages. The proposed composites showed stability after exposure to different dosage of gamma irradiation, as suitable durable engineering applications (Shahdan et al. 2024). Novel composites of natural rubber-cellulose nanofibers-polyaniline complex were proposed for the application of electroconductive, and soft electronic elastomers; one of the prepared composites had ideal dispersion and high aspect ratio through hierarchical 3-D network polymerization. Properties such as ideal conductivity, high sensitivity and stretchability make the green rubber-nanocellulose complex applicable in strain sensors and flexible electrodes. In a similar work, natural rubber was used in sustainable sensors (Han et al. 2019; Wu et al. 2016).

The significant properties of cellulose have motivated researchers to draw on utilizing cellulose and its composites in numerous applications. These applications exploit the important characteristics of cellulose, such as its flexibility, surface activity, hydrophilic nature, free availability, biocompatibility, biodegradability, and good mechanical properties (Mali and Sherje 2022; Torgbo and Sukyai 2020; Guerra et al. 2021). Cellulose provides a fundamental effect on modifying the characteristics of polymer composites, however limited articles studied the related cellulose/natural rubber nanocomposites (Srirachya et al. 2020; Utrera-Barrios 2024; Maciejewska and Sowińska-Baranowska 2022; Zheng et al. 2019).

There are limited studies on antibacterial and mechanical characteristics of cellulose-natural rubber nanocomposites for drug delivery systems. Blending cellulose or natural rubber into nanocomposite forms provides new ways through which medical technology may advance. Such developments are vital towards addressing challenges faced by medical treatment such as managing infections or enhancing bio compatibility of medical implants/devices.

1.9 Applications of Cellulose/Rubber Based Bionanocomposites

Cellulose is a renewable bio based naturally occurring polymer that is widely found in nature, particularly in the form of woody materials and agricultural wastes. The use of natural, renewable, eco-friendly, and sustainable materials has grown in significance in polymer science and engineering during the past few decades (Ali Shah et al. 2013). The processing and application of rubber products based on carbon black may therefore result in significant environmental pollution. Additional drawbacks of using CB-filled rubber composites include the ongoing depletion of petroleum reserves, an increase in greenhouse gas concentrations, and the phenomenon of global warming. The most prevalent natural polymer, cellulose has certain special qualities that make it easy to find, renewable, lightweight, and environmentally friendly. Cellulose has been effectively selected as a non-petroleum based reinforcing filler in rubber technology over the past ten years. Because each anhydroglucopyranose unit contains a hydroxyl group, cellulose is known to have a strong hydrophilic character (Sebastián et al. 2016). Some synthetic rubbers, like ethylene-propylene rubber, are known to have significant swelling resistance against hydrocarbon oil, while NBR is said to have good heat resistance and can withstand service temperatures of up to 150 °C (Samsuri 2010). Nanocelluloses (NC) are being used as an alternative nowadays because of their density, renewability, biodegradability, availability, and lower cost when compared to carbon-based reinforcements (Pasaoglu and Koyuncu 2021). NC can be used to reinforce rubbers, both natural and synthetic. Cellulose could be extracted from natural sources like plants, algae, bacteria, and tunicates by using the right techniques. The cellulose chain's functional groups give it amazing characteristics like chirality, hydrophilicity, chemical change adaptability, and infusibility (Amarasekara 2013).

The degree of polymerization, the length of the cellulose chain, and the source all affect these characteristics. Since cellulose is easily obtained from a variety of natural sources, there are significant differences in its type, morphology, chain length, and lignocellulosic content. It is also possible to attribute variations in celluloses obtained from the same source to pretreatment, isolation methods, and cellulose maturity. Nanocellulose (NC), or cellulose in the form of nanostructures, has emerged as one of the most promising sustainable materials for the twenty-first century. Owing to its higher surface-area-to-volume ratio, high mechanical strength, and high Young's modulus, NC in the nanoscale dimension opens up a new world of possible applications on top of the benefits that are already there.

The process of extracting nanocellulose from Water hyacinth requires several extraction techniques after preliminary pretreatments. It is possible to isolate cellulose nanofibers (CNFs) and cellulose nanocrystals (CNCs) from water hyacinth. It has been discovered that a combination of mechanical and chemical procedures works better to get cellulose to the desired nanoscale (Akter et al. 2023). During the mixing process, cellulose naturally has a tendency to clump together into the rubber matrix. Therefore, in order to achieve the best dispersion, different cellulose surface modification treatments are carried out prior to the mixing process. For the best possible dispersion of cellulose into the rubber matrix, different surface modification treatments are applied to the material prior to mixing (Jiang and Li 2020; Cao et al. 2013). The function of the dual cross-linking system in epoxidized natural rubber (ENR) composites containing CNCs treated with maleic anhydride (MA-CNCs) has been investigated. First, the chemical interaction between the carboxyl groups of MA-CNCs and the epoxy groups of ENR chains can result in the formation of covalent cross-linking. Cellulose nanocrystals (CNCs) have been successfully prepared from Napier grass stems (penny-)Setum purpureum), 24% yield, using the following treatments, viz., alkali-steam explosion and bleaching. The unfilled NR composite has a modulus value of 0.72 MPa (at 100% elongation), while the MNR compatible NR/MCC composite has a modulus value of 1.069 MPa. The MCC surface was coated with in-situ silica that was derived from sol–gel. When mercapto groups were added to the CNC, the tensile strength increased twofold over unmodified samples and roughly five times over neat NR. The increase in elongation at break also implies that the rubber's elasticity was given by this functionalization (Parambath Kanoth et al. 2015). Nagatani (2017) developed a sponge-rubber material based on NR/CNF composites. The material was created using crosslinkers, like dicumyl peroxide, and azodicar-bonamide as a chemical blowing agent. The oleoylation process on the CNF surface gave it hydrophobic qualities and stronger reinforcing effects.

References

Abdul Wahab NM, Samsudin D, Jamaluddin SJS, Ahmad Z, Mustafa MS, Sarip MN, Saiwari S (2022) Characterization and tensile properties of self-healing chitin-devulcanized ethylene propylene rubber (EPDM)—waste natural rubber blend. In: Asian workshop on polymer processing. Springer Nature Singapore, Singapore, pp 265–277

Abitbol T et al (2023) Isolation of mixed compositions of cellulose nanocrystals, microcrystalline cellulose, and lignin nanoparticles from wood pulps. ACS Omega 8:21474–21484

Adak S (2024) Development of nanocellulose-based nanocomposites and its properties. Nanocellulose 113–140

Agrebi F, Ghorbel N, Bresson S, Abbas O, Kallel A (2019) Study of nanocomposites based on cellulose nanoparticles and natural rubber latex by ATR/FTIR spectroscopy: the impact of reinforcement. Polym Compos 40(5):2076–2087

Akter S, Abu S, Haque A, Ashfaqul, Khadem H, Siddiqa F, Masudur ANM, Humayra R, Himu A, Farzana N, Azharul Islam M, Naebe M, Smriti SA, Khadem AH, Rahman ANMM, Haque ANMA, Naebe M, Siddiqa F, Islam MA (2023) Recent developments of the nanocellulose extraction from water hyacinth: a review. Cellulose 30:8617–8641. https://doi.org/10.1007/s10570-023-05374-7

Ali Shah A, Hasan F, Shah Z, Kanwal N, Zeb S (2013) Biodegradation of natural and synthetic rubbers: a review. Int Biodeterior Biodegrad 83:145–157. https://doi.org/10.1016/j.ibiod.2013.05.004

Almeshaal M, Palanisamy S, Murugesan TM, Palaniappan M, Santulli C (2022) Physico-chemical characterization of Grewia Monticola Sond (GMS) fibers for prospective application in biocomposites. J Nat Fibers 19:15276–15290

Amarasekara AS (2013) Acid hydrolysis of cellulose and hemicellulose. In: Handbook of cellulosic ethanol, pp 247–281. https://doi.org/10.1002/9781118878750.ch7

Ates B, Koytepe S, Ulu A, Gurses C, Thakur VK (2020) Chemistry, structures, and advanced applications of nanocomposites from biorenewable resources. Chem Rev 120:9304–9362

Baley C, Perrot Y, Busnel F, Guezenoc H, Davies P (2006) Transverse tensile behaviour of unidirectional plies reinforced with flax fibres. Mater Lett 60(24):2984–2987

Bharagava RN, Goswami L, Kushwaha A, Hussain CM (2023) Bio-based materials and waste for energy generation and resource management: Volume 5 of Advanced zero waste tools: present and emerging waste management practices. Elsevier

Bhat A, Budholiya S, Aravind Raj S, Sultan MTH, Hui D, Md Shah AU et al (2021) Review on nanocomposites based on aerospace applications. Nanotechnol Rev 10:237–253

Bijarimi M, Ahmad S, Rasid R (2013) Mechanical, thermal and morphological properties of poly (lactic acid)/natural rubber nanocomposites. J Reinf Plast Compos 32:1656–1667

Bitinis N, Verdejo R, Bras J, Fortunati E, Kenny JM, Torre L et al (2013) Poly (lactic acid)/natural rubber/cellulose nanocrystal bionanocomposites Part I. Processing and morphology. Carbohydr Polym 96:611–620

Blomfeldt TOJ, Olsson RT, Menon M, Plackett D, Johansson E, Hedenqvist MS (2010) Novel foams based on freeze-dried renewable vital wheat gluten. Macromol Mater Eng 295(9):796–801

Blomfeldt TOJ, Kuktaite R, Johansson E, Hedenqvist MS (2011) Mechanical properties and network structure of wheat gluten foams. Biomacromolecules 12(5):1707–1715

Boonmahitthisud A, Chuayjuljit S (2012) Effects of carbon nanotube on tensile and dynamic mechanical properties of NR/SBR and NR/XSBR nanocomposites prepared by latex compounding. Adv Mater Res 488–489:612–616

Bouthegourd E, Rajisha KR, Kalarical N, Saiter JM, Thomas S (2011) Natural rubber latex/potato starch nanocrystal nanocomposites: correlation morphology/electrical properties. Mater Lett 65:3615–3617. https://doi.org/10.1016/j.matlet.2011.07.069

Cao X, Xu C, Liu Y, Chen Y (2013) Preparation and properties of carboxylated styrene-butadiene rubber/cellulose nanocrystals composites. Carbohydr Polym 92(1):69–76. https://doi.org/10.1016/j.carbpol.2012.09.054

Chang BP, Gupta A, Muthuraj R, Mekonnen TH (2021) Bioresourced fillers for rubber composite sustainability: current development and future opportunities. Green Chem 23:5337–5378

Chazeau L, Cavaille JY, Perez J (2000) Plasticized PVC reinforced with cellulose whiskers. II. Plastic behavior. J Polym Sci Part B: Polym Phys 38(3):383–392

Chen WJ, Gu J, Xu SH (2014) Exploring nanocrystalline cellulose as a green alternative of carbon black in natural rubber/butadiene rubber/styrene-butadiene rubber blends. Express Polym Lett 8:659–668. https://doi.org/10.3144/expresspolymlett.2014.69

Chin SY, Shahruddin S, Chua GK, Samsodin N, Setiabudi HD, Karam Chand NS et al (2021) Palm oil-based chemicals for sustainable development of petrochemical industries in Malaysia: progress, prospect, and challenges. ACS Sustain Chem Eng 9:6510–6533

Cichocki FR, Thomason JL (2002) Thermoelastic anisotropy of a natural fiber. Compos Sci Tech 62(5):669–678
Das M, Lalsangi S, Santra S, Banerjee R (2024) Nanocellulose as a carrier for improved drug delivery: progresses and innovation. J Drug Deliv Sci Technol 97:105743
Ding X, Chen L, Ma T, Ma H, Gu L, Chen T, Ma Y (2019) Laboratory investigation of the recycled asphalt concrete with stable crumb rubber asphalt binder. Constr Build Mater 203:552–557
Diniz J, Gil MH, Castro J (2004) Hornification—its origin and interpretation in wood pulps. Wood Sci Technol 37(6):489–494
Egbujuo WO, Anyanwu PI, Obasi HC (2020) Utilization of chitin powder as a filler in natural rubber vulcanizates: in comparison with carbon black filler. Int Rev Appl Sci Eng
El Hawary O, Boccarusso L, Ansell MP, Durante M, Pinto F (2023) An overview of natural fiber composites for marine applications. J Mar Sci Eng 11:1076
Fan Q, Duan H, Xing X (2024) A review of composite materials for enhancing support, flexibility and strength in exercise. Alex Eng J 94:90–103
Farah S, Anderson DG, Langer R (2016) Physical and mechanical properties of PLA, and their functions in widespread applications—a comprehensive review. Adv Drug Deliv Rev 107:367–392
Galimberti M, Kumar V, Coombs M, Cipolletti V, Agnelli S, Pandini S, Conzatti L (2014) Filler networking of a nanographite with a high shape anisotropy and synergism with carbon black in poly(1,4-cis-isoprene)-based nanocomposites. Rubber Chem Technol 87:197–218
Gea S, Bilotti E, Reynolds CT, Soykeabkeaw N, Peijs T (2010) Bacterial cellulosepoly(vinyl alcohol) nanocomposites prepared by an in-situ process. Mater Lett 64(8):901–904
Gelir A, Yargi O, Yuksel SA (2017) Elucidation of the pore size and temperature dependence of the oxygen diffusion into porous silicon. Thin Solid Films 636:602–607
Gindl W, Keckes J (2004) Tensile properties of cellulose acetate butyrate composites reinforced with bacterial cellulose. Compos Sci Tech 64(15):2407–2413
Gomes ME, Godinho JS, Tchalamov D, Cunha AM, Reis RL (2002) Alternative tissue engineering scaffolds based on starch: processing methodologies, morphology, degradation and mechanical properties. Mater Sci Eng C 20:19–26
Grignard B, Gennen S, Jérôme C, Kleij AW, Detrembleur C (2019) Advances in the use of CO_2 as a renewable feedstock for the synthesis of polymers. Chem Soc Rev 48:4466–4514
Guerra NB et al (2021) Biomedical applications of natural rubber latex from the rubber tree Hevea brasiliensis. Mater Sci Eng C 126:112126
Han J et al (2019) Nanocellulose-templated assembly of polyaniline in natural rubber-based hybrid elastomers toward flexible electronic conductors. Ind Crops Prod 128:94–107
Haris MRHM, Raju G (2014) Preparation and characterization of biopolymers comprising chitosan-grafted-ENR via acid-induced reaction of ENR50 with chitosan. Express Polym Lett 8(2):85–94
Hule RA, Pochan DJ (2007) Polymer nanocomposites for biomedical applications. MRS Bull 32:354–358
Hussain M, Yasin S, Memon H, Li Z, Fan X, Akram MA, Wang W, Song Y, Zheng Q (2020) Rheological and mechanical properties of silica/nitrile butadiene rubber vulcanizates with eco-friendly ionic liquid. Polymers 12. https://doi.org/10.3390/polym12112763
Ifuku S, Saimoto H (2012) Chitin nanofibers: preparations, modifications, and applications. Nanoscale 4:3308–3318
Iguchi M, Yamanaka S, Budhiono A (2000) Bacterial cellulose—a masterpiece of nature's arts. J Mater Sci 35(2):261–270
Isma'ila Mukhtar IM, Zulkiflle Leman ZL, Mohamad Ridzwan Ishak MRI, Edi Syams Zainudin ESZ (2016) Sugar palm fibre and its composites: a review of recent developments
Jabbari F, Babaeipour V (2023) Bacterial cellulose as a potential biopolymer for wound care. A review. Int J Polym Mater Polym Biomater 73:455–477
Jamshidian M, Tehrany EA, Imran M, Jacquot M, Desobry S (2010) Poly-lactic acid: production, applications, nanocomposites, and release studies. Compr Rev Food Sci Food Saf 9:552–571

Jardin JM, Zhang Z, Hu G, Tam KC, Mekonnen TH (2020) Reinforcement of rubber nanocomposite thin sheets by percolation of pristine cellulose nanocrystals. Int J Biol Macromol 152:428–436. https://doi.org/10.1016/j.ijbiomac.2020.02.303

Jasso-Salcedo AB, Caicedo C, Fonseca-García A, Díaz-Cruz CA, Jiménez-Regalado EJ, Aguirre-Loredo RY (2023) Polylactic acid composite materials for packaging and the consumption of food products. Polylactic Acid Compos: Sustain Biocompos 18:31

Jayalakshmy MS (2017) Applications of rubber based biocomposites and bionanocomposites. Rubber Based Bionanocompos Prep 167–176

Jiang W, Li L (2020) Surface modification of nanocrystalline cellulose and its application in natural rubber composites. https://doi.org/10.1002/app.49163

John RP, Anisha GS, Nampoothiri KM, Pandey A (2009) Direct lactic acid fermentation: focus on simultaneous saccharification and lactic acid production. Biotechnol Adv 27:145–152

Johny V, Kuriakose Mani A, Palanisamy S, Rajan VK, Palaniappan M, Santulli C (2023) Extraction and physico-chemical characterization of pineapple crown leaf fibers (PCLF). Fibers 11:5

Jonas R, Farah LF (1998) Production and application of microbial cellulose. Polym Degrad Stabil 59(1–3):101–106

Jose T, Moni G, Salini S, Raju AJ, George JJ, George SC (2017) Multifunctional multi-walled carbon nanotube reinforced natural rubber nanocomposites. Ind Crops Prod 105:63–73

Kamath SS, Sampathkumar D, Bennehalli B (2017) A review on natural areca fibre reinforced polymer composite materials. Ciência Tecnologia Dos Materiais 29:106–128

Kawano A, Yamamoto K, Kadokawa JI (2017) Preparation of self-assembled chitin nanofiber-natural rubber composite sheets and porous materials. Biomolecules 7(3):47

Khui PLN, Rahman MR, Bakri MKB (2023) A review on the extraction of cellulose and nanocellulose as a filler through solid waste management. J Thermoplast Compos Mater 36:1306–1327. https://doi.org/10.1177/08927057211020800

Klemm D, Kramer F, Moritz S, Lindstrom T, Ankerfors M, Gray D, Dorris A (2011) Nanocelluloses: a new family of nature-based materials. Angew Chem Int Ed 50(24):5438–5466

Kodsangma A, Homsaard N, Nadon S, Rachtanapun P, Leksawasdi N, Phimolsiripol Y, Jantanasakulwong K (2020) Effect of sodium benzoate and chlorhexidine gluconate on a bio-thermoplastic elastomer made from thermoplastic starch-chitosan blended with epoxidized natural rubber. Carbohydr Polym 242:116421

Laborel-Préneron A, Aubert J-E, Magniont C, Tribout C, Bertron A (2016) Plant aggregates and fibers in earth construction materials: a review. Constr Build Mater 111:719–734

Lee K-Y, Bharadia P, Blaker JJ, Bismarck A (2012) Short sisal fibre reinforced bacterial cellulose polylactide nanocomposites using hairy sisal fibres as reinforcement. Compos Pt A-Appl Sci Manuf 43(11):2065–2074

Liu Y, Liu M, Yang S, Luo B, Zhou C (2018) Liquid crystalline behaviors of chitin nanocrystals and their reinforcing effect on natural rubber. ACS Sustain Chem Eng 6(1):325–336

Low DYS et al (2021) Recent developments in nanocellulose-reinforced rubber matrix composites: a review. Polymers (Basel) 13:550

Lu DR, Xiao CM, Xu SJ (2009) Starch-based completely biodegradable polymer materials. Express Polym Lett 3:366–375

Maciejewska M, Sowińska-Baranowska A (2022) Bromide and chloride ionic liquids applied to enhance the vulcanization and performance of natural rubber biocomposites filled with nanosized silica. Nanomaterials 12:1209

Magaña I, López R, Enríquez-Medrano FJ, Kumar S, Aguilar-Sanchez A, Handa R et al (2022) Bioelastomers: current state of development. J Mater Chem A 10:5019–5043

Mali P, Sherje AP (2022) Cellulose nanocrystals: fundamentals and biomedical applications. Carbohydr Polym 275:118668

Markovic G, Visakh PM (2017) Applications of chitin-based rubber nanocomposites. Rubber Based Bionanocompos Prep 51–69

Mathew AP, Laborie M-PG, Oksman K (2009) Cross-linked chitosan/chitin crystal nanocomposites with improved permeation selectivity and pH stability. Biomacromol 10:1627–1632

Mathew M, Dominic CM, Neenu KV, Begum PS, Dileep P, Kumar TA, Badawi M (2023) Carbon black and chitin nanofibers for green tyres: preparation and property evaluation. Carbohydr Polym 310:120700

Mekonnen TH, Ah-Leung T, Hojabr S, Berry R (2019) Investigation of the co-coagulation of natural rubber latex and cellulose nanocrystals aqueous dispersion. Colloids Surf Physicochem Eng Asp 583:123949

Moon RJ, Martini A, Nairn J, Simonsen J, Youngblood J (2011) Cellulose nanomaterials review: structure, properties and nanocomposites. Chem Soc Rev 40:3941–3994

Mylsamy B, Shanmugam SKM, Aruchamy K, Palanisamy S, Nagarajan R, Ayrilmis N (2024) A review on natural fiber composites: polymer matrices, fiber surface treatments, fabrication methods, properties, and applications. Polym Eng Sci

Nagalakshmaiah M, Afrin S, Malladi RP, Elkoun S, Robert M, Ansari MA, Svedberg A, Karim Z (2019) Biocomposites: present trends and challenges for the future. https://doi.org/10.1016/B978-0-08-102177-4.00009-4

Nagatani A (2017) Characteristics and applications of cellulose nanofiber reinforced rubber composites. Int Polym Sci Technol 44(7):T1–T8. https://doi.org/10.1177/0307174x1704400701

Narimissa E, Gupta RK, Kao N, Choi HJ, Bhattacharya SN (2015) The comparison between the effects of solvent casting and melt intercalation mixing processes on different characteristics of polylactide-nanographite platelets composites. Polym Eng Sci 55:1560–1570. https://doi.org/10.1002/pen.23996

Neto WPF, Mariano M, da Silva ISV, Silvério HA, Putaux J-L, Otaguro H, Pasquini D, Dufresne A (2016) Mechanical properties of natural rubber nanocomposites reinforced with high aspect ratio cellulose nanocrystals isolated from soy hulls. Carbohydr Polym 153:143–152

Paek SW, Balasubramanian S, Stupples D (2022) Composites additive manufacturing for space applications: a review. Materials 15:4709

Palanisamy S, Murugesan TM, Palaniappan M, Santulli C, Ayrilmis N, Alavudeen A (2024) Selection and processing of natural fibers and nanocellulose for biocomposite applications: a brief review. BioResources 19

Parambath Kanoth B, Claudino M, Johansson M, Berglund LA, Zhou Q (2015) Biocomposites from natural rubber: synergistic effects of functionalized cellulose nanocrystals as both reinforcing and cross-linking agents via free-radical thiol–ene chemistry. ACS Appl Mater Interfaces 7(30):16303–16310. https://doi.org/10.1021/acsami.5b03115

Pasaoglu ME, Koyuncu I (2021) Substitution of petroleum-based polymeric materials used in the electrospinning process with nanocellulose: a review and future outlook. Chemosphere 269. https://doi.org/10.1016/j.chemosphere.2020.128710

Petchnui K, Uwanno T, Phonyiem Reilly M, Pinming C, Treetong A, Yordsri V, Moolsradoo N, Klamcheun A, Wongwiriyapan W (2024) Preparation of chitin nanofibers and natural rubber composites and their triboelectric nanogenerator applications. Materials 17:738

Rahman MR, Khui PLN, Bakri MKB (2022) Chapter 10—Cellulose-reinforced rubber composites. In: Rahman MR (ed) Fundamentals and recent advances in nanocomposites based on polymers and nanocellulose. Elsevier, pp 175–188. https://doi.org/10.1016/B978-0-323-85771-0.00013-0

Rajisha KR, Maria HJ, Pothan LA, Ahmad Z, Thomas S (2014) Preparation and characterization of potato starch nanocrystal reinforced natural rubber nanocomposites. Int J Biol Macromol 67:147–153. https://doi.org/10.1016/j.ijbiomac.2014.03.013

Ranakoti L, Gangil B, Mishra SK, Singh T, Sharma S, Ilyas RA et al (2022) Critical review on polylactic acid: properties, structure, processing, biocomposites, and nanocomposites. Materials 15:4312

Reddy G, Altaf MD, Naveena BJ, Venkateshwar M, Kumar EV (2008) Amylolytic bacterial lactic acid fermentation—a review. Biotechnol Adv 26:22–34

Riyajan S-A, Sasithornsonti Y, Phinyocheep P (2012) Green natural rubber-g-modified starch for controlling urea release. Carbohydr Polym 89:251–258. https://doi.org/10.1016/j.carbpol.2012.03.004

Ross P, Mayer R, Benziman M (1991) Cellulose biosynthesis and function in bacteria. Microbiol Rev 55(1):35–58

Roy K, Pongwisuthiruchte A, Debnath SC, Potiyaraj P (2021) Application of cellulose as green filler for the development of sustainable rubber technology. Curr Res Green Sustain Chem 4:100140

Samsuri A (2010) 3.33 Degradation of natural rubber and synthetic elastomers. Shreir's corrosion, Elsevier, Oxford, pp 2407–2438

Sattayanurak S, Noordermeer JW, Sahakaro K, Kaewsakul W, Dierkes WK, Blume A (2019) Silica-reinforced natural rubber: synergistic effects by addition of small amounts of secondary fillers to silica-reinforced natural rubber tire tread compounds. Adv Mater Sci Eng 2019(1):5891051

Sebastián J, Herrera V, Restrepo SV, Vasquez DG (2016) Mechanical and rheometric properties of gilsonite/carbon black/natural rubber compounds cured using conventional and efficient vulcanization systems. Polym Test. https://doi.org/10.1016/j.polymertesting.2016.09.005

Shahdan D, Chen RS, Zailan FD, Sabtu M, Ahmad S (2024) γ-radiation for improved reinforcement effect of single and hybrid nano-clay and cellulose nanofiber in thermoplastic natural rubber nanocomposite. Radiat Phys Chem 218:111550

Shelma R, Paul W, Sharma CP (2008) Chitin nanofibre reinforced thin chitosan fi lms for wound healing application. Trends Biomater Artif Organs 22(2):107–111

Shokatayeva D, Ignatova L, Savitskaya I, Kistaubaeva A, Talipova A, Asylbekova A, Abdulzhanova M, Mashzhan A (2019) Bacterial cellulose and pullulan from simple and low cost production media. Eurasian Chem Technol J 21:247

Singha NR, Mahapatra M, Karmakar M, Chattopadhyay PK (2019) Processing, characterization and application of natural rubber based environmentally friendly polymer composites. Sustain Polym Compos Nanocompos 855–897

Siro I, Plackett D (2010) Microfibrillated cellulose and new nanocomposite materials: a review. Cellulose 17(3):459–494

Somseemee O, Sae-Oui P, Siriwong C (2022) Bio-based epoxidized natural rubber/chitosan/cellulose nanocrystal composites for enhancing mechanical properties, self-healing behavior and triboelectric nanogenerator performance. Cellulose 29(16):8675–8693

Song Z, Li G, Guan F, Liu W (2018) Application of chitin/chitosan and their derivatives in the papermaking industry. Polymers 10(4):389

Srirachya N, Boonkerd K, Kobayashi T (2020) Effective elongation properties of cellulose–natural rubber composite hydrogels having interconnected domain. J Elastomers Plast 52:337–355

Stevenson A (1992) Engineering properties and applications of rubber. Dev Crop Sci 23:487

Sun D, Saw BLH, Onyianta AJ, Wang B, Wilson C, O'Rourke D, See CH, Popescu C-M, Dorris M, Shyha I, Lu Z (2023) Preparation of elastomeric nanocomposites using nanocellulose and recycled alum sludge for flexible dielectric materials. J Adv Dielectr 13:2242008. https://doi.org/10.1142/S2010135X22420085

Suppanucroa N, Nimpaiboon A, Boonchuay K, Khamkeaw A, Phisalaphong M (2023) Green composite sponge of natural rubber reinforced with cellulose filler using alginate as a dispersing agent. J Mater Res Technol 27:3119–3130

Taherimehr M, Bagheri R, Taherimehr M (2021) In-vitro evaluation of thermoplastic starch/beta-tricalcium phosphate nano biocomposite in bone tissue engineering. Ceram Int 47:15458–15463

Teh TW, Tan CG, Jumaat MZ (2022) Feasibility evaluation of novel high-damping rubbers as energy-dissipation material under axial dynamic load for damper devices. Buildings 12:1917

Thakur R, Pristijono P, Golding JB, Stathopoulos CE, Scarlett C, Bowyer M, Singh SP, Vuong QV (2018) Effect of starch physiology, gelatinization, and retrogradation on the attributes of rice starch-ι-carrageenan film. Starch 70:1700099. https://doi.org/10.1002/star.201700099

Thomas SK, Parameswaranpillai J, Krishn J, Krishnasamy S, Begum PMS, Nandi D, Siengchin S, George JJ, Hameed N, Salim NV, Sienkiewicz N (2021) A comprehensive review on cellulose, chitin, and starch as fillers in natural rubber biocomposites. Carbohydr Polym Technol Appl 2:100095. https://doi.org/10.1016/j.carpta.2021.100095

Torgbo S, Sukyai P (2020) Biodegradation and thermal stability of bacterial cellulose as biomaterial: the relevance in biomedical applications. Polym Degrad Stab 179:109232

Trache D (2018) Nanocellulose as a promising sustainable material for biomedical applications. AIMS Mater Sci 5(2):201–205

Trache D et al (2020) Nanocellulose: from fundamentals to advanced applications. Front Chem 8:392

Urtekin G, Ullah MS, Yildirim R, Ozkoc G, Kodal M (2023) A comprehensive review of the recent developments in thermoplastics and rubber blends-based composites and nanocomposites. Polym Compos 44:8303–8329

Utrera-Barrios S et al (2024) Sustainable composites with self-healing capability: epoxidized natural rubber and cellulose propionate reinforced with cellulose fibers. Polym Compos

Valodkar M, Thakore S (2010) Thermal and mechanical properties of natural rubber and starch nanobiocomposites. Int J Polym Anal Charact 15:387–395. https://doi.org/10.1080/1023666X.2010.500543

Varma M, Chandran S, Vijay Kumar V, Suyambulingam I, Siengchin S (2024) A comprehensive review on the machining and joining characteristics of natural fiber-reinforced polymeric composites. Polym Compos

Vignesh Kumar B, Muthumari B, Kavitha M, John Praveen Kumar JK, Thavamurugan S, Arun A et al (2022) Studies on optimization of sustainable lactic acid production by Bacillus amyloliquefaciens from sugarcane molasses through microbial fermentation. Sustainability 14:7400

Wahba L, D'Arienzo M, Donetti R, Hanel T, Scotti R, Tadiello L, Morazzoni F (2013) In situ sol-gel obtained silica-rubber nanocomposites: influence of the filler precursors on the improvement of the mechanical properties. RSC Adv 3:5832–5844

Walling B, Bharali P, Giridharan B, Gogoi B, Sorhie V, Alemtoshi MSK (2023) Bacterial nanocellulose: a novel nanostructured bio-adsorbent for green remediation technology. Acta Ecol Sin 43:946–967

Wongvasana B, Thongnuanchan B, Masa A, Saito H, Sakai T, Lopattananon N (2022) Comparative structure-property relationship between nanoclay and cellulose nanofiber reinforced natural rubber nanocomposites. Polymers 14(18):3747

Wu Y-P, Ji M-Q, Qi Q, Wang Y-Q, Zhang L-Q (2004) Preparation, structure, and properties of starch/rubber composites prepared by co-coagulating rubber latex and starch paste. Macromol Rapid Commun 25:565–570. https://doi.org/10.1002/marc.200300125

Wu X et al (2016) Cellulose nanowhisker modulated 3D hierarchical conductive structure of carbon black/natural rubber nanocomposites for liquid and strain sensing application. Compos Sci Technol 124:44–51

Xu K (2004) Nonaqueous liquid electrolytes for lithium-based rechargeable batteries. Chem Rev 104:4303–4418

Xu J, Li S, Li Y, Ta X (2016) Preparation, morphology and properties of natural rubber/carbon black/multi-walled carbon nanotubes conductive composites. J Mater Sci Mater Electron 27:9531–9540

Yang KK, Wang X-L, Wang Y-Z (2007) Progress in nanocomposite of biodegradable polymer. J Ind Eng Chem 13(4):485–500

Yasin S, Hussain M, Zheng Q, Song Y (2021) Effects of ionic liquid on cellulosic nanofiller filled natural rubber bionanocomposites. J Colloid Interface Sci 591:409–417. https://doi.org/10.1016/j.jcis.2021.02.029

Yuan S, Shen F, Chua CK, Zhou K (2019) Polymeric composites for powder-based additive manufacturing: materials and applications. Prog Polym Sci 91:141–168

Zhang N, Cao H (2020) Enhancement of the antibacterial activity of natural rubber latex foam by blending it with chitin. Materials 13(5):1039

Zhang G, Wang F, Dai J, Huang Z (2016) Effect of functionalization of graphene nanoplatelets on the mechanical and thermal properties of silicone rubber composites. Materials 9(2):92

Zhang W, Khan A, Ezati P, Priyadarshi R, Sani MA, Rathod NB, Rhim JW (2024) Advances in sustainable food packaging applications of chitosan/polyvinyl alcohol blend films. Food Chem 138506

Zheng D et al (2019) Fabrication and performance of a spherical cellulose nanocrystal-based hydrophobic drug delivery vehicle using rubber wood

Chapter 2
Cellulose-Based Rubber Nanocomposites

Palanisamy Sivasubramanian, Gauri S. Deshmukh, Thambiannan Senthilkumar, Vellaichamy Mageshwaran, and Yalarvathi Nagaraju

2.1 Introduction

Nanotechnology is one of the key factors igniting a new industrial transformation across several industries, from nanocomposites to medical to sensing and biosensing uses (Arof et al. 2019). Compared to mass materials, nanoscale materials, which measure about 100 nm in at least one dimension, have unique physicochemical, optical, magnetic, and biochemical properties (He et al. 2019). The beneficial properties of this nanomaterial include a high surface area to volume ratio, high Young's elasticity and high tensile strength, low coefficient of thermal expansion, hydrogen-bonding capability, biocompatibility, eco-friendliness, renewability, and lack of toxicity (Salimi et al. 2019). Utilizing green, renewable, and viable resources is more crucial than ever for creating various high-quality, environmentally friendly goods. Many academics and business people are interested in this field of study because these materials could be an option for the ever-dwindling non-renewable resources, environmental contamination, global warming, and energy crisis. To be specific, cellulose has emerged as a potential option due to its widespread availability from

P. Sivasubramanian (✉)
Department of Mechanical Engineering, PTR College of Engineering and Technology, Austinpatti, Madurai 625008, Tamil Nadu, India
e-mail: sivaresearch948@gmail.com

G. S. Deshmukh
Department of Applied Chemistry, Yeshwantrao Chavan College of Engineering, Hingna Road, Wanadongri, Nagpur 441110, Maharashtra, India

T. Senthilkumar
ICAR-Central Institute for Research On Cotton Technology, Matunga, Mumbai 400019, Maharashtra, India

V. Mageshwaran · Y. Nagaraju
ICAR-National Bureau of Agriculturally Important Microorganisms, Kushmaur, Mau 275103, Uttar Pradesh, India

© The Author(s), under exclusive license to Springer Nature Switzerland AG 2025
P. M. Visakh (ed.), *Rubber Based Bionanocomposites*, Advanced Structured Materials 227, https://doi.org/10.1007/978-3-031-78557-3_2

various sources, starch, alginate, chitin, chitosan, and gelatin (Trache 2018; Mageshwaran et al. 2023). It has been standard practice to use unusual materials in the compositions and changes of current processing technologies to enhance the properties of polymer matrix materials. Due to their higher aspect ratio and surface energy compared to traditional strands, various nanoparticle types, including natural and synthetic materials, have been shown to improve polymer performance significantly.

The Natural rubbers (NRs), have excellent chemical and physical characteristics like suppleness, elasticity, resilience, abrasion and impact resistance etc. Additionally, NR is extremely compostable and mouldable (Agrebi et al. 2019). The NRs are frequently found in the tires of vehicles, cars, and aeroplanes. Due to their superior cracking resilience and high elasticity, NRs are essential in industrial uses like pneumatic tires and sound separation systems. Despite of several advantages, rubber has low modulus and strength which necessitates the addition of the reinforcing agent to improve the mechanical property and to satisfy the demands of various industrial uses. Typically, silica and carbon black are added to natural rubber during vulcanization to create rubber compounds with improved characteristics (Palanisamy et al. 2021). Despite serving as supports for rubber composites, carbon black and silica are not biodegradable and take much energy to produce (Palaniappan et al. 2024). Being a petrochemical-based substance, carbon black is also non-renewable. As a result, lignocellulosic fibre has attracted interest and momentum in the composite rubber industry as a renewable and ecologically responsible filler because of its inherent qualities. The hydrophilic wood flour or other natural fibres' poor interaction with the hydrophobic rubber matrix is the lignocellulosic fibre-reinforced rubber composite's fundamental flaw, which causes the material's subpar mechanical properties (Karthik et al. 2024). Various methods to change the fillers are frequently used to enhance compatibility between the natural rubber matrix and the filler used. The development of nanotechnology in recent years has made surface-modified nano cellulose a feasible filling for rubber composites.

Nanocellulose has recently shown outstanding promise as a carbon black replacement. In reality, nitrocellulose (NC), which is cellulose in the shape of nanostructures, has emerged in recent years as one of the most promising sustainable materials. NC is a highly preferred reinforcing material due to its excellent properties like renewable nature, higher surface area volume, biocompatible and non-toxic nature, high tensile and elastic modulus (Sharma et al. 2019). Due to their high mechanical strength and adaptable surface chemistry, cellulose nanocrystals (CNC), nanocrystalline cellulose (NCC) and cellulose nano whiskers (CNW) are frequently used and preferred commercially. CNC appears as a rod-like solid structure when magnified. Their liquid crystalline cholesteric (Ch) structures make the most significant input to CNC characteristics (Kim et al. 2019). This capability benefits the industry regarding humidity monitors, optical encryptors, structural pigments, light filters, and blueprints to create inorganic materials. Due to their renewable nature and major use as a reinforcing agent, CNC is quickly becoming an industrial-favoured component.

Additionally, the CNC shows a structural hue that additives can modify in the shape of the solidified film. (Nanoparticles, organic dyes, and surfactants). Bacterial Nano cellulose (BNC) is a highly pure extracellular substance bacteria make. BNC is shaped like a band with 20–100 nm width. BNC was isolated from the bacteria, *Gluconoacetobacter* raised in liquid culture media (Castro et al. 2012). The bacterium culture uses carbon as an energy source, and the expanding culture uses nitrogen. BNC is almost identical in properties to CNC and CNF, but based on the bacterial growth conditions, it is frequently lightweight, non-toxic, and controls microfibril formation. BNC is well-known in biological fields like wound repair and regenerative medicine (Sharma et al. 2023).

The nanocrystals cellulose, or nanocellulose, are emerging green nano reinforcements for polymers. The major advantage for using nanocellulose as a reinforcement for polymers is the possibility of exploiting the high tensile stiffness and strength of the cellulose crystals (Dufresne 2018). Though NC is highly preferred as a reinforcing agent, it has certain drawbacks such as solubility and compatibility with rubber matrix. Therefore, surface modification of NC has become inevitable to improve the interfacial compatibility with the rubber matrix. The different surface modification techniques employed to enhance the compatibility of nanocellulose as fillers in NR composites were silyation, TEMPO mediated oxidation, esterification etc. (Hakimi et al. 2021). This chapter discusses about the impact of addition of modified nanocellulose in the rubber matrix in imparting improved mechanical properties and various characterization techniques to evaluate the same.

2.2 Characterizations Techniques

2.2.1 Tensile Testing

The composite is essentially a mixture of two phases. In this discussion stress will be more upon the composites having reinforcing material as fibres. The fibres are chains of constituent polymers and belong to the category of visco-elastic materials. When a visco-elastic material is strained, it shows instantaneous extension (which is completely recoverable on removal of load) followed by gradual extension with time known as "creep". Time given for the rate of creep to occur, decides the ultimate strength. If the rate of straining is slow, there is more time for creep to advance and consequently, the breaking extension is reached at a lower load. The strength of the visco-elastic fibres is a function of the rate of straining (Iyyadurai et al. 2023). Extensive studies carried out by researchers to ascertain the influence of rate of straining on tensile characteristics reveal that a rapid test produces a higher breaking load than a slow test. It is observed that as time to break decreases (due to faster rate of extension), the breaking strength gets enhanced. Breaking load is reported to vary inversely with log (time to break). The tensile strength of composite materials I is characterized by Instron.

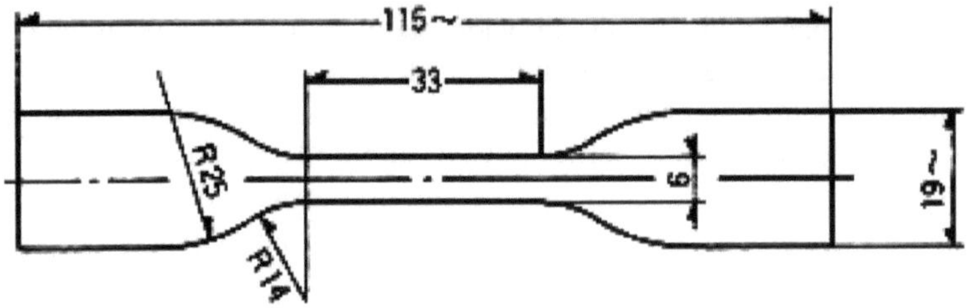

Fig. 2.1 The tensile test specimen as per the ASTM D638 (Ertugrul et al. 2023)

Instron Tensile Tester

This instrument works based on the principal of constant rate of extension (CRE) and used to determine properties such as tensile and flexural strength of the materials. By using this instrument, it can calculate the tensile value range from 2.5 N to 10 kN. It can also be used for compression tests. A wide spectrum of mechanical properties of fibres, yarns, fabrics and composites can be studied by using this instrument. The test specimens are cut in to the shape given in Fig. 2.1 according to ASTM D638 Type IV standard.

The tested specimens are place between two jaws as shown in the Fig. 2.2. The applied load between the jaws are calculated by load cells which convert the mechanical energy into electrical signal. The tensile strength of the composites reported in MPa.

The polymer composite's tensile strength is also decided by the properties of reinforcement and the proportion of the reinforcement and resin. When the proportion of the reinforcement and resin is low, the reinforcement cannot support the redistributed load, and therefore it breaks. When the proportion of the reinforcement and resin is high, the redistributed load can be supported by the bridging fibre bundles, and the composite tensile failure reaches up to the fracture strength of the fibre. According to Karthik et al. (2024), Ku et al. (2011) it was reviewed that factors influencing the strength of the composite were found to be resin and reinforcement compatibility, chemical modifications on fibre, and the fibre content. It analysed the effects of stacking sequence on tensile and interlaminar shear strength of the produced composite and reported that fibre alignment and stacking sequence have affected the selected properties of the composite materials (Ahmed and Vijayarangan 2008; Govindarajan et al. 2024). The number of reinforcement layers and their directions has a significant result on interlaminar shear strength and flexural strength than on the mechanical strength. Authors in Hargitai et al. (2008), Palanisamy et al. (2023) analyzed the fibre content and anisotropy of non-woven mats on the strength of composites. They reported that optimal mechanical performance was achieved in composites made from 40 to 50% of fibre content. Authors in Anuar and Zuraida (2011), Sumesh et al. (2024) developed the natural fibre composite with Kenaf fibre reinforcement and reported that tensile and impact properties were higher for alkali treated reinforcement than the raw form. Authors in Ramakrishna and Hull

Fig. 2.2 Broken tensile test specimen

(1994) studied the knitted fabric reinforced epoxy composites and concluded that knitted configuration has greatly affected the tensile strength and modulus of the produced composites. Authors in Bhambure et al. (2023) produced kenaf/kapok hybrid composite and concluded that kenaf/kapok hybrid composites increased the tensile and impact strength by 38% and 29% respectively. Authors in Wu and Huang (2020) synthesized graphene from raw poplar leaves and modified poplar leaves, and reinforced with silicone rubber composites and found that the tensile strength and wear resistance increased 36.2% and 57.1% respectively. Authors in Maiti et al. (2016) studied the effect of mesoporous silica, silica and blend of mesoporous nanosilica with carbon black reinforcement in polybutadiene rubber. They have found the significant improvement in tensile strength of 250% when compared to virgin matrix.

Composites strengthened with carbon nanofiller are appropriate for this study. Dampening parameters and contact parameters were recalculated. Regarding how well conventional particles engage with rubbers, nanoparticles were contrasted with conventional particles. Rubber was combined with CB and single-wall carbon nanotubes (SWNT), reducing dampening parameters and raising the glass transition temps. SWNT displayed more B than CB, indicating improved interaction between rubber and SWNT (Zhang et al. 2016). The nanocrystalline cellulose (NCC) was obtained from different natural sources such as kenaf, ramie, Napier grass, cotton cellulose etc. The NCC thus prepared was surface modified by silylation, TEMPO mediated oxidation, esterification etc. to increase the compatibility with natural rubber. The modified nanocrystalline cellulose and unmodified monocrystalline cellulose were reinforced with natural rubber and produced by nanocomposites. Usually, 3–10% addition of NCC was done to prepare the nanocomposites. The different mechanical properties were evaluated. The reports showed that the NR nanocomposites prepared by addition of modified nanocrystalline had better mechanical properties than unmodified nanocrystalline cellulose (Xu et al. 2012; Nechyporchuk et al. 2016; Singh et al. 2020; Noguchi et al. 2021). In one of the study the nanocellulose was extracted from dried leaves of rubber (Dittanet et al. 2019). The tensile strength recorded was 15.04 MPa in 2% nanocellulose reinforced NR composites and 12.30 in NR which was 22.2% higher in composites than the control (natural rubber). The elongation (%) in nanocellulose reinforced composites was 807.23 ± 15.79 while in NR, 723.95 ± 10.54.

2.2.2 Differential Scanning Calorimeter

Studying numerous phenomena, including melting, crystallization, healing rates, and glass transition, that occur during a thermal scan of nanofillers and PN has extensively used differential scanning calorimeters (DSC) (Gopinath et al. 2021). When dispersion at the nanoscale is accomplished, these properties undergo an odd shift. Diverse processes, including glass transition (Tg), freezing, crystallization, and curing, occur during organoclays' thermal heating, and polymer/clay nanocomposites or nanotubes have been extensively studied using differential scanning calorimetry (Bitinis et al. 2011). One of the most popular methods used to study the transition in polymers and their mixtures is the DSC approach. The transition is connected to the corresponding dipoles' easing and the major chains' Brownian motion at the change from the glassy to the rubbery state. Using clay nanocomposites as an example, DSC method can show how nanosized montmorillonite in many polymers results in noticeable improvements in Tg (Pradhan et al. 2015).

2.2.3 Dynamical Mechanical Analysis

This instrument used to determine the viscoelastic behaviour of the composite materials. Through a dynamic mechanical analyzer, the composite specimen's stress/strain behaviour is measured between different ranges of temperature. The storage modulus, loss factor and loss modulus are the three important parameters that have been analysed. Bashir (2021) studied that viscoelastic behaviour of neat and filled polymers and reported that to know the effect of fillers in matrix and their interaction was important for parameters which will be determined by the dynamic mechanical analyser. Rajesh et al. (2021) studied the fibre volume fractions, effect of coupling agents for nylon 6 reinforced in acrylonitrile butadiene rubber with different temperatures and different frequencies by using a dynamic mechanical analyser. Bakošová (2018) studied the interaction of carbon nanotubes in rubber composites and studied its dynamical mechanical analysis. Vijay and Menon (2022) studied the interaction of organo-modified fillers as a reinforcement in rubber matrix using dynamic mechanical analysis and reported that DMA is the instruments, which was useful to study the interaction of surface modified reinforcement in matrix. Researchers also have looked into the dynamic tensile properties of NR/partially reduced graphene oxide (PRGO) nanocomposites over a various temperature, from 280 to 40 °C. Rubber's storage elasticity rises linearly with PRGO concentration, confirming the PRGO's significant incorporation impact on NR. However, in the glass transition zone between 250 and 0 °C, PRGO1 s E0 is lower than NR's (Song et al. 2015; Mat Ali and Ahmad 2016).

2.2.4 *Thermo Gravimetric Analysis*

The thermal stability of reinforcement, resin and composites materials can be predicted by using Thermo gravimetric analysis (TGA) (Nam et al. 2015). This analysis can be done based on the rate of change in weight of the material with respect to the temperature in an inert atmosphere. The inert atmosphere maintains by using nitrogen gas (N_2) to avoid oxidation effect. The known quantity of samples approximately 5–10 mg was taken for the analysis and the temperature from 25 °C to maximum decomposing temperature at a heating rate of 10 °C per min in the inert atmosphere (Bhambure et al. 2022). The weight of the material is decreased due to decomposition or volatilization during the thermal analysis. The TGA can provide the information of the evaluated material such as composition and thermal stability of polymeric materials. The thermogravimetric analysis is one of the effective studies for determining the decomposition of the materials (Motiee et al. 2013). This analysis will help to compare between different materials' nature and decide the final application. The thermal stability of polymeric composites, depends on the type of reinforcement, fillers and matrix materials used (Dickens 1980).

Mamanpush and Tabatabaei (2019), studied the isothermal thermogravimetric analysis (TGA) of recycled carbon fibre reinforced with epoxy and vinyl ester composites (CFCs). Gurukarthik Babu et al. (2019) reported that the material is heating up to its maximum decomposition temperature, and the mass reduction during the heating process is measured. Mohanasundaram and Sivakumar (2021) studied the thermal behaviour of the Palmyra fibre, alkali treated Palmyra fibre and glass fibre and its composite with epoxy resin were studied. They have reported that the thermal behaviour of the alkali treated Palmyra and raw Palmyra fibre shows higher thermal stability. Intapun et al. (2021) investigated the thermal stability of the rubber composites by adding the Klason lignin as a filler. The thermal durability of NR/SWCNT nanocomposites created through latex stage mixing is largely unaffected at smaller SWCNT concentrations. MWCNT-reinforced NR nanocomposites were created by Jose et al. (2017) using the two-roll mixture, and their temperature behavior in nitrogen was investigated. Compared to the gum sample, the matching thermal profiles demonstrated the superior thermal behaviour of NR/MWCNTs. Most likely, the MWCNTs' spread in the NR grid stops the NR from thermally degrading. This implied that 3.5 phr loaded MWCNTs in NR have better temperature stability than pure NR. This is attributed to MWCNTs' capacity to maintain the NR matrix regarding temperature (Song et al. 2015).

2.2.5 Scanning Electron Microscopy

The filler size, form, dispersion, and matrix-filler interaction all affect the characteristics of polymer-matrix composites. Transmission electron microscopy (TEM) is frequently used for the morphological and structural study of nanocomposites, where comprehensive imaging is necessary to guarantee a fair picture of the substance (Yamauchi et al. 2005). Carbon-based nanofillers like carbon nanotubes, thin graphitic platelets, carbon black, etc., also have very poor TEM contrast when incorporated in a polymer matrix. At the nanometer scale, comprehensive information on the morphology, conductivity, and even the modes of fracture of the nanocomposite can be acquired by merging several scanning electron microscopy (SEM) methods. The area of the electron beam spot, which is smaller at greater acceleration voltages, has historically been linked to SEM sharpness (Sivasubramanian et al. 2022; Srisuwan et al. 2014). The sample volume from which secondary electrons are generated and gathered and the effectiveness of the low-energy electron detectors are intrinsically related to resolution and information-passing capability for high-resolution imaging of nanostructures. The carbon nanotube contrast has a distinct sign at various AVs, and surface characteristics are primarily visible at low AVs. Like the charge contrast imaging method, their contrast/brightness values depend on the electron flux per unit time and discharge processes. The mix of topographic imaging by atomic force microscopy and scanning Kelvin probe microscopy is comparable to low AV imaging. However, the sub-surface imaging method is more helpful in analyzing filler particles' distribution and structure and evaluating nanocomposites' conductivity (Li et al. 2013). SEM is used to analyse any surface changes in the material. The earlier reports showed that the addition of nanocellulose in rubber composites resulted in increase in roughness on the surface of the composite material. The addition of cellulose nanowhiskers at the rate of 2.5 and 5.5% addition in natural rubber composites had no micro-scale aggregates which was observed through SEM micrograph (Visakh et al. 2012).

The compatibility of matrix and fibres are mainly depending on the surface morphological features such as roughness and porosity which can be analysed by SEM. Reinforcement with increasing surface roughness showed an increase in surface properties. However, the roughness of the fibre surface improved the bonding behaviour with the matrix component (Hashemifard et al. 2015). From this SEM analysis higher resolution images can be taken, which will provide the clear information about the bonding nature between the reinforcement and matrix (Ravindran et al. 2020). They were coated with gold/palladium using a sputter coater to prevent the accumulation of electrical charges (Manimaran et al. 2018).

Senthilkumar et al. (2020) studied the mechanical and morphological analysis of raw coir fibre (RCF) and Fibrillated coir fibre (FCF) reinforced with the epoxy composite. Figure 2.3 clearly shows the surface morphology of the raw coir fibre, fibrillated coir fibre and their fracture surfaces of RCF and FCF composites with epoxy resin. Boominathan et al. (2022) studied the fractured surface analysis of raw *Acacia Concinna* fibre (ACF) and alkali treated ACF reinforced polyester composites. They have reported that raw ACF reinforced fibre reinforced composites show poor bonding between polyester resin than alkali treated ACF. The alkali treated ACF shows more surface roughness and better bonding with resin. The raw fibres are having natural coating with wax, pectin and other foreign particles which was covered with microvoids of the fibres, due to alkali treatment the wax, pectins and other foreign particles were removed and the fibre surface roughness increased so that it enhanced the composite properties (Ilangovan et al. 2020; Bessa et al. 2017).

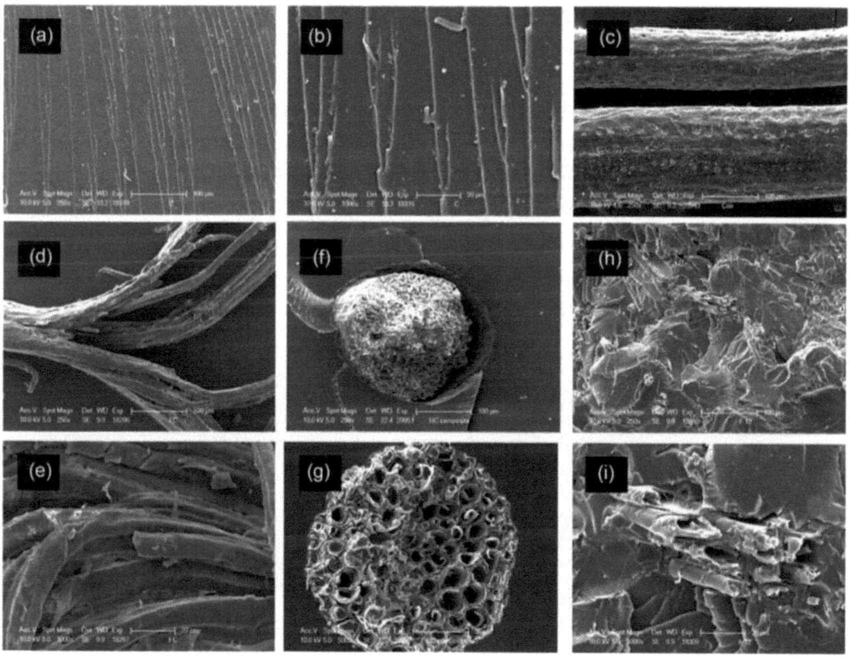

Fig. 2.3 Fractural studies of raw coir fibre and fibrillated coir fibre reinforced with epoxy composites (Senthilkumar et al. 2020)

2.2.6 Transmission Electron Microscopy

In recent years, structural and chemical analysis at the nanometer scale has become crucial for all materials. This chapter provides an overview of the transmission electron microscope (TEM), the ideal tool for this task. Transmission electron micrographs can determine particle size, grain size, lattice type, morphological data, crystallographic specifics, chemical makeup, phase type, and dispersal. Quantitative data such as size, phase identity, orientation relationship, crystal flaws in the lattice structure, etc., can be obtained from the electron diffraction images of nanomaterials. Typical electron diffraction, high-resolution transmission, and scanning transmission electron microscope imaging in materials study are covered in this volume.

Strong acid decomposition can easily produce high aspect ratio nanoparticles called cellulose nanocrystals (CNCs) from cellulose microfibrils. They have special qualities brought about by the chemical of their surfaces, the crystallinity, and the three-dimensional structure. Aerogels, nanocomposite materials, optically active films, and other uses have all made use of CNCs. Determining the size and form of CNCs is a significant task, and transmission electron microscopy (TEM) is one of the most effective methods for accomplishing this. CNCs need special consideration due to TEM imaging's unique requirements. They have a low density, are readily aggregated, and are very vulnerable to harm from electron beams. Over the past few decades, specific imaging and collection methods have been created (Srivastava and Mittal 2017).

The TEM analysis showed that the average length of cellulose nanocrystals (CNC) obtained from rubber leaves was 375. 4 nm while the CNC obtained from commercial cellulose was 502.3 nm as shown in Fig. 2.4. The aspect ratio of CNC obtained from both the sources was more than 10 (Dittanet et al. 2019).

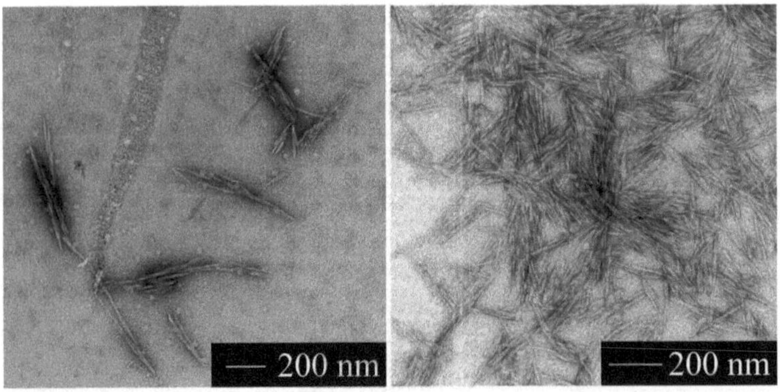

a. b.

Fig. 2.4 TEM images of cellulose nanocrystals (CNC-R) obtained from rubber leaves (**a**) and cellulose nanocrystals (CNC-C) obtained from commercial cellulose (**b**) (Dittanet et al. 2019)

2.2.7 *Atomic Force Microscopy*

Atomic force microscopy examines the surface roughness and structure (AFM). The surface roughness is determined by quantifying the height of the peaks in the out-of-plane orientations (Palanisamy et al. 2022; Kumar et al. 2022). Due to its capability to generate three-dimensional (3D) topographic pictures with a precision in the order of nanometers) for polymeric nanocomposites, atomic force microscopy (AFM) is one of the most potent characterization methods for the morphological and structural study of polymeric nanocomposites. Due to these unusual properties, it is possible to map the heights and locations of tiny features on the sample surface and thus learn more about the anatomy. The regulated contact between the microscopic probes and the sample surface allows various surface characterizations to be performed using the AFM. The AFM is also beneficial because it requires little sample processing and can be used in ambient circumstances; the sample does not need to be conductive or undergo the metallization process before being subjected to morphological analysis. This method is an exceptional instrument for directly characterizing various samples with complex morphological organizations because of these capabilities (Ma et al. 2010; Liu et al. 2013).

The academic community and businesses have recently shown great interest in nanostructured polymeric composites. One can create multifunctional materials with entirely novel characteristics thanks to the ability to modify these materials on a nanometric scale. For instance, a tried-and-true method for enhancing polymeric composites' mechanical, electrical, and thermal performance is the filling use of carbonaceous nanomaterials like carbon nanotubes, carbon nanofibers, and graphene nanoplatelets (Bisht et al. 2017).

AFM is extensively used to analyse nanosized cellulosic particles. Visakh et al. (2012) used AFM for characterization of nanowhiskers (CNWs). The author used bamboo cellulose pulp for the synthesis of nanowhiskers. The size of the CNW described was in the range of 4–14 nm.

2.2.8 *FT-IR Spectroscopy*

One of the most widely used spectroscopic methods by both organic and inorganic chemists is Fourier transform infrared (FTIR) spectroscopy. A fundamental goal of IR spectroscopic study is to identify the chemical functional groups in the material since these groups absorb distinctive IR radiation frequencies. All atoms in molecules constantly vibrate with respect to one another above temperature of absolute zero. When the frequency of a given vibration equals a particular IR radiation frequency, the radiation is absorbed.

IR spectroscopy is a crucial method for identifying and understanding rubber nanocomposites. The FTIR method is utilized in rubber nanocomposites to find the functional groups present in polymers and nanofillers. By comparing the FTIR spectra with the standard vibrational modes of common functional groups it is possible to determine the surface interaction of natural rubber with different nanofillers.

Fourier Transform Infrared Spectroscopy (FTIR) is a direct method which is used to recognize specific reactions taking place within matrix, nanofiller surface, or amongst nanofiller and matrix and thus it can be used to characterize the modification of interface in rubber nanocomposites. Cellulose nanofillers are generally modified by grafting reaction using organic molecules that may contain various functional groups which are capable of molecular entanglements or forming chemical linkages with the polymer matrix. The grafting of organic molecules on nanocellulose exhibits distinct absorption peaks on FTIR.

In a very comprehensive study Agrebi and coworkers investigated the chemical bonding taking place between natural rubber (NR) reinforced with cellulose nanowhiskers (CNW) and nanofibrillated cellulose (NFC) by means of FTIR in order to better understand potential chemical interactions that occur between the reinforcement and the matrix (Agrebi et al. 2019). They analyzed the composites in four spectral regions of 3500–2600, 1800–1500, 1500–1200, and 1200–600 cm^{-1} and reported that when cellulose nanoparticles were added to the NR matrix, several vibrational modes were discovered which were found to be significant indicators of structural changes. The main vibrational modifications were revealed by way of shifts in IR frequency, notable fluctuations in intensity, and disappearance and development of certain IR vibrational modes with respect to a particular amount of reinforcement in matrix. The main changes were observed by way of evolution in the intensity of stretching mode of v(O–H) bond and peaks that corresponded to asymmetric stretching vibration in case of =C–H bond that were ascribed to intramolecular and intermolecular hydrogen bonding between nanocellulose and NR. In addition, these peaks shifted towards higher IR frequencies indicating significant impacts of the cellulose nanofibers through hydrogen bonding on the NR matrix. In general, it was observed that compared to the NR/CNW samples, the NR/NFC were considerably more vulnerable to vibrational shifts depending on the proportion of cellulose in the rubber matrix.

This was mainly ascribed due to the presence of hemicelluloses and residual lignin on NFC surface whereas, CNW surface did not have these constituents and included solely cellulose chains giving it a hydrophilic character which was also confirmed by the appearance of the IR peak corresponding to C=O stretching mode in the IR spectrum for the NR/NFC composite filled with 15% of the filler.

Furthermore, analyzing the intensity ratio of Iv(=C–H)/Iv (C–OH) peaks in the range of 3500–2600 cm^{-1} revealed a drop in this ratio. However, this decrease occurred at different ratios according to the cellulose percentage in NR/CNW and NR/NFC composites, wherein it occurred at 7.5% in case of NR/CNW while this decrease began at lower cellulose percentage for NR/NFC nanocomposites. This phenomenon was explained on the basis of different structural characteristics of the two nanoparticles, particularly due to the presence of hemicellulose and residual lignin on the surface of NFC, imparting it hydrophobic properties. By way of FTIR spectroscopy the authors exhibited that interfacial adhesion in case of NR/NFC nanocomposites was higher as compared to that of the NR/CNW due to the residual lignin presence on

NFC surface, which acted as a compatibilizer to improve and enhance the adhesion between NFC and rubber matrix. Overall, the research was successful in demonstrating that vibrational characteristics of cellulose-rubber nanocomposite materials are affected by the introduction of cellulose nanoparticles in different forms and mass fraction.

Similar results have also been reported by Thomas et al. in their study on natural rubber nanocomposites that were reinforced with nanocellulose extracted from jute fibers (Thomas et al. 2015). They also reported nanocellulose and natural rubber interact through hydrogen bond to produce a network-like structure that limits the motion of rubber chains that are trapped inside the network of cellulose nanofibers. The observed enhancement in mechanical properties of nanocellulose filled nanocomposites was attributed to these chains of immobilized rubber chains.

Surface functionalization of nanocellulose is widely studied in order to improve its compatibility and reinforcing efficiency in the rubber matrix. Kanoth et al. studied the reinforcing efficiency of functionalized cellulose nanocrystals (CNC) in NR (Parambath Kanoth et al. 2015). CNC functionalized by mercapto groups was crosslinked with NR by UV irradiation and this reaction was examined by real time FTIR analysis. The authors were able to show that modified CNC, was not only evenly distributed in the NR matrix but it also promoted covalent crosslinking at the filler matrix interface by way of free radical coupling reaction viz. thiol-ene between the mercapto groups present on CNC surface and the alkene groups of natural rubber. The remarkable enhancement in the mechanical characteristics of the composites was the result of network structure developed in the nanocomposites.

A similar study was also performed by Kato et al. wherein CNF functionalized by sulphur and oleic acid and unsaturated fatty acid chains were used to reinforce natural rubber (Kato et al. 2015). The FTIR analysis demonstrated that both sulphur and unsaturated fatty acid improved the degree of crosslinking in NR which considerably enhanced the mechanical properties including tensile strength and Young's modulus. However, the unsaturated fatty acid modified CNF resulted in a greater improvement in tensile strength due to the additional crosslinking reactions that occurred at the C=C bonds.

2.2.9 NMR Spectroscopy

Nuclear magnetic resonance spectroscopy (NMR) is a technique which analyses the chemical environment of a nucleus with nuclear spin by measurement of its free induction decay by using electromagnetic frequency that is characteristic of the magnetic field of the nucleus. NMR spectroscopy is generally utilized to study both crystalline as well as noncrystalline materials. Atoms in a molecule or a group of similar molecules may be identified by analyzing the peaks from NMR. However, characterizing rubber nanocomposites by NMR is challenging due to the presence of structural disorder and dynamic heterogeneity. Nevertheless, currently solid-state

NMR spectroscopy receives increased attention as it provides valuable information on the dynamic molecular interaction in polymer nanocomposites.

Solid state NMR is generally perceived as a low resolution technique and hence significant efforts has been made over the last few decades to transform it into a fundamental tool for structural as well as surface dynamic analysis of a variety of materials in various physical states (Borisov et al. 2010). Solid state NMR spectra are often characterized by very broad lines and weak intensity that are usually ascribed to interactions of the magnetic fields with the active nuclear spins. In order to suppress and reduce broadness in solid state NMR spectra, several solid state NMR approaches, including newly designed pulse sequences, have been devised (Polenova et al. 2015). Moreover, the amalgamation of solid-state NMR along with magic angle spinning (MAS) proved to be the most successful technique in reducing the dipolar and anisotropic interactions prevalent in solid state NMR.

Both ^{1}H and ^{13}C NMR techniques are used for structural analysis of cellulose-based rubber nanocomposites. Characterization of the main chain units or characterization of the terminal units and crosslinking junctions of the rubbers are the two approaches that are typically used for structural analysis of rubbers. However, because of the physical and chemical interactions that limit the mobility of rubber chains, it is challenging to characterize the terminal and crosslinking junctions by NMR (Kawahara et al. 2019).

Presently nanocellulose is used as a reinforcement in rubber and is largely been investigated from a commercial point of view. It has been established that nanocellulose forms a network kind of structure affecting the mobility of rubber chains and which ultimately influences its reinforcement effect in the rubber matrix. With a view to understand the surface dynamics of the solid–solid interface in nanocomposites, NMR has been used and it provides useful data in terms of the interfacial interactions that occurs at the nanocellulose-rubber interface. NMR can also be used to understand the behavior of the functional groups on the surface of modified nanocellulose and the surface dynamics of nanocellulose in rubber nanocomposites. NMR also provides transverse and longitudinal relaxation times that provide valuable information on the surface dynamics of polymer chains adsorbed on nanofiller surface. In particular, in nanocellulose filled rubber nanocomposites, the transverse nuclear magnetic relaxation time, T2, is influenced by chemical and physical crosslinking on the particle surface which is used in calculating the fraction of polymer chains with limited mobility, on the adsorbed polymer layer that surrounds the filler particles.

Nunes employed 13C NMR which is a high resolution solid state NMR method with magic angle spinning and cross polarization to investigate natural rubber/cellulose-II nanocomposites (Nunes 2017). 1H relaxation periods in the rotating frame (T1p) were determined by means of a conventional approach, which involved a spin lock period of delayed contact time followed by a constant CP time. Figure 2.5 shows that the plot of stress at break versus T1ρH values at 127 ppm for cellulose nanocomposites. The study demonstrated that with 15 phr cellulose-II loading, the nanocomposite exhibited decrease in molecular mobility than unfilled natural rubber, as well as the stress at break is also the highest. These finding were consistent with the cellulosic filler's excellent dispersion all through the rubber matrix, and it indicated

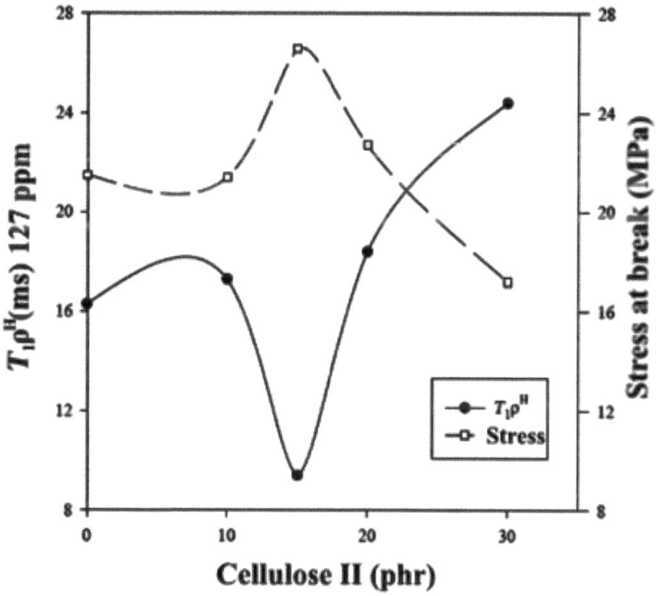

Fig. 2.5 $T_{1p}{}^{H}$ versus break for natural rubber-cellulose II composites (Nunes 2017)

a strong interaction between the rubber and nanofiller creating a network with stiff molecular dynamics. 13C NMR also revealed the presence of a physical interaction between the rubber and nanofiller, which correlated very well with the results of the mechanical study.

Figure 2.6 displays that NMR spectroscopy was used by Ogunsona et al. to study the structure property changes in cured acrylonitrile-butadiene rubber (NBR) that were triggered by cellulose nanocrystals (CNC) (Ogunsona et al. 2020). The change in the crosslink densities in case of NBR and its corresponding nanocomposites filled with 0.5, 1 and 2 phr nanocellulose were observed using H-NMR. They reported that signals approximately at 5 ppm, corresponding to the protons of NBR double bonds, were the most informative (Table 2.1). With the inclusion of CNC, these signals were substantially decreased, suggesting that the percentage of NBR having double bonds declined with the addition of CNC leading to greater efficiency of crosslinking. The double bonds were disassembled during crosslinking to allow short chains to bridge across the rubber chains. Additionally, the decreasing trend in the double to single bond proton ratio indicated a strong correlation between increased crosslinking on increasing CNC loading in the rubber matrix. The study also demonstrated that the combined impact of an enhancement in crosslinking and the presence of CNC resulted in manyfold improvements in mechanical properties viz. toughness tensile strengths and elongation at break.

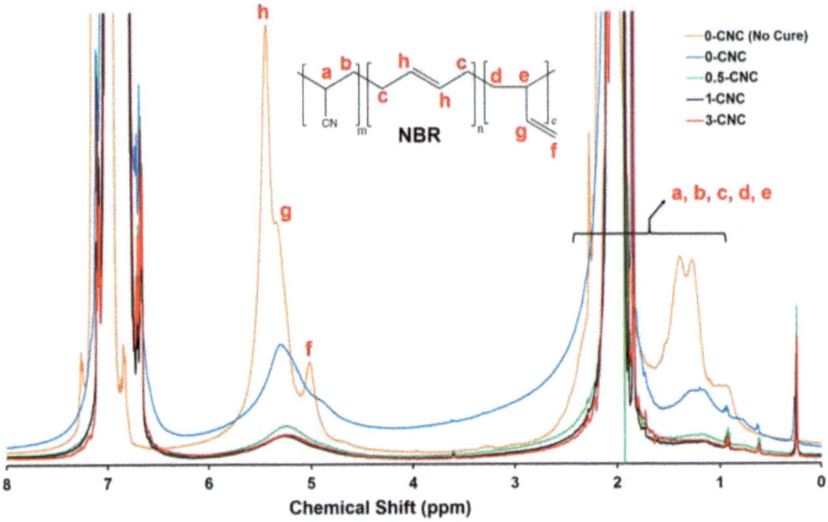

Fig. 2.6 ^{1}H NMR spectra of acrylonitrile-butadiene rubber-CNC composites (Ogunsona et al. 2020)

Table 2.1 Results of NMR quantification (Ogunsona et al. 2020)

CNC (phr)	Double/Single bond protons signal ratio	
	Signal in toluene	Signal in absence of toluene
0 (no cure)	0.110	0.184
0	0.020	0.071
0.5	0.004	0.049
1.5	0.005	0.036
3	0.005	0.041

2.2.10 *Raman Spectroscopy*

Raman spectroscopy is commonly used to examine the structural as well as the chemical characteristics of nanofillers and their composites. In this spectroscopic technique monochromatic light in the form of laser light is generally used to illuminate the sample. The interaction between the vibrational energy levels of the sample and the radiation gives rise to characteristic peaks in Raman Spectrum and its shift is termed as Raman shift. As compared to incoming light, the energies of the Raman scattered light might be lower or greater, and variation in this energy is equivalent to the sample's vibrational energies. Raman spectroscopy is a sophisticated characterization method that examines the inelastic scattering of light across a substance.

The Raman spectrum provides information regarding the chemical composition as well as crystallite features like orientation of crystal and strain in lattice. The spectral characteristics associated with the Raman spectra may be assigned to a variety of physicochemical features of the sample in such a manner, that the distinctive Raman frequency provides information about the material's composition. Intensity and line width of the Raman peak indicates the quality and amount of the constituent materials in the sample, while the polarisation the Raman peak provides indicates orientation and crystal symmetry.

Figure 2.7 shows that in cellulose reinforced nanocomposites the degree of crosslinking may be determined using Raman spectroscopy. Raman spectroscopic studies have been used to assess the crosslink density of the nanocomposites. Recently, Krainoi and Boonkerd used Raman spectroscopy to understand the interaction in natural rubber hybrid nanocomposites reinforced with carbon nanotubes (CNT) and cellulose nanofibers (CNF) in proportion of 1:0.05, 1:0.5, and 1:1 (Krainoi and Boonkerd 2022). The Raman spectrum exhibited two bands: G and D which corresponded to C–C bond's in-plane oscillation and the disorder of the carbon systems, respectively. Raman spectroscopy was able to confirm the interaction at the interface between CNF and CNT as evidenced by an upshift of the D and G bands along with a drop in the ID/IG ratio as the amount of CNF increased. The Raman results were interpreted in terms of noncovalent interaction between CNF and CNT by way of hydrogen bonding.

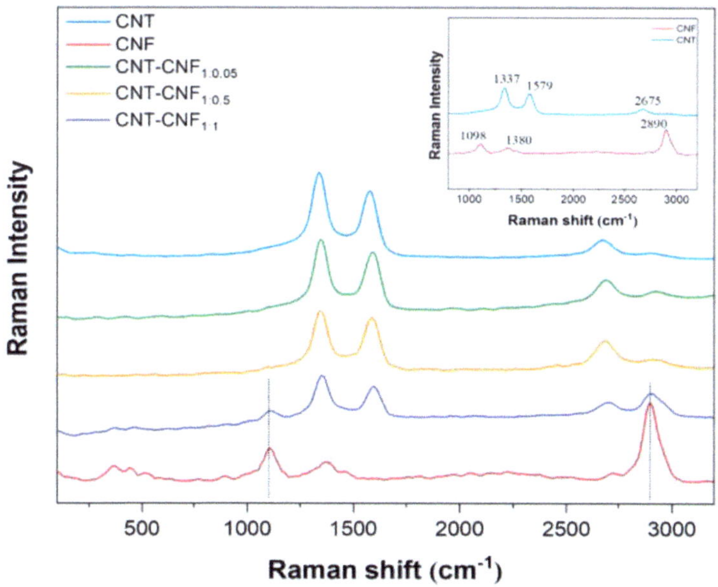

Fig. 2.7 Raman spectra of carbon nanotubes (CNT), cellulose nanofibers and corresponding composites (Krainoi and Boonkerd 2022)

2.2.11 UV–Vis Spectroscopy

UV–Vis spectroscopy is a type of spectroscopy that uses the visible and ultraviolet portions of the electromagnetic spectrum to investigate the materials. When UV–Visible radiations pass through the sample, it absorbs energy from radiation, causing the excitation of electrons from lower molecular orbitals to higher molecular orbitals producing an excited state. The output signal lacks the wavelengths associated with the absorbed energy and as a result, the peaks for the absorbed wavelength are specific. The chromophore can be determined by identifying the corresponding peaks formed by the absorption of certain wavelengths in a graph between absorptivity and wavelength.

Mariano et al. used cellulose nanocrystals to reinforce oxidized natural rubber nanocomposites (Mariano et al. 2016). Figure 2.8 shows that initially natural rubber was subjected to oxididation in order to encourage the introduction of polar hydroxyl groups on the surface of NR chains that would encourage chemical connections among the polar hydroxyl groups of oxidized N R and those of CNC by hydrogen bonding. The degree of oxidation using $KMnO_4$ as oxidant was studied by using UV–Visible spectroscopy. UV–Vis spectra of oxidized natural rubber suspension n and the $KMnO_4$ solution that was used for oxidation is shown in the above figure. $KMnO_4$ oxidizes the double bonds present on natural rubber and in the process gets converted from manganese (VII) to manganese (IV) state. The lack of the manganese (VII) signal at 525 nm in the NR solution suggested that it had been completely consumed during oxidation. The additional hydroxyl groups created on oxidation by permanganate in the natural rubber chain altered the compatibility of rubber matrix with CNC due to potential interfacial interactions in the form of hydrogen bonding at the interface.

2.2.12 EPR or ESR Spectroscopy

Electron spin resonance (ESR), is a spectroscopic method commonly referred to as electron paramagnetic resonance (EPR) that identifies paramagnetic centers having unpaired electrons. It is a powerful technique for examining molecular movements and structural characteristics in a variety of polymers and their nanocomposites. However, like with the bulk of polymers, rubber nanocomposite materials often lack inherent para magnetism and do not produce an ESR spectrum. Nonetheless, this methodology may be used, though indirectly, to analyze the rubber matrix by introducing stable free radicals. ESR spectroscopy can be used to analyze rubber nanocomposites in two ways: (i) the use of metallic ions and other agents present in nanocomposite materials that give an ESR signal and (ii) spin probe or spin label methods. In both situations, the ESR results are related to modifications in material nanostructure. The nitroxides are the most prevalent free radicals utilized for this purpose. Nitroxide radicals have a free electron and occupy the nitrogen atom's *pz*

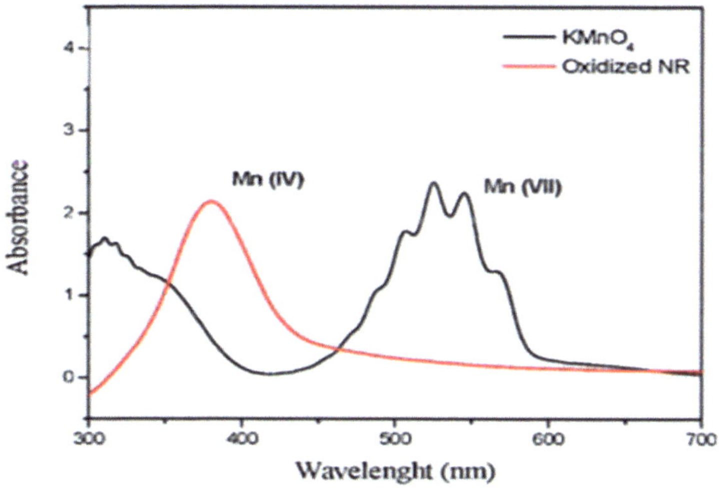

Fig. 2.8 UV–visible spectra of natural rubber oxidized by KMnO4 (Mariano et al. 2016)

orbital. Four substituted methyl groups are frequently present in the nitrogen group's immediate vicinity. These groups sterically shield the paramagnetic centre from potential attacking reagents, which helps to maintain the stability of this radical. Nitroxides can be injected into the polymer matrix by means of diffusion termed as spin probe or connected to polymer chains via covalent bonding termed as the spin label method. The benefit of spin labelling is that the nitroxide position in a rubber chain is exactly known. However, this approach requires additional synthetic steps and this method cannot be used to investigate polymers that do not have an appropriate functional group.

The ESR spectroscopy has been applied for the investigation of nanocellulose, cured and noncured rubber and its nanocomposites but not yet to nanocellulose based rubber nanocomposites. ESR has been successfully used for investigating natural rubber nanocomposites containing silica (Valić 2010), organically modified clay nanoparticles (Wang et al. 2015), organically modified montmorillonite (Ivanoska-Dacikj et al. 2017), multiwalled carbon nanotubes (Neelesh et al. 2020) and grapheme oxide. The common observation in all these studies was the hindrance offered to the molecular mobility of rubber chains in the presence of a nanofiller. This was explained on the basis of an unusually strong connection between the rubber chains and the nanoparticles. Moreover, with higher filler loadings, nanoparticles agglomerated, resulting in a decreased contact between the nanofiller and the rubber matrix which also resulted in lowering the mechanical properties.

2.2.13 X-Ray Diffraction Analysis

Rubbers or elastomers primarily due to their amorphous nature do not appear to be particularly amenable to X-ray diffraction studies. However, it should be remembered that many rubbers can indeed crystallize when stretched. Furthermore, the production of rubber-based nanocomposites involves the addition of fillers that are crystalline or produce materials of varying crystallinity on addition to elastomeric matrices. The use of nanocellulose largely investigated for reinforcing rubbers has necessitated the development of methodologies that can precisely describe the degree of nanocellulose dispersion in the rubber matrix and invariably the structural morphology of the resulting composites. X-ray scattering investigations in the form of wide-angle X-ray diffraction (WAXD) and small angle X-ray scattering (SAXS) can contribute to understand these aspects and provide for a highly detailed picture of the structural morphology of cellulose-based rubber nanocomposites. These techniques have an especially intriguing benefit in that they can scan the entire sample providing a more comprehensive image of its morphology that helps in understanding of the ultimate performance of the nanocomposites.

Generally, the physicochemical properties of rubber nanocomposites are directly affected by the nanofillers specific crystal structure and crystallite plane orientation, which promotes crystallization in matrix. XRD analysis may be used to determine the nanofillers crystal structure along with average crystallite size. Particularly, it is possible to infer the fillers nanodimensional characteristics from their geometrical behavior. The width of the peaks in XRD patterns is used to calculate the average crystallite size, wherein broadening of peaks is a direct indication of reduction of crystallite size.

The crystallization in natural rubber during deformation, known as strain-induced crystallization is responsible for crystallites formation in the rubber network that increases its strength by providing rubber with a self-reinforcing characteristic is a well-established theory (Masa et al. 2021). In addition, common nanofillers such as silica, nanoclay, carbon nanotubes and carbon black, were also found to stimulate early crystallization and increase total crystallization of rubber under uniaxial deformation. Nunes and co-workers were the first to study the structure–property relationship of natural rubber nanocomposites filled with *cellulose II* in the range of 0–30 phr (Nunes et al. 2004). It assessed the structure of natural rubber and its nanocomposites during tensile testing at various elongations by WAXD. The results of the WAXD study revealed that in natural rubber the strain induced crystallization occurred at elongations above 300% and the amorphous halo persisted even at large strains indicating that most of the polymer chains could not orient to form crystallites. However, in the nanocomposites the presence of nanocellulose promoted the strain-induced crystallization that was apparent from the rise in the intensity of peaks and suppression of amorphous halo with the optimum structural arrangement observed for 15 phr of nanocellulose in WAXD patterns. The results of WAXD study were also in alignment with mechanical properties with the 15 phrnanocomposites exhibiting maximum tensile strength.

Figure 2.9 shows that in yet another recent interesting study Wongvasana and co-workers analyzed the reinforcement mechanism of chemically unmodified cellulose nanofibers in natural rubber nanocomposites (Wongvasana et al. 2023). They studied rubber composites filled with 1, 3 and 5 phr cellulose nanofibers and tried to correlate the structure property relationship by using TEM, tensile testing, WAXD and dynamic mechanical analysis. WAXD was used to analyze the effects of cellulose nanofibers on the structure of rubber network and also to evaluate degree of crystallinity X_c. The X_c for rubber and 1 and 3 phr nanocomposites were comparable and observed at strain of 300% whereas for 5 phr nanocomposite the crystallization onset started at 175% strain value but the sample failed at 300% strain. WAXD results of the study revealed that at 1 and 3 phr the cellulose nanofibers did not accelerate strain induced crystallization but even though 122% enhancement in tensile strength was observed which was attributed to their uniform dispersion and shear stress transfer at the cellulose nanofibers and rubber interface. However, 5 phr cellulose nanofibers loading promoted strain-induced crystallization in nanocomposites but substantially reduced the strain and stress on account of aggregation of nanofibers which acted as local stress concentration points.

In the literature it is well acknowledged that the tendency of crosslinked natural rubber to exhibit strain induced crystallization is the primary reason for its superior properties and fracture resistance. The strain-induced crystallization process in case of vulcanized natural rubber during deformation has been well explained on the basis of X-ray diffraction study which postulated that stretched rubber chains undergo

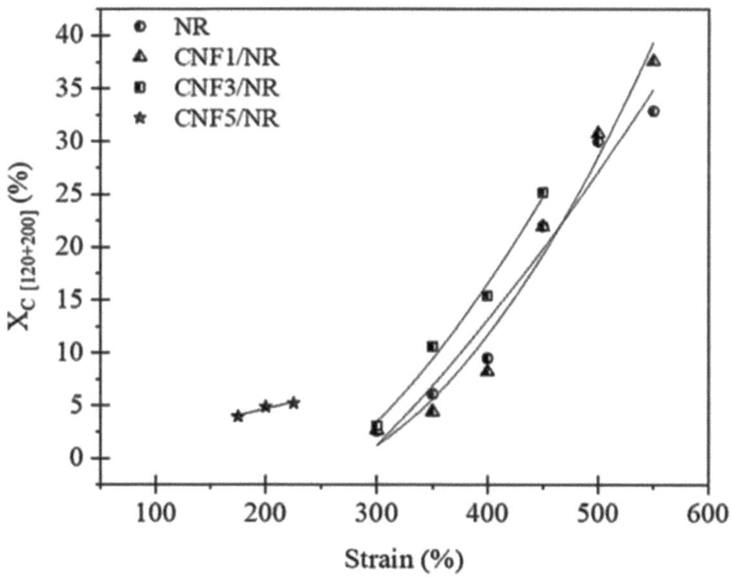

Fig. 2.9 Degree of crystallinity (X_c) versus applied strain graph of natural rubber and natural rubber/CNF composites (Wongvasana et al. 2023)

molecular alignment between dense network points, which eventually produced crystallites (Sainumsai et al. 2017). Besides this, it has been reported that the addition of the nanofillers to natural rubber changes its crystallization behavior under tensile stretching. This structural property relationship was addressed in research by Wongvasana and colleagues that compared nanoclay (NC) and cellulose nanofibers (CNF) reinforced natural rubber nanocomposites (Wongvasana et al. 2022). The study demonstrated that reinforcement effect of CNF was distinct from that NC on the basis of WAXD observations and developed a model explaining the reinforcement effect of CNF. The high values of degree of crystallinity at low strain levels evaluated from WAXD results exhibited that CNF did accelerate strain stimulated crystallization in natural rubber due to the mutual entanglement of CNF aggregate and natural rubber, but at the same time immobilized natural rubber chains that resulted in local stress concentration sites and eventual failure of composites.

2.2.14 Neutron Scattering

Neutron scattering probes the elastic scattering by neutrons in small angles to analyze structures and aggregates/inhomogeneities with dimensions between 1 and 1000 nanometers. As the scattering angles investigated in this technique are much smaller than XRD it is termed as small angle neutron scattering (SANS). Most investigations are based on the difference in scattering length densities between isotopes, more specifically between hydrogen and deuterium atoms. Neutron scattering is used to study the structure of various polymers including polymer chain dimensions, conformations, miscibility of blends, structures of block copolymers, semicrystalline polymers and biomacromolecules. The unique property of neutrality makes neutron an excellent probing tool for nanocomposites. Interactions of neutrons with matter can produce both elastic and inelastic scattering.

Inelastic neutron scattering experiments used for studying the structure and shape of cellulose nanowalkers have demonstrated that major disordered chains are present on the surface but they maintain a preferential parallel orientation to the microfiber axis. SANS method is also used for determining the dispersion of the nanocellulose into the matrix. Chazeau utilized this approach to analyze the dispersion behavior of tunicate whiskers in plasticized poly(vinyl chloride) and reported an isotropic dispersion without the formation of any aggregates (Chazeau et al. 2000). Yamauchi et al. used SANS to study the structure of natural rubber thermoplastic elastomer composites filled with carbon black (Yamauchi et al. 2005). Besides the structure, the authors successfully evaluated the thickness of interface of the blends and composite by sing SANS. Since the blends' measured thickness values ranged between 5 and 6 nm, it indicated that NR and HDPE were strongly immiscible, whereas in composites interface thickness was found to be 2.4 nm that corresponded to a layer of rubber bound on carbon black. The study demonstrated that SANS can be was very helpful in determining the thickness of the interface in case of the natural rubber and thermoplastic elastomers and its miscibility with polyethylene.

2.2.15 Rheology Measurements

For over a century, it has been well understood that addition of fillers into rubber composites has a significant impact on the overall properties of the materials. From a rheological standpoint a major change in the behavior of constant shear viscosity and viscoelastic characteristics are a direct result of the inclusion of nanofillers in rubber. One such phenomenon is the Payne effect which is related to the dependence of the dynamic viscoelastic properties on the strain in filled polymers. In general, rubbers containing nanoparticles exhibit behavior similar to that of solids, which comprises a nonterminal zone of relaxation and a viscosity dependency on shear thinning, with respect to concentration of nanofillers and their dispersion. The presence of a rubber chain network structure causes this rheological behavior.

Rheology offers unique ways to evaluate the nature of dispersion of nano-fillers in nanocomposites. It offers a unique way and to analyze the effects of dispersion of nanofillers on flow conditions as the melt rheological properties of polymer nanocomposites are sensitive not only to the concentration, structure and particle size but also to the surface characteristics of the fillers. Nanofillers are generally prone to aggregation and thus easily form three dimensional networks in molten rubber matrix on account of their small particle size and high specific surface area. In fact, the nature of the developed network structure, polymer-filler interaction, has fuelled the contentious debate over the Payne effect.

Figure 2.10 shows that the rheology study of rubber filled with nanocellulose can help to understand and quantify the matrix-filler interaction. Abraham et al. investigated the rheological properties of natural rubber reinforced with 0, 2.5, 5, 7.5 and 10 wt% of nanocellulose (Abraham et al. 2013). The effect of nanocellulose on the rheological characteristics of natural rubber was investigated by examining the variation in viscosity as a function of frequency as shown in above figure. The degree of compatibility between the nanocellulose and the rubber matrix determines nanocomposites' flow. It was reported that with an increase in the cellulose weight %, the nanocomposite's viscosity increased indicating that the viscosity of the resulting composite was significantly influenced by the interface between the nanocellulose and the rubber matrix. The study proposed the formation of a strong interconnecting network structure between the crosslinked rubber and three-dimensional network of Zn-cellulose on vulcanization that resulted in the formation of cavities. These cavities were crucial in the immobilisation of the trapped rubber chains and, as a result, altered the overall viscosity properties of the nanocomposites. The level of interfacial interaction between the nanofiller and polymer increased as the % of filler increased. As a result, the viscosity of the nanocomposite increased as the nanofiller content increased, and this increase in viscosity was attributable to the twofold network established during the pre-vulcanization stage between cellulose and cellulose and rubber-rubber.

In a recent study Yasin et al. investigated the impact of an ionic liquid ([EMIM]OAc) on cellulose nanofiber (CNF) and CNF filled natural rubber (NR) composites and vulcanizates (Yasin et al. 2021). The study exhibited that the presence

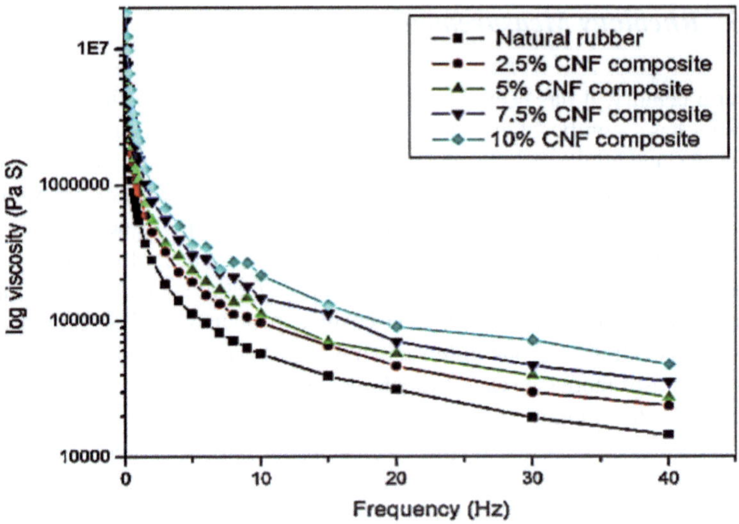

Fig. 2.10 Frequency versus log viscosity of natural rubber/nanocellulose composites with varying % of nanocellulose (Abraham et al. 2013)

of the ionic liquid enhanced the dispersion of cellulose nanofillers in rubber matrix and boosted the density of crosslinking and thereby enhanced the mechanical strength of the vulcanized natural rubber. The researchers also evaluated the viscoelastic rheological behavior of nanocellulose and NR composites to that of vulcanizates. The study demonstrated that with increasing % of cellulose nanofillers and addition of ionic liquid, the magnitude of the Payne effect reduced in the composites but rose in the vulcanized compounds.

2.3 Conclusion

NR is the gift of nature having excellent physical properties and is most suitable for industrial applications. However, the major limitations of NR for its extensive application in industry are low strength and modulus. The addition of a reinforcing agent significantly improves the mechanical property. Recently, the surface modified nano-cellulose has gained much attention as the reinforcing agent in rubber composites due to multi-fold improvement of the mechanical property. The different forms of nanocellulose used in the composites are nanocrystals, nanofibrils, nanowhiskers, bacterial cellulose etc. In this chapter, the characterization of cellulose-based rubber nanocomposites for different mechanical properties has been discussed. The cellulose-based nanocomposites have variety of industrial applications that include paper and pulp, paint and adhesive, filtration, electronic sensors, food, hygiene, cosmetic and medical products. It is obvious that the type of nanomaterial used and the processing technique adopted decides the mechanical property of composites.

References

Abraham E, Thomas MS, John C et al (2013) Green nanocomposites of natural rubber/nanocellulose: membrane transport, rheological and thermal degradation characterisations. Ind Crops Prod 51:415–424

Agrebi F, Ghorbel N, Bresson S et al (2019) Study of nanocomposites based on cellulose nanoparticles and natural rubber latex by ATR/FTIR spectroscopy: the impact of reinforcement. Polym Compos 40:2076–2087

Ahmed KS, Vijayarangan S (2008) Tensile, flexural and interlaminar shear properties of woven jute and jute-glass fabric reinforced polyester composites. J Mater Process Technol 207:330–335

Anuar H, Zuraida A (2011) Improvement in mechanical properties of reinforced thermoplastic elastomer composite with kenaf bast fibre. Compos B Eng 42:462–465

Arof AK, Nor NAM, Aziz N et al (2019) Investigation on morphology of composite poly(ethylene oxide)-cellulose nanofibers. Mater Today: Proc 17:388–393

Bakošová D (2018) Dynamic mechanical analysis of rubber mixtures filled by carbon nanotubes. Manuf Technol 18:345–351

Bashir MA (2021) Use of dynamic mechanical analysis (DMA) for characterizing interfacial interactions in filled polymers. Solids 2:108–120

Bessa J, Matos J, Mota C et al (2017) Influence of surface treatments on the mechanical properties of fibre reinforced thermoplastic composites. Procedia Eng 200:465–471

Bhambure SS, Rao AS, Senthilkumar T (2022) Characterization of control and chemically modified Kenaf fiber. J Nat Fibers 19:10320–10330

Bhambure SS, Rao AS, Senthilkumar T (2023) Analysis of mechanical properties of kenaf and kapok fiber reinforced hybrid polyester composite. J Nat Fibers 20:2156964

Bisht A, Srivastava M, Kumar RM et al (2017) Strengthening mechanism in graphene nanoplatelets reinforced aluminum composite fabricated through spark plasma sintering. Mater Sci Eng A 695:20–28

Bitinis N, Hernández M, Verdejo R et al (2011) Recent advances in clay/polymer nanocomposites. Adv Mater 23:5229–5236

Boominathan SK, Amutha V, Senthamaraikannan P et al (2022) Investigation of mechanical, thermal, and moisture diffusion behavior of Acacia concinna fiber/polyester matrix composite. J Nat Fibers 19:13495–13510

Borisov AS, Hazendonk P, Hayes PG (2010) Solid-state nuclear magnetic resonance spectroscopy: a review of modern techniques and applications for inorganic polymers. J Inorg Organomet Polym Mater 20:183–212

Castro C, Zuluaga R, Álvarez C et al (2012) Bacterial cellulose produced by a new acid-resistant strain of Gluconacetobacter genus. Carbohydr Polym 89:1033–1037

Chazeau L, Cavaille JY, Perez J (2000) Plasticized PVC reinforced with cellulose whiskers. II. Plastic behavior. J Polym Sci Part B: Polym Phys 38:383–392

Dickens B (1980) Thermal degradation and oxidation of polystyrene studied by factor-jump thermogravimetry. Polym Degrad Stab 2:249–268

Dittanet P, Somphol W, Lampang TN et al (2019) Natural rubber reinforced by nanocellulose extracted from dried rubber leaves. In: AIP conference proceedings. AIP Publishing LLC, p 30008

Dufresne A (2018) Cellulose nanomaterials as green nanoreinforcements for polymer nanocomposites. Philos Trans R Soc A: Math Phys Eng Sci 376:20170040

Ertugrul I, Ulkir O, Ersoy S, Ragulskis M (2023) Additive manufactured strain sensor using stereolithography method with photopolymer material. Polymers 15:991

Gopinath R, Billigraham P, Sathishkumar TP (2021) Investigation of physico-chemical, mechanical, and thermal properties of new cellulosic bast fiber extracted from the bark of Bauhinia purpurea. J Nat Fibers. https://doi.org/10.1080/15440478.2021.1990180

Govindarajan PR, Shanmugavel R, Subramanian K et al (2024) Effect of stacking sequence on mechanical and water absorption characteristics of jute/banana/basalt fabric aluminium fibre laminates with diamond microexpanded mesh. Int J Polym Sci 2024:3835788

Gurukarthik Babu B, Prince Winston D, SenthamaraiKannan P et al (2019) Study on characterization and physicochemical properties of new natural fiber from Phaseolus vulgaris. J Nat Fibers 16:1035–1042

Hakimi NMF, Lee SH, Lum WC et al (2021) Surface modified nanocellulose and its reinforcement in natural rubber matrix nanocomposites: a review. Polymers 13:3241

Hargitai H, Rácz I, Anandjiwala RD (2008) Development of hemp fiber reinforced polypropylene composites. J Thermoplast Compos Mater 21:165–174

Hashemifard SA, Matsuura T, Ismail AF et al (2015) Characterization of partial pore wetting in hollow fiber gas absorption membrane contactors: an EDX analysis approach. Chem Eng J 281:970–980

He X, Deng H, Hwang H (2019) La aplicación actual de la nanotecnología en la alimentación y la agricultura. J Food Drug Anal 27:1–21

Ilangovan M, Guna V, Prajwal B et al (2020) Extraction and characterisation of natural cellulose fibers from Kigelia africana. Carbohydr Polym 236:115996

Intapun J, Rungruang T, Suchat S et al (2021) The characteristics of natural rubber composites with Klason lignin as a green reinforcing filler: thermal stability, mechanical and dynamical properties. Polymers 13:1109

Ivanoska-Dacikj A, Bogoeva-Gaceva G, Valić S et al (2017) Benefits of hybrid nano-filler networking between organically modified montmorillonite and carbon nanotubes in natural rubber: experiments and theoretical interpretations. Appl Clay Sci 136:192–198

Iyyadurai J, Arockiasamy FS, Manickam T et al (2023) Experimental investigation on mechanical, thermal, viscoelastic, water absorption, and biodegradability behavior of Sansevieria ehrenbergii fiber reinforced novel polymeric composite with the addition of coconut shell ash powder. J Inorg Organomet Polym Mater 33:796–809

Jose T, Moni G, Salini S et al (2017) Multifunctional multi-walled carbon nanotube reinforced natural rubber nanocomposites. Ind Crops Prod 105:63–73

Karthik A, Bhuvaneshwaran M, Senthil Kumar MS et al (2024) A review on surface modification of plant fibers for enhancing properties of biocomposites. ChemistrySelect 9:e202400650

Kato H, Nakatsubo F, Abe K, Yano H (2015) Crosslinking via sulfur vulcanization of natural rubber and cellulose nanofibers incorporating unsaturated fatty acids. RSC Adv 5:29814–29819

Kawahara S, Yamamoto Y, Saito T (2019) Rubber-state NMR Spectroscopy. NMR Methods Character Synth Nat Polym 20:156

Kim H, Lee JA, Ambulo CP et al (2019) Intelligently actuating liquid crystal elastomer-carbon nanotube composites. Adv Func Mater 29:1905063

Krainoi A, Boonkerd K (2022) Novel hybrid natural rubber nanocomposites with carbon nanotube and cellulose nanofiber for strain-sensitive sensor. Ind Crops Prod 187:115455

Ku H, Wang H, Pattarachaiyakoop N, Trada M (2011) A review on the tensile properties of natural fiber reinforced polymer composites. Compos Part B: Eng 42:856–873. https://doi.org/10.1016/j.compositesb.2011.01.010

Kumar R, Sivaganesan S, Senthamaraikannan P et al (2022) Characterization of new cellulosic fiber from the Bark of Acacia nilotica L. plant. J Nat Fibers 19:199–208. https://doi.org/10.1080/15440478.2020.1738305

Li H, Wang H, Wu Y et al (2013) Preparation and characterization of poly(dimethylsiloxane) grafted carbon nanotubes/silicone rubber nanocomposites. Sci Adv Mater 5:107–115

Liu J, Lu X, Wu C, Zhao C (2013) Effect of preparation conditions on the morphology, polymorphism and mechanical properties of polyvinylidene fluoride membranes formed via thermally induced phase separation. J Polym Res 20:1–10
Ma P-C, Siddiqui NA, Marom G, Kim J-K (2010) Dispersion and functionalization of carbon nanotubes for polymer-based nanocomposites: a review. Compos A Appl Sci Manuf 41:1345–1367
Mageshwaran V, Sivasubramanian P, Kumar P, Nagaraju Y (2023) Antibacterial response of nanostructured chitosan hybrid materials. Chitosan Nanocompos Bionanomech Appl 161–179
Maiti M, Basak GC, Srivastava VK, Jasra RV (2016) Mesoporous silica reinforced polybutadiene rubber hybrid composite. Int J Ind Chem 7:131–141
Mamanpush SH, Tabatabaei AT (2019) Dynamic thermogravimetric analysis of recycled carbon fiber composites: mini review
Manimaran P, Senthamaraikannan P, Sanjay MR et al (2018) Study on characterization of Furcraea foetida new natural fiber as composite reinforcement for lightweight applications. Carbohydr Polym 181:650–658
Mariano M, El Kissi N, Dufresne A (2016) Cellulose nanocrystal reinforced oxidized natural rubber nanocomposites. Carbohydr Polym 137:174–183
Masa A, Hayeemasae N, Soontaranon S et al (2021) Effect of stretching rate on tensile response and crystallization behavior of crosslinked natural rubber. Malays J Fundam Appl Sci 17:217–225
Mat Ali A, Ahmad SH (2016) Thermal analysis of polylactic acid/liquid natural rubber reinforced with multi-walled carbon nanotubes. In: Advanced materials research. Trans Tech Publ, pp 481–485
Mohanasundaram S, Sivakumar M (2021) Thermal properties of natural fiber reinforced hybrid composites. Int J Eng Res Technol 9:125–128
Motiee F, Taghvaei-Ganjali S, Malekzadeh M (2013) Investigation of correlation between rheological properties of rubber compounds based on natural rubber/styrene-butadiene rubber with their thermal behaviors. Int J Ind Chem 4:1–8
Nam TH, Ogihara S, Kobayashi S, Goto K (2015) Effects of surface treatment on mechanical and thermal properties of jute fabric-reinforced poly(butylene succinate) biodegradable composites. Adv Compos Mater 24:161–178
Nechyporchuk O, Pignon F, Do Rego AMB, Belgacem MN (2016) Influence of ionic interactions between nanofibrillated cellulose and latex on the ensuing composite properties. Compos B Eng 85:188–195
Neelesh A, Vidhyashree S, Meera B (2020) The influence of MWCNT and hybrid (MWCNT/nanoclay) fillers on performance of EPDM-CIIR blends in nuclear applications: mechanical, hydrocarbon transport, and gamma-radiation aging characteristics. J Appl Polym Sci 137:49271
Noguchi T, Niihara K, Kurashima A et al (2021) Cellulose nanofiber-reinforced rubber composites prepared by TEMPO-functionalization and elastic kneading. Compos Sci Technol 210:108815
Nunes RCR (2017) Rubber nanocomposites with nanocellulose. In: Progress in rubber nanocomposites. Elsevier, pp 463–494
Nunes RCR, Martins AF, Visconte LLY et al (2004) Nanocomposites of natural rubber and cellulose II: comparison between different processes of preparation. J Rubber Res 7:1–13
Ogunsona E, Hojabr S, Berry R, Mekonnen TH (2020) Nanocellulose-triggered structural and property changes of acrylonitrile-butadiene rubber films. Int J Biol Macromol 164:2038–2050
Palaniappan M, Palanisamy S, Mani Murugesan T et al (2024) Influence of washing with sodium lauryl sulphate (SLS) surfactant on different properties of ramie fibres. BioResources 19:2609–2625
Palanisamy S, Mayandi K, Palaniappan M et al (2021) Mechanical properties of phormium tenax reinforced natural rubber composites. Fibers 9:11
Palanisamy S, Kalimuthu M, Palaniappan M et al (2022) Characterization of Acacia caesia bark fibers (ACBFs). J Nat Fibers 19:10241–10252

Palanisamy S, Kalimuthu M, Santulli C et al (2023) Tailoring epoxy composites with Acacia caesia bark fibers: evaluating the effects of fiber amount and length on material characteristics. Fibers 11:63

Parambath Kanoth B, Claudino M, Johansson M et al (2015) Biocomposites from natural rubber: synergistic effects of functionalized cellulose nanocrystals as both reinforcing and cross-linking agents via free-radical thiol–ene chemistry. ACS Appl Mater Interfaces 7:16303–16310

Polenova T, Gupta R, Goldbourt A (2015) Magic angle spinning NMR spectroscopy: a versatile technique for structural and dynamic analysis of solid-phase systems

Pradhan B, Roy S, Srivastava SK, Saxena A (2015) Synergistic effect of carbon nanotubes and clay platelets in reinforcing properties of silicone rubber nanocomposites. J Appl Polym Sci 132

Rajesh C, Divia P, Dinooplal S et al (2021) Dynamic mechanical analysis of nylon 6 fiber-reinforced acrylonitrile butadiene rubber composites. Polym Polym Compos 29:S1328–S1339

Ramakrishna S, Hull D (1994) Tensile behaviour of knitted carbon-fibre-fabric/epoxy laminates—Part II: Prediction of tensile properties. Compos Sci Technol 50:249–258

Ravindran D, Sundara Bharathi SR, Indran S (2020) Characterization of surface-modified natural cellulosic fiber extracted from the root of Ficus religiosa tree. Int J Biol Macromol 156:997–1006

Sainumsai W, Toki S, Amnuaypornsri S et al (2017) Dependence of the onset of strain-induced crystallization of natural rubber and its synthetic analogue on crosslink and entanglement by using synchrotron X-ray. Rubber Chem Technol 90:728–742

Salimi S, Sotudeh-Gharebagh R, Zarghami R et al (2019) Production of nanocellulose and its applications in drug delivery: a critical review. ACS Sustain Chem Eng 7:15800–15827

Senthilkumar T, Bharimalla AK, Sundaramoorthy C et al (2020) Fibrillation of coconut fibers by mechanical refining to enhance its reinforcing potential in epoxy composites. Fibers Polym 21:2111–2117

Sharma A, Thakur M, Bhattacharya M et al (2019) Commercial application of cellulose nanocomposites—a review. Biotechnol Rep 21:e00316

Sharma P, Mittal M, Yadav A, Aggarwal NK (2023) Bacterial cellulose: nano-biomaterial for biodegradable face masks—a greener approach towards environment. Environ Nanotechnol Monit Manag 19:100759

Singh S, Dhakar GL, Kapgate BP et al (2020) Synthesis and chemical modification of crystalline nanocellulose to reinforce natural rubber composites. Polym Adv Technol 31:3059–3069

Sivasubramanian P, Mayandi K, Santulli C et al (2022) Effect of fiber length on curing and mechanical behavior of pineapple leaf fiber (PALF) reinforced natural rubber composites. J Nat Fibers 19:4326–4337

Song Y, Yu J, Yu L et al (2015) Enhancing the thermal, electrical, and mechanical properties of silicone rubber by addition of graphene nanoplatelets. Mater Des 88:950–957

Srisuwan S, Prasoetsopha N, Suppakarn N (2014) The effects of alkalized and silanized woven sisal fibers on mechanical properties of natural rubber modified epoxy resin. Energy Procedia 56:19–25. https://doi.org/10.1016/j.egypro.2014.07.127

Srivastava SK, Mittal V (2017) Recent developments on elastomer/hybrid filler nanocomposites. In: Srivastava SK, Mittal V (eds) Hybrid nanomaterials: developments in energy, environments and polymer nanocomposites, pp 423–490

Sumesh KR, Palanisamy S, Khan T et al (2024) Mechanical, morphological and wear resistance of natural fiber/glass fiber-based polymer composites. BioResources 19:3271–3289

Thomas MG, Abraham E, Jyotishkumar P et al (2015) Nanocelluloses from jute fibers and their nanocomposites with natural rubber: preparation and characterization. Int J Biol Macromol 81:768–777

Trache D (2018) Nanocellulose as a promising sustainable material for biomedical applications. AIMS Mater Sci 5:201–205

Valić S (2010) Electron spin resonance in studying nanocomposite rubber materials. In: Rubber nanocomposites: preparation, properties, and applications, pp 391–406

Vijay R, Menon ARR (2022) Reinforcing effect of organo-modified fillers in rubber composites as evidenced from DMA studies. J Mater Sci Eng A 12:13–23

Visakh PM, Thomas S, Oksman K, Mathew AP (2012) Crosslinked natural rubber nanocomposites reinforced with cellulose whiskers isolated from bamboo waste: processing and mechanical/thermal properties. Compos A Appl Sci Manuf 43:735–741

Wang L, Hu L, Gao S et al (2015) Bio-inspired polydopamine-coated clay and its thermo-oxidative stabilization mechanism for styrene butadiene rubber. RSC Adv 5:9314–9324

Wongvasana B, Thongnuanchan B, Masa A et al (2022) Comparative structure-property relationship between nanoclay and cellulose nanofiber reinforced natural rubber nanocomposites. Polymers 14:3747

Wongvasana B, Thongnuanchan B, Masa A et al (2023) Reinforcement behavior of chemically unmodified cellulose nanofiber in natural rubber nanocomposites. Polymers 15:1274

Wu W, Huang H (2020) Silicone rubber composites fabricated using KH550-modified poplar leaves graphene. Bull Mater Sci 43:16

Xu SH, Gu J, Luo YF, Jia DM (2012) Effects of partial replacement of silica with surface modified nanocrystalline cellulose on properties of natural rubber nanocomposites. Express Polym Lett 6

Yamauchi K, Akasaka S, Hasegawa H et al (2005) Structural study of natural rubber thermoplastic elastomers and their composites with carbon black by small-angle neutron scattering and transmission electron microscopy. Compos A Appl Sci Manuf 36:423–429

Yasin S, Hussain M, Zheng Q, Song Y (2021) Effects of ionic liquid on cellulosic nanofiller filled natural rubber bionanocomposites. J Colloid Interface Sci 591:409–417

Zhang G, Wang F, Dai J, Huang Z (2016) Effect of functionalization of graphene nanoplatelets on the mechanical and thermal properties of silicone rubber composites. Materials 9:92

Chapter 3
Cellulose-Based High-Performance Rubber Micro- or Nano-composites

Sohail Yasin, Wang Zhongzhen, Asif Mahmood, Munir Hussain, Li Yang, Chen Qi, MaioMiao Yang, Nasir M. Ahmad, Song Yihu, and Jianfeng Shi

Abstract This chapter discusses the process of preparation, various applications, and the potential uses of cellulose-reinforced rubber micro- and nanocomposites. Cellulose has recently been recognized for its efficiency as a reinforcement filler and has gained prominence across diverse applications, particularly in material engineering and science. As a polysaccharide abundant in plant cell walls, cellulose-filled rubber composites as environmentally friendly due to more eco-conscious alternatives to traditional fillers such as silica and carbon black and provide vast applications for polymer products. This aligns with the growing emphasis on sustainable practices and materials in various industries, reflecting a broader societal shift toward environmentally responsible choices. The chapter centers on the general preparation process and modification of rubbery materials through the incorporation of cellulose,

S. Yasin · L. Yang
Hydrogen Energy Institute, College of Energy Engineering, Zhejiang University, Hangzhou 310000, China

S. Yasin (✉) · W. Zhongzhen · C. Qi · M. Yang · J. Shi (✉)
Institute of Process Equipment, Zhejiang University, Hangzhou 310027, China
e-mail: soh.yasin@gmail.com

J. Shi
e-mail: shijianfeng@zju.edu.cn

A. Mahmood
School of Mathematical and Physical Sciences, Faculty of Science, Center for Clean Energy Technology, University of Technology Sydney, Sydney, Australia

M. Hussain · S. Yihu (✉)
Department of Polymer Science and Engineering, Zhejiang University, Hangzhou 310027, China
e-mail: s_yh0411@zju.edu.cn

N. M. Ahmad
School of Chemical and Materials Engineering, National University of Sciences and Technology, Islamabad 44000, Pakistan

M. Hussain
College of Textile Science and Engineering, Zhejiang Sci-Tech University, Hangzhou 310018, China

© The Author(s), under exclusive license to Springer Nature Switzerland AG 2025
P. M. Visakh (ed.), *Rubber Based Bionanocomposites*, Advanced Structured Materials 227, https://doi.org/10.1007/978-3-031-78557-3_3

resulting in enhanced properties such as morphological, viscoelastic, mechanical, and biodegradability.

3.1 Introduction to the Cellulose-Based Rubber Composites

Enhancement of the mechanical properties of rubbers has always been a topic of significant interest, and nanofillers have served as a promising avenue for achieving this goal (Song and Zheng 2016; Yasin et al. 2021a). Even in relatively small quantities, nanofillers can bring about crucial modifications in macroscopic behaviors (Yasin et al. 2024a), with their effectiveness influenced by factors such as topology, surface nature, and dispersion within the rubber matrix (Song and Zheng 2016; Hussain et al. 2022a). Traditionally, carbon black (CB) has been the primary reinforcing nanofiller in the rubber industry due to its excellent reinforcing properties and cost efficiency. However, the extensive use of CB has raised significant environmental concerns, due to its sourcing from nonrenewable petroleum, which contributes to greenhouse gas emissions during combustion and increases its carbon footprint. CB is known to be carcinogenic (Yasin et al. 2022a), consequently, various alternative reinforcing nanofillers have emerged, including clay, lignin, and non-black metal oxides, aiming to address these environmental issues. Among these alternatives, silica (Sil) stands out as a promising option among inorganic oxides, having successfully replaced CB in various applications. Sil offers enhanced resistance to cutting, tear strength, chip resistance, low rolling resistance, and reduced heat buildup in rubber compounds when properly incorporated (Singh et al. 2024).

Substitution of CB with Sil as a reinforcing nanofiller is a strategic choice aimed at reducing rolling resistance, leveraging the polar groups present on the surface of Sil (Hussain et al. 2019). The environmental implications of Sil production and the substantial use of fuel resources for CB have triggered a pressing need for the rubber industry to explore better biodegradable nanofillers (Yasin et al. 2021b). Some of the unique characteristics of cellulose biomolecules are displayed in Fig. 3.1. Among the range of cellulose-based products, cellulose nanocrystals (CNC) and cellulose nanofibrils (CNF) have emerged as promising candidates for reinforcing rubbers (Mariano et al. 2014, 2016). Various incorporation techniques, including melt compounding, coagulation of rubber latex, and masterbatch techniques, have been employed to integrate cellulosic nanofillers into rubbers (Nunes 2014).

Rubber products primarily rely on natural rubber (NR) extracted from latex, making latex mixing a preferred method over in situ emulsion polymerization. Achieving a stable latex aqueous dispersion of fillers and additives remains a challenge despite advancements in latex blending technology, typically while incorporating nanofillers. The use of cellulosic nanofillers for elastomers is a relatively new area of research, as NR has traditionally been the preferred choice for investigating their characteristics. NR is a type of polyisoprene derived from latex. In its uncured state, it exhibits weak mechanical properties and limited uses. However, vulcanization enhances these characteristics by crosslinking the long chains of rubber

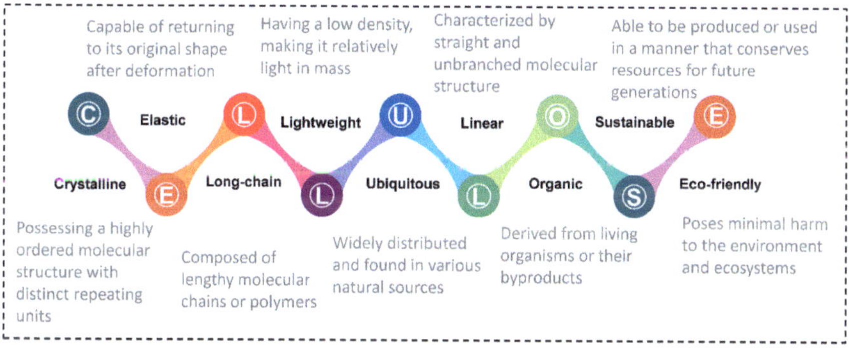

Fig. 3.1 The notable properties of the cellulose biomolecule and its special features

molecules, thereby improving its mechanical strength, fatigue resistance, and elasticity. Meanwhile, industries like tire and thread manufacturing, where NR properties alone may not meet high-performance requirements, require incorporating CB fillers, which is a common practice. CB improves the mechanical and physical properties of rubber matrices, enhancing volume and strengthening vulcanization. However, recent research suggests that cellulose fillers may offer a viable alternative to CB and Sil fillers, showing potential for improving mechanical properties in tire and thread manufacturing applications. The incorporation of cellulose fillers into rubber matrices holds promise for enhancing mechanical properties and performance, offering a greener option for tire and thread manufacturing while reducing reliance on synthetic fillers. Cellulose naturally consists of both amorphous and crystalline regions. By removing a significant portion of the amorphous areas, the cellulose becomes stronger and stiffer, thereby enhancing its potential as a reinforcing filler for different applications.

Microcrystalline cellulose forms large particles ranging from 20 to 90 μm and is highly polar, which limits its effectiveness in reinforcing nonpolar polymer matrices. High polarity and a tendency to agglomerate can make dispersion within the rubber matrix challenging. Chemical treatments, including esterification, etherification, grafting, and salinization of the hydroxyl groups of microcrystalline cellulose, significantly improve interfacial interactions and compatibility with the rubber matrix. While chemical treatments can significantly improve the compatibility and bonding between microcrystalline cellulose and the rubber matrix, they are not ideal for large-scale production due to the extensive use of chemicals, unsafe reaction conditions, and high costs.

3.2 Processing of Cellulose-Based Rubber Composites

General preparation processes for cellulose-reinforced rubber nanocomposites involve solvent casting, blending, in situ polymerization, and latex compounding (Nunes 2014). The two main cellulose nanofillers, derived primarily from natural plant sources, yield CNC and CNF, depending on the extraction processes (Khui et al. 2023). The polysaccharide made up of linear chains of D-glucose units plays a crucial role in many plants, such as trees, hemp, wheat straw, and bamboo. In these plants, the basic units align to form bundles of microfibrils, which create a cellulosic fiber network. This network is further complexed with other polymers, including hemicellulose and lignin, which are attached to the cellulose surface and embedded within the microfibrils (Sun et al. 2023). Common morphology is depicted in Fig. 3.2. However, as mentioned earlier, the main bottleneck of these sustainable nanofillers is their dispersion in rubber matrices (Yasin et al. 2021a, b). In this regard, solution blending involves dispersing nanofillers and blending rubber in an appropriate solvent with an optimized formulation. Meanwhile, CNC or CNF can be dispersed in a separate solvent solution, allowing swelling to occur before blending with the dissolving rubber (Azeredo et al. 2017). Techniques like shear mixing, magnetic stirring, and ultrasonication are frequently used for solution blending, aiding in the incorporation of nanofillers into the matrix (Rahman et al. 2022). The melt blending process involves mixing rubber and cellulose to create a masterbatch before subjecting it to heating and mechanical processing, a technique also referred to as dry mixing. Dry mixing helps disperse the cellulose within the matrix, impacting the melt blending process and facilitating the intercalation of cellulose and rubber (Narimissa et al. 2015). Inadequate heating and mechanical processing may result in filler agglomeration, poor matrix-filler intercalation, and consequently, adverse effects on the final product's properties (Rahman et al. 2022).

Latex compounding or blending technology originates from NR in the form of latex, with latex mixing emerging as a pragmatic choice over in situ emulsion polymerization (Yasin et al. 2021b). Despite advancements in latex blending technology, challenges persist in achieving a stable aqueous dispersion of fillers (Rahman et al. 2022). Cellulose-reinforced NR or silicon micro- and nanocomposites are prepared from cured latex matrices in a liquid state, with fillers typically incorporated and mechanically processed as a masterbatch (Yasin et al. 2021b; Hussain et al. 2020). Masterbatch mixing incorporates several mechanical techniques, including ultrasonication, shear blending, and homogenization, which promote the exfoliation and intercalation of cellulose fillers into the latex matrix. The coagulation process, using acids such as acetic and formic acid, triggers the coagulation of latex, allowing it to bond with the dispersed cellulose fillers (Rahman et al. 2022; Mekonnen et al. 2019). All the processes involved in cellulose-reinforced rubber nanocomposites are coupled with vulcanization, a cross-linking procedure for both natural and synthetic rubbers that uses sulfur and heat. Sulfur acts as the cross-linking agent, while various additives like activators, accelerators, coagulants, antioxidants, color pigments, surfactants, softeners, anti-foaming agents, and antitack agents can alter the physical and

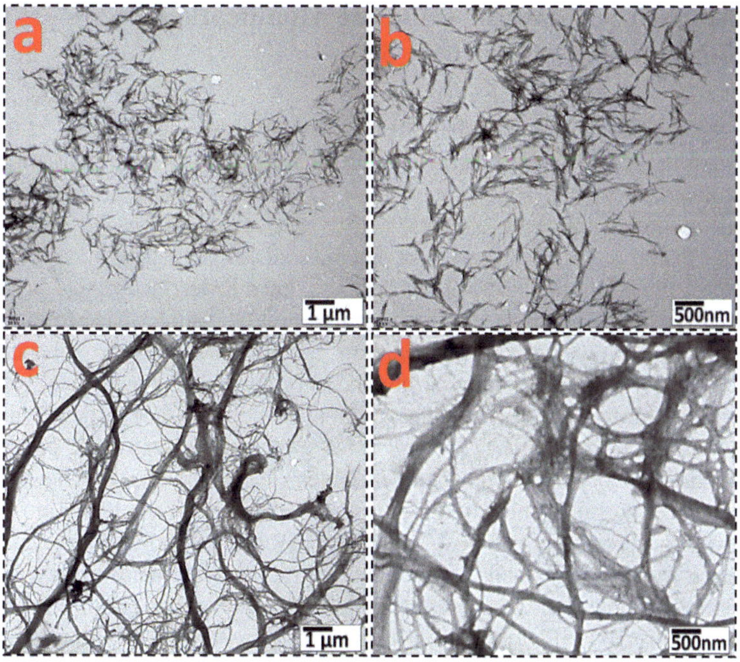

Fig. 3.2 Transmission electron microscopy images of CNC (**a**, **b**) and CNF (**c**, **d**) derived from cellulose at different magnifications (Adapted from Yasin et al. 2021b)

mechanical characteristics of the rubber. The temperature during vulcanization typically ranges from 120 to 200 °C, with heating promoting quicker and complete cross-linking (Nunes 2014; Rahman et al. 2022; Hussain et al. 2020; Coran 2005).

The addition of CNC also accelerated vulcanization and enhanced the mechanical properties of NR composites. When combined with epoxidized natural rubber (ENR), CNC forms two types of cross-linking networks. The introduction of carboxyl groups onto the CNC surface through a maleic anhydride grafting process (MAH) creates both covalent and physical cross-links, improving interfacial adhesion and ensuring a more uniform dispersion of CNC within the composites (Cao et al. 2018). Similarly, a multifunctional biomass rubber composites incorporating a three-dimensional heterostructure of zinc oxide (ZnO), integrated onto the surface of cellulose fibers (Li et al. 2019).

3.3 Modern Solvents for Cellulose Modifications for Rubber Composites

The global interest in natural fibers has witnessed a significant surge in the last decade, with production reaching 32 million metric tons in 2018 (Townsend 2020). Natural fibers, especially vegetable fibers, offer abundant renewable resources that are easily biodegradable, non-toxic, low density, and cost-effective. They also exhibit resistance to corrosion, alkali, and most organic acids. Despite their numerous benefits, natural fibers, particularly vegetable fibers, have lower mechanical properties compared to synthetic fibers due to their hydrophilic nature. Therefore, surface treatments for cellulose-based modifications are often necessary to enhance their adhesion and coupling to polymeric phases, improving their mechanical response in polymeric matrices. Moreover, polar cellulosic nanofillers face further challenges in rubber matrices, particularly in terms of agglomeration, restricting processing, vulcanization kinetics, and mechanical attributes. These bottlenecks hinder the full potential of cellulose in rubber nanocomposites. However, new solvents are used to assist in the processing of cellulose in rubbers. NR (cis-1,4-polyisoprene) is found to enhance stretchability using unsaturated fatty acids (oleic acid) to crosslink the polyisoprene double bonds incorporated on the CNF surfaces (Kato et al. 2015). Cellulose-reinforced NR through phase inversion, prepared with lithium chloride in N,N-dimethylacetamide, is found to exhibit optimal reinforcement, high elasticity, and mechanical properties, forming an interconnection layer between cellulose and NR components (Srirachya et al. 2018). In terms of modern solvents, ionic liquids (ILs) and deep eutectic solvents (DES) have emerged as a versatile solution (Yasin et al. 2021b, 2024a; Hussain et al. 2020, 2022a; Song et al. 2022), serving as solvents, dispersing agents, and dissolution facilitators for cellulose (Yasin et al. 2021b, 2022b). Figure 3.3 reveals morphological improvement with the addition of ILs in different rubbers. ILs have demonstrated the capability to enhance filler dispersity, filler-rubber interactions, thermal stability, and overall mechanical properties when incorporated into rubber composites (Yasin et al. 2024a; Liu et al. 2023).

3.4 Payne Effect and Dynamic Viscoelastic Properties of Cellulose-Based Rubber Composites

To understand the material behavior under dynamic loading conditions, providing information on material's response for both elastic (recoverable) and viscous (non-recoverable) is crucial for elastomers. It provides insights into the mechanical properties and performance of materials under real-world conditions, including those subjected to oscillatory or cyclic stresses. Rubbery materials deform and flow in response to applied strains (γ) that vary with time or frequency. Dynamic viscoelastic properties are characterized by storage modulus (G'), which indicates material's elastic behavior, and loss modulus (G'') indicating the material's viscous behavior.

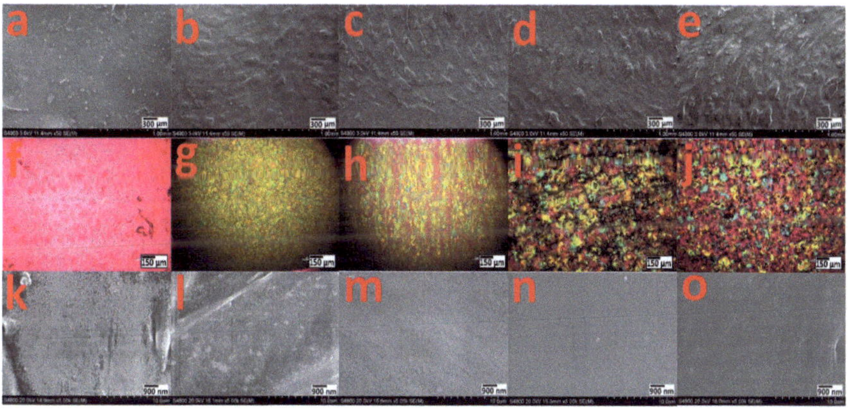

Fig. 3.3 Scanning electron microscope images of silicon rubber (SR) (**a**) and its microcomposites (**b–e**) with randomly dispersed (**b** and **d**) or aligned (**c** and **e**) CNC (**b** and **c**) or CNC graphene nanoplatelets fillers (**d** and **e**) (Adapted from Hussain et al. 2022b). Polarized optical microscopies of SR (**f**) and its composites (**g–j**) with randomly dispersed (**g** and **h**) or aligned (**i** and **j**) CNC (**g** and **h**) or CNC graphene nanoplatelets (**i** and **j**) (Adapted from Hussain et al. 2022b). Scanning electron microscope images of natural rubber (NR) vulcanizate (**k**) and 10 phr CNF (**l–m**) or CNC filled (**n–o**) without (**l** and **n**) and with IL (**m** and **o**) (Adapted from Yasin et al. 2021b)

Nitrile butadiene rubber (NBR) with varying CNC loadings, under different γ at a set temperature and frequency show a gradual increase in the G' of uncured nanocomposites, suggesting the formation of rigid filler-filler structures as the content of CNC goes higher. The linear viscoelastic region of the CNC-filled nanocomposites at lower γ demonstrates G' is independent, while, a significant decrease in the G' with higher applied γ, particularly in high CNC content rubbers. This phenomenon is called the Payne effect, suggesting results from the breakdown and reformation of physical bonds between nanocellulose domains. The Payne effect of sole and secondary nanocellulose fillers NR nanocomposites is shown in Fig. 3.4. The stronger rebuilding ability of filler-filler networks leads to higher strain amplitudes required for the breakdown of these networks. This robust three-dimensional filler-filler network in CNC nanocomposites is attributed to the presence of hydroxyl groups in the biofibers' chemical structure and the high aspect ratio of nanocellulose. The unique characteristics of cellulose nanofillers extend their influence beyond mere mechanical reinforcement. Cellulose's interaction with rubber matrices also influences the nonlinear mechanical behavior described earlier as the Payne effect. The crosslinked networks impose restrictions on the mobility of rubber chains, while the interaction between filler and elastomer matrix acted as physical crosslink points, contributing to higher G' values for rubber nanocomposites. Understanding and manipulating these behaviors are critical in tailoring rubber components for specific biomedical or other applications. By enhancing elastic nonlinearity and weakening viscous nonlinearity (Zhu et al. 2021), cellulose nanofillers contribute to the overall performance and

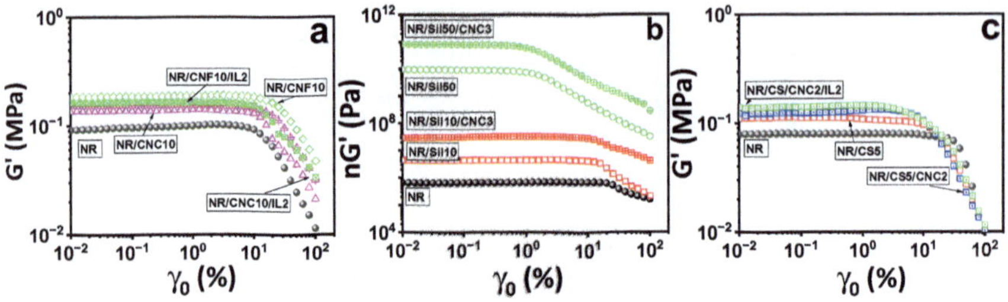

Fig. 3.4 The Payne effect (decline of storage modulus G' along strain amplitude γ_0) of cellulose nanocrystal (CNC) and cellulose nanofiber (CNF) and ionic liquid (IL) as a sole filler (**a**), and with silica (Sil) (**b**) and candle soot (CS) (**c**) as a secondary filler in natural rubber (NR) nanocomposites (Adapted from Yasin et al. 2021a, b, 2022b)

response of rubber materials (Yasin et al. 2021a). CNC in carboxylated nitrile butadiene rubber (XNBR) combined with clay nanofillers also demonstrates significant nonlinear viscoelastic behavior. This suggests that increased filler content results in reduced interaggregate distances and a higher likelihood of forming a filler network. These alterations in viscoelastic properties, along with the Payne effect observed in nanocellulose nanocomposites, clearly indicate enhancements in mechanical performance. These improvements are mainly influenced by the nanocellulose material's structure and size, as well as the interfacial interactions between the nanocellulose and the elastomer matrix (Paran 2024). Similarly, in SBR, Payne effect increases with the addition of CNC, bonds with the non-rubber components in rubbers forming intermolecularly hydrogen-bonded networks, thus increasing the overall mechanical properties (Yasin et al. 2024b).

Furthermore, the viscoelastic properties of cellulose-reinforced rubbers can be determined by using dynamic oscillatory shear rheological tests, where dynamic stress (σ) is decomposed into elastic and viscous stresses (σ' and σ'') (Ewoldt et al. 2008). The resulting Lissajous plots can be seen in Fig. 3.5. Cellulosic nanofillers change the Lissajous shapes when added isotopically, which is a qualitative indicator of the transformative properties of rubbers. This change is enhanced when a secondary filler is added, graphene nanoplatelets (GP) in silicone rubber, where σ and σ' against dynamic γ as well as σ and σ'' against strain rate ($\dot{\gamma}$) for the elastic and viscous perspective (Fig. 3a, b). Changing plots from rectangular and ellipse shapes to others with increasing γ and demonstrates the ability of cellulose to transform the intracycle shear thickening behavior of elastomers.

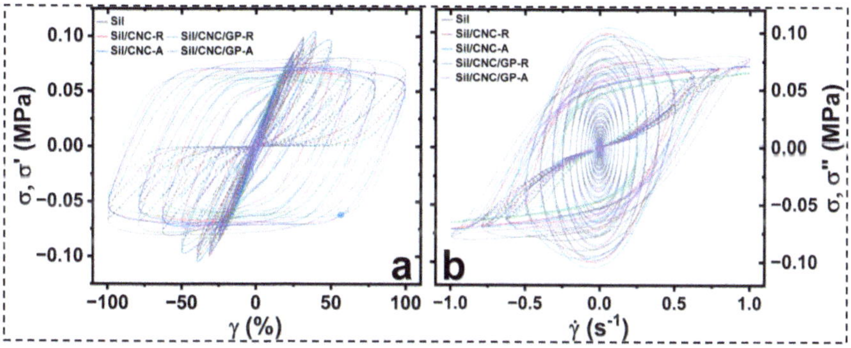

Fig. 3.5 Elastic (**a**) and viscous Lissajous plots (**b**) for silicone rubber and its microcomposites with CNC, Sil, and graphene nanoplatelets. Dynamic stress (σ; lines) and elastic stress (σ'; dotted lines) as a function of dynamic strain (γ) and σ (lines) and viscous stress (σ''; dotted lines) as a function of strain rate ($\dot{\gamma}$) for silicone rubber and microcomposites with randomly dispersed (R) or aligned (A) CNC, Sil, and graphene nanoplatelets (Adapted from Hussain et al. 2022b)

3.5 Mechanical Properties and Sustainability of Cellulose-Based Rubber Composites

With the growing demand for sustainable and green materials, industries have started utilizing life cycle assessment (LCA) to streamline and compare alternatives in the process of product/service development recently, especially in rubber tires. LCA is standardized according to ISO 14040 (2014—revised edition), which details the entire process, including defining the objectives and scope, conducting inventory analysis, performing an impact assessment, and interpreting the results (Ferreira et al. 2024). In terms of the biodegradability and sustainability of rubber waste, the accumulation of rubber particles raises environmental concerns. The dispersion of micro-waste in environments is challenging; for instance, 79% of 359 million tons of plastic waste is either landfilled or lost to the environment (Schwarz et al. 2021). In contrast, rubber tires wear off particles, and road markings are a significant source of microplastics in the oceans (Boucher and Friot 2017). The susceptibility of synthetic rubbers to complete degradation in the environment exceeds 100 years, and large accumulations in landfills pose a fire hazard (Katarzyna et al. 2020). Non-biodegradable rubber waste can negatively impact ecosystems, and the production and incorporation of synthetic nanofillers have diverse environmental impacts (Yasin et al. 2024d).

Cellulose-based nanofillers present a unique reinforcing opportunity due to their abundant hydroxyl groups (Yasin et al. 2021b), with inexhaustible bioresources availability. As mentioned before, cellulosic nanofillers have been found to enhance tearing, abrasion resistance, and tensile properties of natural and synthetic rubbers, such as NBR and styrene butadiene rubber (SBR) nanocomposites (Mekonnen et al. 2019; Chen et al. 2014; Jardin et al. 2020; Neto et al. 2016). NBR and SBR are commonly used for tires, hoses, footwear, seals, adhesives, and conveyor belts (Jardin et al. 2020; McKeen 2013). CNC being a biodegradable nanofiller, its susceptibility toward natural weathering, compared to CB and Sil, has been explored with soil burial aging and LCA (Yasin et al. 2022a). Table 3.1 shows the tensile strength and soil burial aging of cellulose-reinforced SBR or NBR with partial replacement of CB or Sil, where the rubbers with CNC showed improved degradation properties in the natural environment.

An LCA study on rubber recycling is deemed sufficient, avoiding virgin production (Yasin et al. 2022a). This results in negative values on the environmental impact balance, considered environmentally beneficial (Yasin 2017; Yasin et al. 2014, 2018; Yasin and Sun 2018, 2019a). End-of-life (EOL) scenarios for recycling are limited compared to other sustainable EOL choices (Yasin et al. 2017, 2020, 2024c; Yasin and Sun 2019b). Waste-to-energy in LCA literature is often unclear for similar waste types, where anaerobic digestion is a preferable EOL choice (Khoo et al. 2010), while incineration is found to be more optimal EOL in other cases (Schwarz et al. 2021; Yasin and Sun 2019a; Yasin et al. 2020; Manfredi et al. 2011). These substantial limitations in LCA EOL usability can be minimized with experimental results supporting EOL choices, including the thermal and biodegradation of materials, which can potentially be applied to existing waste stream boundaries. In this regard, cellulose-based composites can assist in eliminating environmental burdens by offering ease in biodegradation of products.

In a gate-to-grave analysis, comprising the manufacturing phase without raw materials extraction, transportation, and use or maintenance phases, with system boundaries shown in Fig. 3.6, the functional unit was one kg rubber system with the addition of CB, Sil, CNC, and curing agents. EOL modeling is mostly carried out with a choice of scenarios (Yasin et al. 2018), where the discarded product or a system is completely or partially directed to waste management options. Recycling, incineration, or landfills are commonly modeled EOL scenarios (Mansor et al. 2019; Yasin et al. 2015, 2016a), while some involve varying percentages of waste management options. The authors considered ten midpoint-level environmental impact categories

Table 3.1 Percentage retention of elongation at break and tensile strength for nanocomposites after soil burial aging (Adapted from Yasin et al. 2022a)

Sample no.	Retention %		Tensile strength (MPa)		Elongation at break (%)	
	30d	60d	30d	60d	30d	60d
NBRCB50	86.96	53.43	14	10	392	241
NBRSil50	85.57	49.46	15	11	401	232
NBRCB25/CNC25	96.02	31.99	22	9	600	199
NBRSil25/CNC25	91.05	26.33	19	8	629	182
SBRCB50	84.04	50.08	11	9	421	251
SBRSil50	82.59	54.23	10	8	440	289
SBRCB25/CNC25	84.78	32.5	20	7	576	221
SBRSil25/CNC25	81.94	27.48	19	6	599	201

for the life cycle impact assessment (LCIA) of the manufacturing phase and EOL. The environmental impacts of 1 kg rubber nanocomposite CB, Sil, or CNC at consistent ingredients composition are compared in Table 3.2. High energy use during NBR or SBR compounding and materials source are the major contributors to the environmental impacts, while partial replacement of CNC in CB or Sil-filled NBR or SBR nanocomposites, resulted in higher AA, AT, RMD, WD, and water-related impact categories. Cotton is one of the most water-consumptive crops, utilizing 2.6% of global water use, resulting in high environmental impacts due to energy consumption, land occupation, fertilizer, and pesticide (Baydar et al. 2015; Fidan et al. 2021; Yasin et al. 2016b). This implies a need for mass production of cellulosic nanofillers using low energy-intensive techniques (Bevilacqua et al. 2014). Similarly, substituting CB with Sil reduces energy and crude oil resource consumption, although it is known to have a negative impact on wastewater-related environmental categories (Krömer et al. 1999).

The average lifespan of a product is based on its functionality, generally ~4 years are considered functional for the rubber product (tires) and then are discarded (Mohajerani et al. 2020). Environmental impacts of landfill and incineration as EOL are shown in Table 3.3, where environmental impact values for CB and Sil-filled rubber, are demonstrated higher in landfills compared to CNC. Similarly, lower or negative values represent the calorific values of the incinerated materials compensated in the

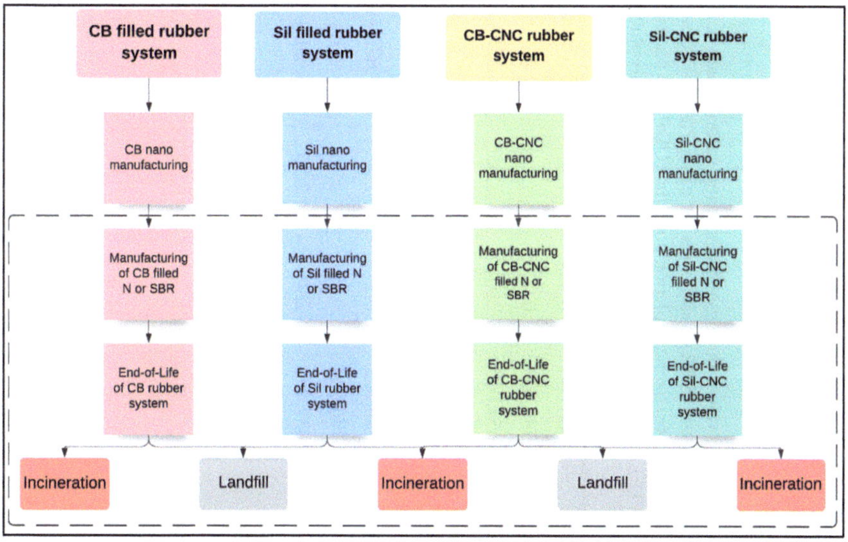

Fig. 3.6 System boundary (dashed line) used for carbon black (CB), silica (Sil) or cellulose nanocrystals (CNC) filled nitrile butadiene rubber (NBR) and styrene butadiene rubber (SBR) nanocomposites

Table 3.2 Environmental impact values during manufacturing carbon black (CB), silica (Sil), or cellulose nanocrystals (CNC) filled nitrile butadiene rubber (NBR) and styrene butadiene rubber (SBR) composites for categories: (a) air acidification (AA), (b) air toxicity (AT), (c) freshwater ecotoxicity (FWE), (d) global warming potential (GWP), (e) ozone depletion potential (ODP), (f) photochemical ozone creation potential (POCP), (g) raw material depletion (RMD), (h) terrestrial ecotoxicity (TE), (i) water depletion (WD), and (j) water eutrophication (WE) (Adapted from Yasin et al. 2022a)

Samples	AA	AT	FWE	GWP	ODP	POCP	RMD	TE	WD	WE
NBRCB50	0.012	606,000	0.398	1.32	6.03E−6	0.0153	5.53E−4	2.52E−5	38.8	3.52E−5
NBRSil50	0.0117	588,000	0.0777	1.26	4.4E−6	0.0108	5.28E−4	1.79E−5	38.8	3.49E−5
NBRCB25/ CNC25	0.0238	654,000	286	1.65	5.22E−6	0.0131	5.42E−4	0.0441	293	0.00664
NBRSil25/ CNC25	0.0237	645,000	286	1.62	4.41E−6	0.0108	5.29E−4	0.0441	293	0.00664
SBRCB50	0.00839	389,000	0.521	2.22	1.67E−6	0.00653	8.55E−5	7.47E−5	11.7	4.54E−6
SBRSil50	0.00807	370,000	0.201	2.16	3.91E−8	0.00207	6.02E−5	6.74E−5	11.7	4.17E−6
SBRCB25/ CNC25	0.0202	437,000	286	2.55	8.63E−7	0.00432	7.42E−5	0.0441	266	0.00661
SBRSil25/ CNC25	0.02	427,000	286	2.52	4.54E−8	0.00209	6.15E−5	0.0441	266	0.00661

AA [AE]; AT [m^3]; FWE [CTUe]; GWP [kg CO_2 eq.]; ODP [kg CFC-11 eq.]; POCP [kg NMVOC eq.]; RMD [person reserve]; TE [kg 1,4-DB eq.]; WD [dm^3] and WE [kg $PO_3{}^{-4}$-eq.]

Table 3.3 Environmental impact values of carbon black (CB), silica (Sil), or cellulose nanocrystals (CNC) filled rubber at landfill EOL for categories: (a) air acidification (AA), (b) air toxicity (AT), (c) freshwater ecotoxicity (FWE), (d) global warming potential (GWP), (e) ozone depletion potential (ODP), (f) photochemical ozone creation potential (POCP), (g) raw material depletion (RMD), (h) terrestrial ecotoxicity (TE), (i) water depletion (WD) and (j) water eutrophication (WE) (Adapted from Yasin et al. 2022a)

Samples	AA	AT	FWE	GWP	ODP	POCP	RMD	TE	WD	WE
At landfill EOL										
CB50	4.12E−4	21,500	8.76E−4	0.702	2.31E−9	5.99E−4	8.06E−7	1.25E−5	0.242	3.94E−4
Sil50	3.56E−4	18,500	7.31E−4	0.62	1.9E−9	5.3E−4	6.71E−7	1.15E−6	0.209	3.23E−4
CB25/ CNC25	3.98E−4	20,700	8.75E−4	0.639	2.31E−9	5.58E−4	8.06E−7	1.14E−5	0.229	3.91E−4
Sil25/ CNC25	2.85E−4	14,700	5.86E−4	0.475	1.5E−9	4.19E−4	5.36E−7	1.24E−6	0.162	2.49E−4
At incineration EOL										
CB50	−0.00195	−54,000	−0.0138	−0.114	−8.97E−9	−2.65E−4	−1.05E−4	−9.46E−6	3.08	−3.06E−7
Sil50	−0.00249	−77,500	−0.0244	−0.229	−2.97E−8	−7.14E−4	−2.83E−4	−1.57E−5	3.9	−6.45E−7
CB25/ CNC25	−0.00244	−67,700	−0.0155	−0.146	−2.73E−8	−3.28E−4	−1.09E−4	−1.06E−5	3.05	−4.36E−7
Sil25/ CNC25	−0.00271	−79,500	−0.0315	−0.279	−3.77E−8	−0.00104	−4.55E−4	−1.96E−5	3.51	−7.24E−7

AA [AE]; AT [m^3]; FWE [CTUe]; GWP [kg CO_2 eq.]; ODP [kg CFC-11 eq.]; POCP [kg NMVOC eq.]; RMD [person reserve]; TE [kg 1,4-DB eq.]; WD [dm^3] and WE [kg PO_3^{-4}-eq.]

manufacturing phase. The corresponding lower values of CNC-filled nanocomposites are associated with their fractional percentage and calorific values without a waste recovery process.

3.6 Summary

Cellulosic nanofillers have emerged as a promising biofiller for rubber composites due to their reinforcing capabilities, alongside their abundance, biodegradability and sustainability factors. The effectiveness of nanocellulose as a filler depends on various factors such as its size, shape, surface energy, dispersion, and processing techniques. Despite the potential benefits, achieving uniform dispersion of cellulosic nanofillers in hydrophobic rubber matrices remains a challenge, which is currently being researched with various strategies, including surface modification of nanocellulose and structural modification of rubbers with modern solvents. These modification studies have highlighted nanocellulose potential applications in diverse fields such as automotive, sports goods, baby care products, biomedical devices, electronics, and wearable sensors. Continuous academia and industry collaborative work is expected to yield novel and sustainable nanocellulose rubber materials in the coming years.

Acknowledgements The authors gratefully acknowledge the financial support given under the National Natural Science Foundation of China (Grant No. 5217050434) and National Key R&D Program of China (Grant No. 2022YFB4003002).

References

Azeredo HMC, Rosa MF, Mattoso LHC (2017) Nanocellulose in bio-based food packaging applications. Ind Crops Prod 97:664–671. https://doi.org/10.1016/j.indcrop.2016.03.013

Baydar G, Ciliz N, Mammadov A (2015) Life cycle assessment of cotton textile products in Turkey. Resour Conserv Recycl 104:213–223. https://doi.org/10.1016/j.resconrec.2015.08.007

Bevilacqua M, Ciarapica FE, Mazzuto G, Paciarotti C (2014) Environmental analysis of a cotton yarn supply chain. J Clean Prod 82:154–165. https://doi.org/10.1016/j.jclepro.2014.06.082

Boucher J, Friot D (2017) Primary microplastics in the oceans: a global evaluation of sources, Iucn Gland, Switzerland. https://doi.org/10.2305/IUCN.CH.2017.01.en

Cao L, Huang J, Chen Y (2018) Dual cross-linked epoxidized natural rubber reinforced by tunicate cellulose nanocrystals with improved strength and extensibility. ACS Sustain Chem Eng 6:14802–14811. https://doi.org/10.1021/acssuschemeng.8b03331

Chen WJ, Gu J, Xu SH (2014) Exploring nanocrystalline cellulose as a green alternative of carbon black in natural rubber/butadiene rubber/styrene-butadiene rubber blends. Express Polym Lett 8:659–668. https://doi.org/10.3144/expresspolymlett.2014.69

Coran AY (2005) 7—Vulcanization. In: Mark JE, Erman B, Eirich FR (eds) Science and technology of rubber, 3rd edn. Academic Press, Burlington, pp 321–366. https://doi.org/10.1016/B978-012464786-2/50010-9

Ewoldt RH, Hosoi AE, McKinley GH (2008) New measures for characterizing nonlinear viscoelasticity in large amplitude oscillatory shear. J Rheol 52:1427–1458

Ferreira RR, Barbosa RFS, Camani PH, Marciano JS, Rosa DS (2024) 24—Life cycle assessment from laboratory to industrial scale—case of elastomeric composite development. In: Thomas S, Mohamed PK, Kim J, Tom M (eds) Elastomeric nanocellulose composites. Woodhead Publishing, pp 535–557. https://doi.org/10.1016/B978-0-443-18608-0.00019-X

Fidan FŞ, Aydoğan EK, Uzal N (2021) An integrated life cycle assessment approach for denim fabric production using recycled cotton fibers and combined heat and power plant. J Clean Prod 287:125439. https://doi.org/10.1016/j.jclepro.2020.125439

Hussain M, Yasin S, Adnan Akram M, Xu H, Song Y, Zheng Q (2019) Influence of ionic liquids on structure and rheological behaviors of silica-filled butadiene rubber. Ind Eng Chem Res 58:18205–18212. https://doi.org/10.1021/acs.iecr.9b03494

Hussain M, Yasin S, Memon H, Li Z, Fan X, Akram MA, Wang W, Song Y, Zheng Q (2020) Rheological and mechanical properties of silica/nitrile butadiene rubber vulcanizates with eco-friendly ionic liquid. Polymers 12. https://doi.org/10.3390/polym12112763

Hussain M, Yasin S, Ali A, Li Z, Fan X, Song Y, Zheng Q, Wang W (2022a) Synergistic impact of ionic liquid on interfacial interaction and viscoelastic behaviors of silica filled nitrile butadiene rubber nanocomposites. Compos Part Appl Sci Manuf 163:107202. https://doi.org/10.1016/j.compositesa.2022.107202

Hussain M, Yasin S, Uddin A, Lu M, Qiang Z, Song Y (2022b) Nonlinear rheology of silicone rubber composites with tailored mechanical and dielectric properties. Compos Commun 35:101328. https://doi.org/10.1016/j.coco.2022.101328

Jardin JM, Zhang Z, Hu G, Tam KC, Mekonnen TH (2020) Reinforcement of rubber nanocomposite thin sheets by percolation of pristine cellulose nanocrystals. Int J Biol Macromol 152:428–436. https://doi.org/10.1016/j.ijbiomac.2020.02.303

Katarzyna P, Izabela P, Patrycja B-W, Weronika K, Andrzej T (2020) LCA as a tool for the environmental management of car tire manufacturing. Appl Sci 10. https://doi.org/10.3390/app10207015

Kato H, Nakatsubo F, Abe K, Yano H (2015) Crosslinking via sulfur vulcanization of natural rubber and cellulose nanofibers incorporating unsaturated fatty acids. RSC Adv 5:29814–29819. https://doi.org/10.1039/C4RA14867C

Khoo HH, Lim TZ, Tan RBH (2010) Food waste conversion options in Singapore: environmental impacts based on an LCA perspective. Sci Total Environ 408:1367–1373. https://doi.org/10.1016/j.scitotenv.2009.10.072

Khui PLN, Rahman MR, Bakri MKB (2023) A review on the extraction of cellulose and nanocellulose as a filler through solid waste management. J Thermoplast Compos Mater 36:1306–1327. https://doi.org/10.1177/08927057211020800

Krömer S, Kreipe E, Reichenbach D, Stark R (2006) Life cycle assessment of a car tire, Cont. Hann. Ger. Accessed 20 April 1999

Li Y, Sun H, Zhang Y, Xu M, Shi SQ (2019) The three-dimensional heterostructure synthesis of ZnO/cellulosic fibers and its application for rubber composites. Compos Sci Technol 177:10–17. https://doi.org/10.1016/j.compscitech.2019.04.012

Liu B, Hu Z, Zhong X, Yang L, Jiang X, Zuo M, Du M, Song Y, Zheng Q, Wang D (2023) Effect of deep eutectic solvents on vulcanization and rheological behaviors of rubber vulcanizates. J Clean Prod 429:139426. https://doi.org/10.1016/j.jclepro.2023.139426

Manfredi S, Tonini D, Christensen TH (2011) Environmental assessment of different management options for individual waste fractions by means of life-cycle assessment modelling. Resour Conserv Recycl 55:995–1004. https://doi.org/10.1016/j.resconrec.2011.05.009

Mansor MR, Mastura MT, Sapuan SM, Zainudin AZ (2019) 11—The environmental impact of natural fiber composites through life cycle assessment analysis. In: Jawaid M, Thariq M, Saba N (eds) Durability and life prediction in biocomposites, fibre-reinforced composites and hybrid composites. Woodhead Publishing, pp 257–285. https://doi.org/10.1016/B978-0-08-102290-0.00011-8

Mariano M, El Kissi N, Dufresne A (2014) Cellulose nanocrystals and related nanocomposites: review of some properties and challenges. J Polym Sci Part B Polym Phys 52:791–806. https://doi.org/10.1002/polb.23490
Mariano M, El Kissi N, Dufresne A (2016) Cellulose nanocrystal reinforced oxidized natural rubber nanocomposites. Carbohydr Polym 137:174–183. https://doi.org/10.1016/j.carbpol.2015.10.027
McKeen LW (2013) 12—Elastomers and rubbers. In: McKeen LW (ed) The effect of UV light and weather on plastics and elastomers, 3rd edn. William Andrew Publishing, Boston, pp 299–370. https://doi.org/10.1016/B978-1-4557-2851-0.00012-8
Mekonnen TH, Ah-Leung T, Hojabr S, Berry R (2019) Investigation of the co-coagulation of natural rubber latex and cellulose nanocrystals aqueous dispersion. Colloids Surf Physicochem Eng Asp 583:123949
Mohajerani A, Burnett L, Smith JV, Markovski S, Rodwell G, Rahman MT, Kurmus H, Mirzababaei M, Arulrajah A, Horpibulsuk S, Maghool F (2020) Recycling waste rubber tyres in construction materials and associated environmental considerations: a review. Resour Conserv Recycl 155:104679. https://doi.org/10.1016/j.resconrec.2020.104679
Narimissa E, Gupta RK, Kao N, Choi JH, Bhattacharya SN (2015) The comparison between the effects of solvent casting and melt intercalation mixing processes on different characteristics of polylactide-nanographite platelets composites. Polym Eng Sci 55:1560–1570. https://doi.org/10.1002/pen.23996
Neto WPF, Mariano M, da Silva ISV, Silvério HA, Putaux J-L, Otaguro H, Pasquini D, Dufresne A (2016) Mechanical properties of natural rubber nanocomposites reinforced with high aspect ratio cellulose nanocrystals isolated from soy hulls. Carbohydr Polym 153:143–152
Nunes RCR (2014) 10—Natural rubber (NR) composites using cellulosic fiber reinforcements. In: Kohjiya S, Ikeda Y (eds) Chemistry, manufacture and applications of natural rubber. Woodhead Publishing, pp 284–302. https://doi.org/10.1533/9780857096913.2.284
Paran SMR (2024) 12—Nonlinear viscoelasticity and Payne effect of nanocellulose reinforced elastomer composites. In: Thomas S, Mohamed PK, Kim J, Tom M (eds) Elastomeric nanocellulose composites. Woodhead Publishing, pp 301–317. https://doi.org/10.1016/B978-0-443-18608-0.00006-1
Rahman MR, Khui PLN, Bakri MKB (2022) Chapter 10—Cellulose-reinforced rubber composites. In: Rahman MR (ed) Fundamentals and recent advances in nanocomposites based on polymers and nanocellulose. Elsevier, pp 175–188. https://doi.org/10.1016/B978-0-323-85771-0.00013-0
Schwarz AE, Ligthart TN, GoBizarro D, De Wild P, Vreugdenhil B, van Harmelen T (2021) Plastic recycling in a circular economy; determining environmental performance through an LCA matrix model approach. Waste Manag 121:331–342. https://doi.org/10.1016/j.wasman.2020.12.020
Singh S, Maji PK, Das C (2024) 9—Nanocellulose as a reinforcing filler in natural rubber composites. In: Thomas S, Mohamed PK, Kim J, Tom M (eds) Elastomeric nanocellulose composites. Woodhead Publishing, pp 205–230. https://doi.org/10.1016/B978-0-443-18608-0.00022-X
Song Y, Zheng Q (2016) Concepts and conflicts in nanoparticles reinforcement to polymers beyond hydrodynamics. Prog Mater Sci 84:1–58. https://doi.org/10.1016/j.pmatsci.2016.09.002
Song Y, Wu G, Peng J, Zhang C, Wang D, Zheng Q (2022) Vulcanization kinetics of natural rubber and strain softening behaviors of gum vulcanizates tailored by deep eutectic solvents. Polymer 263:125504. https://doi.org/10.1016/j.polymer.2022.125504
Srirachya N, Boonkerd K, Nakajima L, Kobayashi T (2018) Bio-composite hydrogels of cellulose and vulcanized natural rubber with nanointerconnected layers for reinforced water-retaining materials. Polym Bull 75:5493–5512. https://doi.org/10.1007/s00289-018-2341-y
Sun D, Saw BLH, Onyianta AJ, Wang B, Wilson C, O'Rourke D, See CH, Popescu C-M, Dorris M, Shyha I, Lu Z (2023) Preparation of elastomeric nanocomposites using nanocellulose and recycled alum sludge for flexible dielectric materials. J Adv Dielectr 13:2242008. https://doi.org/10.1142/S2010135X22420085

Townsend T (2020) 1B—World natural fibre production and employment. In: Kozłowski RM, Mackiewicz-Talarczyk M (eds) Handbook of natural fibres, 2nd edn. Woodhead Publishing, pp 15–36. https://doi.org/10.1016/B978-0-12-818398-4.00002-5
Yasin S (2017) Eco-design for end-of-life phase of flame retardant textiles. Université Lille 1—Sciences et Technologies. https://hal.science/tel-02086728/
Yasin S, Sun D (2018) Life cycle assessment of technical textile waste: pilot evaluation of environmental impacts, Tianjin, China. https://hal.science/hal-02133644v1
Yasin S, Sun D (2019a) Propelling textile waste to ascend the ladder of sustainability: EOL study on probing environmental parity in technical textiles. J Clean Prod 233:1451–1464. https://doi.org/10.1016/j.jclepro.2019.06.009
Yasin S, Sun D (2019b) Critical review on the waste-to-energy techniques for flame retardant treated textiles, Bali, Indonesia, pp 31–33. https://hal.science/hal-02105423/
Yasin S, Perwuelz A, Behary N (2014) A case study of life cycle inventory of cotton curtain, Bursa, Turkey. https://hal.science/hal-01984356/
Yasin S, Nemeshwaree B, Giraud S, Anne P, Rovero G (2015) Degradation of phosphorus based flame retardant by IR radiation. In: Autex. http://porto.polito.it/2670416/
Yasin S, Behary N, Giraud S, Perwuelz A (2016a) In situ degradation of organophosphorus flame retardant on cellulosic fabric using advanced oxidation process: a study on degradation and characterization. Polym Degrad Stab 126:1–8. https://doi.org/10.1016/j.polymdegradstab.2015.12.005
Yasin S, Behary N, Rovero G, Kumar V (2016b) Statistical analysis of use-phase energy consumption of textile products. Int J Life Cycle Assess 21:1776–1788. https://doi.org/10.1007/s11367-016-1129-2
Yasin S, Massimo C, Rovero G, Behary N, Perwuelz A, Giraud S, Migliavacca G, Chen G, Guan J (2017) An alternative for the end-of-life phase of flame retardant textile products: degradation of flame retardant and preliminary settings of energy valorization by gasification. Bioresource 12(3). https://doi.org/10.15376/biores.12.3.5196-5211
Yasin S, Behary N, Perwuelz A, Guan J (2018) Life cycle assessment of flame retardant cotton textiles with optimized end-of-life phase. J Clean Prod 172:1080–1088. https://doi.org/10.1016/j.jclepro.2017.10.198
Yasin S, Curti M, Rovero G, Hussain M, Sun D (2020) Spouted-bed gasification of flame retardant textiles as a potential non-conventional biomass. Appl Sci 10. https://doi.org/10.3390/app10030946
Yasin S, Hussain M, Zheng Q, Song Y (2021a) Large amplitude oscillatory rheology of silica and cellulose nanocrystals filled natural rubber compounds. J Colloid Interface Sci 588:602–610. https://doi.org/10.1016/j.jcis.2020.10.094
Yasin S, Hussain M, Zheng Q, Song Y (2021b) Effects of ionic liquid on cellulosic nanofiller filled natural rubber bionanocomposites. J Colloid Interface Sci 591:409–417. https://doi.org/10.1016/j.jcis.2021.02.029
Yasin S, Hussain M, Zheng Q, Song Y (2022a) Thermo-soil weathering and life cycle assessment of carbon black, silica and cellulose nanocrystal filled rubber nanocomposites. Sci Total Environ 835:155521. https://doi.org/10.1016/j.scitotenv.2022.155521
Yasin S, Hussain M, Zheng Q, Song Y (2022b) Influence of ionic liquid on rheological behaviors of candle soot and cellulose nanocrystal filled natural rubber nanocomposites. Compos Commun 33:101214. https://doi.org/10.1016/j.coco.2022.101214
Yasin S, Shi J, Yihu S, Qi C, Yang M, Peng W, Zheng C, Li Y, Gu C (2024a) Impacts of deep eutectic solvent on silica/nitrile rubber nanocomposites for high-pressure hydrogen storage applications. Compos Commun 46:101820. https://doi.org/10.1016/j.coco.2024.101820
Yasin S, Hafeez M, Hussain M, Mahmood A, Yihu S, Qiang Z, Gheewala SH, Li X, Li Y, Shi J, Gu C (2024b) Cellulose nanocrystal filled synthetic rubber: sustainable production, environmental assessment and end-of-life biodegradation by polyphagous pests. Sustain Mater Technol e00947. https://doi.org/10.1016/j.susmat.2024.e00947

Yasin S, Hussain M, Uddin A, Zheng Q, Shi J, Song Y (2024c) Recycling of binary polymer (PET/SBR) carpet into microfibrillar composites: a life cycle perspective with microplastics quantification. Sustain Mater Technol e00988. https://doi.org/10.1016/j.susmat.2024.e00988

Yasin S, Shi J, Ye S, Yihu S, Ullah A, Li G, Peng W, Gu C (2024d) Sustainable rubber nanocomposites for hydrogen sealings: impact of carbon nano-onions and bio-based plasticizer. Polym Degrad Stab 230:111051. https://doi.org/10.1016/j.polymdegradstab.2024.111051

Zhu F, Yasin S, Hussain M (2021) Viscoelastic rheological behaviors of polypropylene and LMPP blends. Polymers 13. https://doi.org/10.3390/polym13203485

Chapter 4
Natural Rubber-Based Biocomposites and Bionanocomposites

Nadras Othman and Muhamad Sharan Musa

Abstract This chapter reviews various fillers from natural resources as alternatives to conventional fillers in natural rubber composites and nanocomposites. The aim is to provide a comprehensive overview of the mechanical, morphological, and thermal properties of natural rubber-based composites and nanocomposites. The drawbacks of most natural fillers are discussed, including the influence of hydroxyl groups on the surface of the natural fillers and their compatibility issues with the rubber matrix. Additionally, the characterization techniques used to assess the impact of natural fillers on mechanical properties, along with the mechanisms of their performance reduction, are reviewed. This review is expected to give progress in minimizing the problem between the natural filler and natural rubber matrix. It can be done by reviewing the most recent study in this work.

Keywords Natural rubber · Natural filler · Mechanical properties · Morphological properties · Thermal properties

4.1 Introduction

Natural rubber (NR), derived from the *Hevea brasiliensis* plant, is one of the most important elastomers used in various applications. Its superior and unique mechanical properties make it widely utilized in industries such as automotive and engineering, including in tires, mounts, gaskets, and seals (Ismail et al. 2011). The advantages of natural rubber include high elasticity, stiffness, toughness, and hardness (Xuan et al. 2024). However, natural rubber also has limitations, such as being soft in normal weather conditions, brittle in cold temperatures, and exhibiting poor resistance to oils, hydrocarbons, greases, and fats (Whba et al. 2024). In practical applications, NR often requires reinforcing fillers to enhance its performance characteristics, which are otherwise limited without such fillers. Using reinforcement filers in NR

N. Othman (✉) · M. S. Musa
School of Materials and Mineral Resources Engineering, Universiti Sains Malaysia, Engineering Campus, 14300 Nibong Tebal, Penang, Malaysia
e-mail: srnadras@usm.my

© The Author(s), under exclusive license to Springer Nature Switzerland AG 2025
P. M. Visakh (ed.), *Rubber Based Bionanocomposites*, Advanced Structured Materials 227, https://doi.org/10.1007/978-3-031-78557-3_4

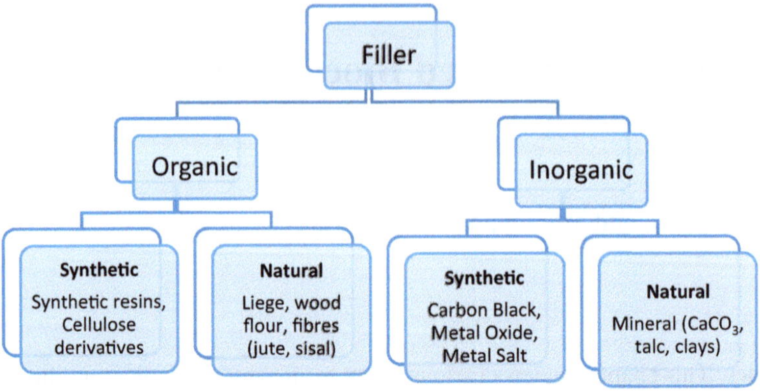

Fig. 4.1 Classification of fillers by type: organic versus inorganic (Leblanc 2009)

can improve mechanical performance such as tensile properties, hardness, abrasion, friction, swelling, and others (Mogy and Lawandy 2024).

Reinforcing fillers can be categorized as organic or inorganic, as shown in Fig. 4.1. Organic fillers include synthetic resins and natural materials such as lignin, wood flour, and fibers. In traditional rubber compounding processes, inorganic fillers like carbon black are commonly used as the primary reinforcement to improve the strength, modulus, stiffness, and weather resistance of rubber composites (Leblanc 2009). However, carbon black presents several issues, including being non-biodegradable, non-renewable, energy-intensive to produce, and costly (George 2023). As a result, many studies have been conducted to identify new reinforcing fillers that provide similar properties to those of inorganic fillers (Chandran et al. 2024; Kumar et al. 2024). The growing interest in bio-based fillers, such as hemp, kenaf, jute, wood flour, rice husks, sisal, coconut coir, and banana trunk fiber, has encouraged increasing research on natural rubber composites filled with these materials (Dave and Macwan 2024). Consequently, fillers from natural resources offer a promising alternative to traditional inorganic fillers.

4.2 Natural Rubber-Based Composites

In filled NR composite systems, foreign rigid materials such as powdered minerals, fibers, and other additives are used to enhance the mechanical performance of NR. Various natural fillers can be considered for reinforcing natural rubber, including cellulose (Roy et al. 2021), chitin (Thomas et al. 2021), polylactic acid (PLA) (Masek et al. 2021), starch (Songtipya et al. 2024), and others. These natural fillers can serve as alternatives to carbon black in NR composites. Therefore, many studies have been conducted to identify new reinforcing fillers that offer similar reinforcement characteristics to those of inorganic fillers (Chang et al. 2021; Dutta and Ray 2024; Mathew et al. 2024). The advantages of using natural resource fillers include their renewability, biodegradability, environmental benefits, and cost-effectiveness (Kumar et al. 2024). However, a key disadvantage is the incompatibility between hydrophilic natural fillers and the hydrophobic natural rubber matrix (Wong et al. 2024). To address this issue, treatments are often required to improve the compatibility between the natural rubber matrix and the natural resource fillers.

4.2.1 Natural Rubber/Cellulose-Based Composites

Cellulose is a bio-organic material that is produced from green plant resources such as palm oil empty fruit bunches (Cholilie et al. 2024), banana leaves (Dáger-López et al. 2024), coconut husk (Chand et al. 2025), bamboo (Suzuki et al. 2024), soybean hull (Anandkumar et al. 2024), and others. Cellulose makes up to 40–44% of wood (Yapar et al. 2024). Cellulose has been studied as a reinforcement filler in rubber composites over the past five years (Cholilie et al. 2024; Chand et al. 2025; Anandkumar et al. 2024). Cellulose is a renewable resource, low cost, easy availability, and easy physical and chemical modification (Riseh et al. 2024).

NR/cellulose composites have excellent elasticity, flexibility, electrical properties, tear strength, and resistance in water and most fluid chemicals (Zhou et al. 2015). The main disadvantage between rubber and cellulose is an interaction between the cellulose filler and NR. Cellulose fillers contain hydroxyl groups in their structure, making them hydrophilic, whereas natural rubber (NR) is composed of hydrocarbon chains, giving it hydrophobic characteristics (Whba et al. 2024; Chang et al. 2021). Many studies show that unmodified NR/cellulose composites exhibit poor compatibility between the cellulose and natural rubber matrix, resulting in inadequate mechanical properties of the composites (Sajith and Sambhudevan 2023). According to some studies (Manimaran et al. 2024; Jantachum et al. 2023), modifying cellulose fillers with a coupling agent can enhance the miscibility and compatibility between NR and cellulose fillers. A coupling agent helps establish molecular networking at the interface between the cellulose surface and natural rubber, improving the distribution of the cellulose filler within the rubber matrix. By using a coupling agent, the hydroxyl content in the cellulose structure can be reduced, enhancing its compatibility with natural rubber (Jantachum and Phinyocheep 2024).

The mechanical performance of NR/cellulose composites depends on several factors, including the amount of cellulose added to the natural rubber, the treatment process, the dispersion of cellulose, and the adhesion between the two components (Kazemi et al. 2022). Table 4.1 shows the scorch, curing time, and maximum torque of modified and unmodified NR/Cellulose composites. By using a coupling agent as a modification on the cellulose, the cure time and scorch time improve drastically as compared to unmodified composites. The presence of a coupling agent is known to exact a catalytic effect on accelerating the curing process of NR compounds (Tom et al. 2022). In summary, coupling agents modify cellulose to enhance its compatibility with natural rubber and optimize the curing process. This modification reduces

Table 4.1 Scorch time, curing time, and maximum torque of unmodified and modified NR/cellulose composites with silane coupling agent (Daud et al. 2016)

Sample code	Scorch time (min)	Cure time (min)	Maximum torque (dNm)
Gum NR	2.6	4.91	7
NR/PKS 5	2.66	4.88	7.15
NR/PKS 10	2.67	4.86	7.57
NR/PKS 15	2.65	4.82	7.98
NR/PKS 20	2.62	4.80	8.31
NR/SPKS 5[a]	1.74	4.75	7.4
NR/SPKS 10	1.46	3.5	7.87
NR/SPKS 15	1.40	3.4	8.4
NR/SPKS 20	1.28	2.99	8.89

[a] NR/SPKS composite with a silane coupling agent

both cure and scorch times, thereby improving the efficiency and performance of NR/cellulose composites. For example, modifications that enhance the interaction between cellulose and NR can lead to a more effective distribution of crosslinking agents, resulting in a faster and more uniform cure process (Parambath Kanoth et al. 2015).

Figure 4.2 shows that the mechanical properties of NR/cellulose composites with and without silane coupling agent modification exhibit a similar pattern of decreasing mechanical performance, except for hardness, which increases with higher filler loading. NR/cellulose composites modified with a silane coupling agent demonstrate greater improvements compared to the unmodified composites. The reduction in tensile strength in both composites is attributed to the geometry of the filler and the polarity of the natural rubber and cellulose filler. The irregular shape of cellulose fillers is insufficient to support stress transfer from the rubber matrix, leading to uneven stress distribution (Hakimi et al. 2021).

Additionally, the hydrophilic nature of cellulose fillers tends to cause aggregation and agglomeration within the natural rubber matrix, further reducing the composite's uniaxial strength. The toughness decreases due to increased restriction of molecular movement, resulting in reduced elongation at break. However, cellulose, being a hard and rigid particle, can enhance the hardness of the composites (Mahadi et al. 2024). The addition of rigid cellulose introduces additional resistance to the rubber composite's flow, contributing to a more rigid structure.

In summary, the addition of filler in both conditions leads to a decrease in mechanical performance (except for hardness). However, the introduction of a silane coupling agent improves the mechanical properties compared to composites without it, due to better filler attachment between the filler and the rubber matrix.

To better understand how the filler interacts with the rubber matrix, SEM micrographs of surface fractures in modified and unmodified NR/cellulose composites (shown in Fig. 4.3a, b) are observed. These micrographs reveal a reduction in voids

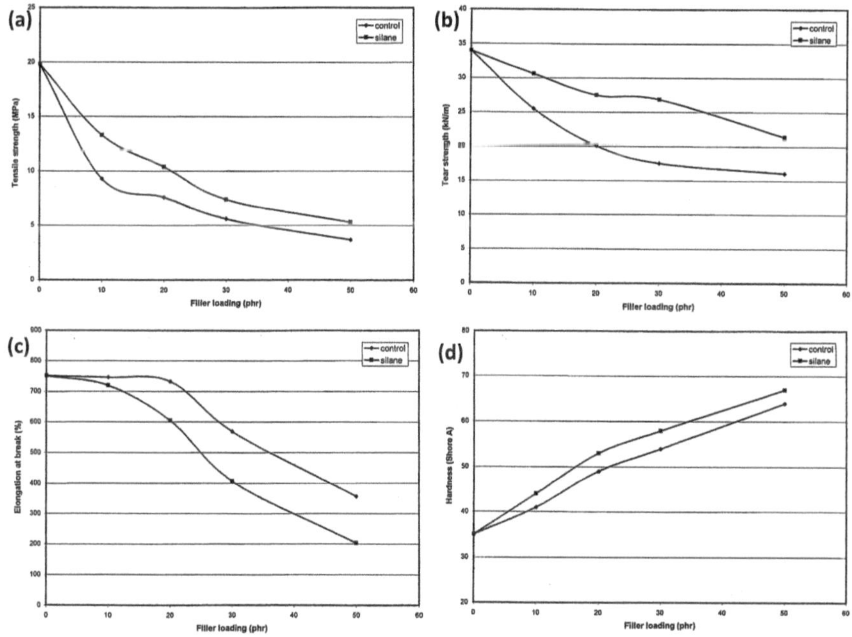

Fig. 4.2 **a** Tensile strength, **b** tear strength, **c** elongation at break (%), and **d** hardness of NR/cellulose composites with and without a modification (Ismail et al. 2002)

compared to unmodified composites. The fewer voids suggest that the modification process has enhanced the integration of the filler within the NR matrix, leading to a more uniform dispersion and better interaction between the filler and the rubber. This improved attachment contributes to the overall performance and stability of the composite material. This indicates improved homogenization of the filler within the NR matrix. The fracture surfaces exhibit rough zones and deeper tearing lines, indicating high stress during rupture. In contrast, Fig. 4.3c, d display poor adhesion and less effective interfacial bonding between the filler and the NR matrix, resulting in voids around the filler detachment areas. This lack of adhesion leads to reduced stress distribution within the rubber composites.

4.2.2 Natural Rubber/Chitosan-Based Composites

Chitosan is obtained through the N-deacetylation of chitin and consists of (1/4)-linked 2-amino-2-deoxy-β-D-glucopyranose units. It is a linear polymer and is classified as a natural polymer due to the presence of the degradable enzyme chitosanase (Ali et al. 2024). The primary sources of chitosan are crustaceans such as shrimp, crabs, and lobster (Rani et al. 2024). This biomass of chitosan usually can be obtained from

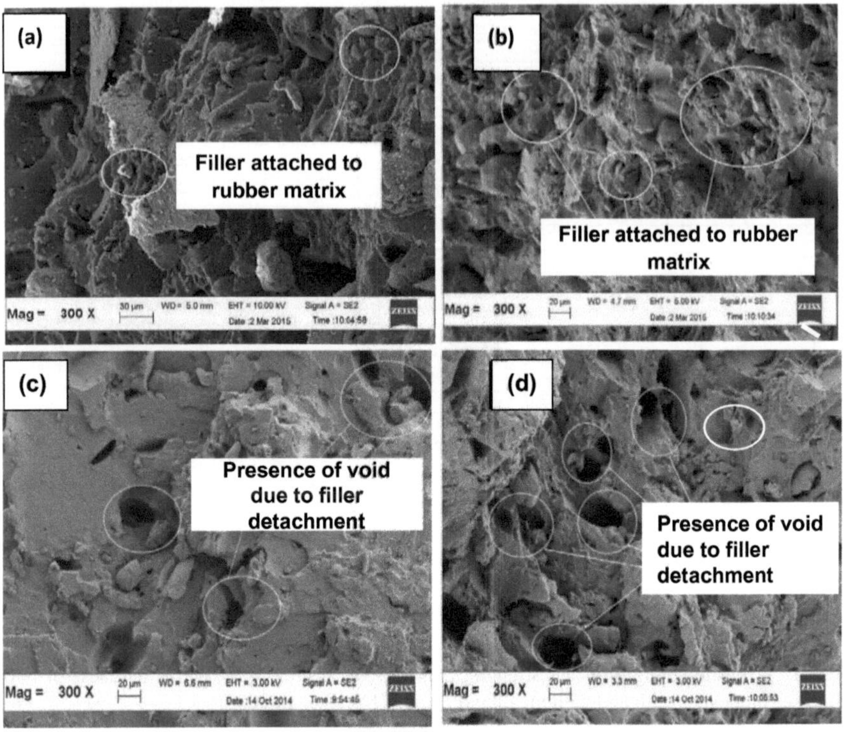

Fig. 4.3 SEM images of cellulose powder-filled NR composites: **a, b** 10 phr and 20 phr with silane coupling agent modification, and **c, d** 10 phr and 20 phr unmodified (Daud et al. 2016)

the marine and food industries. Due to its chemical and physical properties, many researchers focused on chitosan as a functional biopolymer and it is being used in a wide range of applications such as cosmetic products, water treatment, biomedical engineering, plant protection, and pharmaceutical (Yadav et al. 2024). Chitosan is a natural-based resource and it has many useful features such as antimicrobial properties, biodegradability, non-toxicity, biocompatibility, and good mechanical strength (Ul-Islam et al. 2024).

Blending two or more polymers can result in new polymer properties with a better balance of characteristics than the original polymers. In some cases, mechanical properties are enhanced based on filler parameters such as particle size, moisture content, surface area, surface reactivity, and the presence of metal oxides (Wawrzyńczak et al. 2024). The addition of chitosan powder to natural rubber reduces the scorch time and curing time of the composites. However, increasing the chitosan content in NR/chitosan composites further reduces curing time and increases maximum torque due to the stronger interaction between chitosan and NR (Rani et al. 2024).

Chitosan contains a high number of functional groups, such as hydroxyl (–OH) and amide (–NH_2), within its complex chemical structure (Pandey and Mathur 2024).

The amide group forms strong interaction bonds between chitosan and NR. However, as the amount of chitosan increases, there is a reduction in the tensile strength of NR/chitosan composites. The increased filler loading restricts the movement of the natural rubber coil entanglements (Shaari et al. 2011). Figure 4.4 illustrates the morphology of NR/chitosan composites with increasing chitosan content. In Fig. 4.4b, the rubber composite exhibits a smooth surface with 0% chitosan addition. At lower chitosan loading, the distribution is uniform, as shown in Fig. 4.4c. The circle in Fig. 4.4d highlights the increased surface roughness of NR/chitosan composites due to the higher chitosan content.

Figure 4.5 illustrates the tensile properties of NR/chitosan composites. A reduction in tensile strength and elongation at break, along with an increase in modulus at 100% elongation, is observed in the NR/chitosan composites. The increasing size and geometry of the chitosan filler contribute to a reduction in composite strength. At higher chitosan loadings, the filler struggles to form a uniform distribution, reducing the stress transfer from the rubber matrix (Chang et al. 2021). Larger filler particles create a smaller surface area, weakening the interaction between the filler and the rubber matrix (Wang 1998). The increase in modulus indicates that the presence of chitosan reduces the elasticity and flexibility of the rubber chains within the composites (Ismail et al. 2011). At high filler loadings, the interaction between rubber and filler weakens, reducing filler interaction and, consequently, reducing the tensile strength of the composites.

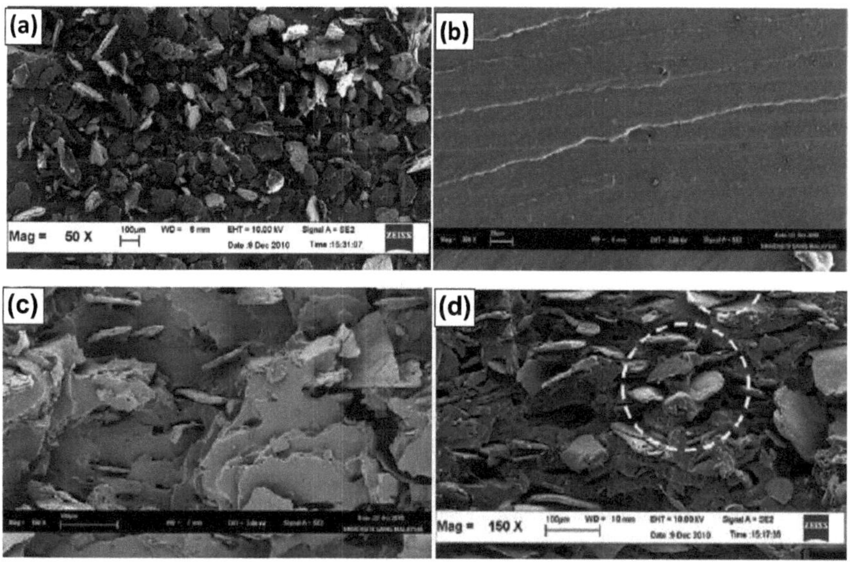

Fig. 4.4 Morphology of NR/chitosan composites with increasing chitosan loading: **a** chitosan particles, **b** compound with 0% chitosan, **c** compound with 10% chitosan, **d** compound with 40% chitosan (Shaari et al. 2011)

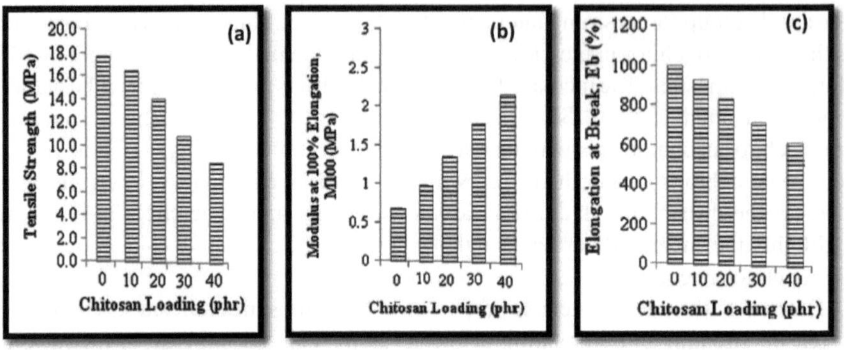

Fig. 4.5 Mechanical properties of NR/chitosan composites **a** tensile strength, **b** modulus, and **c** elongation at break (Shaari et al. 2011)

The reduction in elongation at break is caused by the increasing brittleness of the composites. This brittleness results from the transition within the rubber matrix, which becomes unstable with increased chitosan loading (Seidi et al. 2021). As filler loading increases, the rubber vulcanization fails at lower elongation values. At higher chitosan levels (phr), the elongation at break decreases further due to the dominance of filler-filler interactions.

Chitosan also affects the thermal stability of NR/chitosan composites. Table 4.2 shows that chitosan exhibits poorer thermal stability compared to NR. Rao and Johns (2008) mention that chitosan undergoes thermal degradation between 200 °C and 270 °C, while NR degrades between 287 and 400 °C. The degradation of NR/chitosan composites begins with the breakdown of chitosan crosslinking at temperatures above 350 °C, followed by the destruction of amino groups. As the chitosan content increases in the composite system, mass loss also rises, likely due to the degradation of the amino groups.

4.2.3 Natural Rubber/Starch-Based Composites

Cereal grains and tubers are rich in starch, which is a major component used in various industries, including textiles, food, and paper. Starch offers several advantages, including being biodegradable, inexpensive, renewable, and environmentally friendly (Jiang et al. 2020). Research has explored different types of starch, such as sago starch (Izmar et al. 2012), cassava starch (Ruamcharoen et al. 2023), and corn starch (Sowińska-Baranowska et al. 2022). Goodyear Tire Co., for instance, uses modified corn starch as a base filler in their natural rubber composites, branded as BioTRED. This modification aims to replace traditional fillers like carbon black and silica, thereby reducing rolling resistance, tire weight, and energy consumption during production. The result is reduced fuel consumption and lower noise levels

Table 4.2 Thermal stability of NR/chitosan composites (Johns and Rao 2008)

Sample code	Mass loss %								Activation energy, Ea/ $kJmol^{-1}$	
	100 °C	200 °C	300 °C	350 °C	400 °C	500 °C	600 °C	700 °C	I stage	II stage
CS_0NR_{100}	0.7	2.9	6.5	15.2	74.1	93.2	94.2	94.5	44.3	–
CS_5NR_{95}	1.0	3.9	8.6	18.4	77.2	97.3	97.7	98.0	43.1	–
$CS_{10}NR_{90}$	1.4	4.8	12.9	34.4	77.0	91.6	99.4	99.5	52.5	12.3
$CS_{15}NR_{85}$	1.4	8.2	15.7	25.8	72.7	89.6	91.1	92.4	55.9	27.0
$CS_{20}NR_{80}$	1.8	8.5	15.8	25.6	72.3	89.9	91.1	92.2	42.0	26.7
$CS_{35}NR_{65}$	3.0	12.4	21.9	32.0	73.7	88.0	89.2	90.4	44.2	24.1
$CS_{50}NR_{50}$	3.4	14.3	30.9	48.9	68.1	78.0	91.5	99.9	30.8	22.8
$CS_{70}NR_{30}$	6.9	18.5	37.0	50.2	71.6	79.7	81.8	83.5	35.3	7.3
$CS_{100}NR_0$	9.5	22.5	9.0	57.1	61.6	65.8	68.2	69.4	28.5	6.0

(Rødland et al. 2024). Research on NR/starch composites can be approached in several ways, including the direct addition of starch powder, gelatinization, and modified gelatinization (coagulant method) (Misman et al. 2018). Figure 4.6 presents the morphology of these three starch modifications. The images show that while starch powder and gelatinized starch remain in a rigid form, the modified gelatinization (coagulant method) results in a softer, more phase-like filler.

Several factors influence the mechanical properties of starch-filled NR composites, including interfacial strength, dispersion level, and the type of starch used (Kamal and Clarke 2013). However, starch presents several challenges as a reinforcing filler, such as its large particle size, highly polar surface (due to numerous hydroxyl groups), high cohesion energy, and relatively high softening temperature (156 °C), which make it difficult to disperse finely within the NR matrix (Liu et al. 2008). Figure 4.7 shows the mechanical properties result of NR/starch composites. It shows that directly adding starch powder to NR results in deteriorated mechanical properties, including reduced tensile strength, tear strength, and elongation at break. This degradation is attributed to poor dispersion and weak interfacial interaction between the starch and NR.

Starch in its gelatinized phase (without the coagulant method) also fails to significantly improve the mechanical properties of the composites. This is because temperatures below 156 °C are insufficient to break down or soften the starch particles, which are hindered by their high cohesion and softening temperature (Kamal and Clarke 2013). Figure 4.8 illustrates the small fractures between starch and natural rubber in both direct filling and gelatinization scenarios, highlighting the weak interfacial interaction. This weakness decreases the compatibility between hydrophilic starch and hydrophobic NR, which fails to effectively prevent starch aggregation, re-crystallization, and hydrogen bonding. The strong interaction between hydroxyl

Fig. 4.6 SEM micrographs of starch **a** natural starch powder (without modification), **b** gelatinized starch (without modification), **c** modified starch (coagulant method) (Liu et al. 2008)

groups in starch particles leads to self-aggregation, resulting in poor dispersion within the natural rubber matrix (Wang et al. 2009).

Modified NR/starch composites show improved mechanical properties compared to unmodified composites. The graft chains in modified starch are designed to prevent hydrogen bonding, reduce starch crystallization, and enhance compatibility between starch and NR (Amaraweera et al. 2021). Starch modification techniques include plasticization, mechanical smashing, gelatinization (combined with co-coagulation with natural latex), and acid hydrolysis. The modified starch exhibits better properties than unmodified starch, whether gelatinized or in powder form. Modification methods include using butyl acrylate (Yimmut et al. 2018), carbon disulfide combined with sodium hydroxide (Lee and Azura 2013), and dimethylaminoethyl methacrylate (DMAEMA) (Zhou et al. 2024).

The advantages of modifying starch in NR/starch composites include finer dispersion, reduced starch particle size, and improved interfacial strength between starch and NR. Modified starch also lowers the glass transition temperature from 156 to 84 °C (Yimmut et al. 2018) making it easier to soften during the mixing process and improving dispersion within the NR matrix.

Figure 4.9 shows the crystalline structure of both original and modified NR/starch composites, analyzed using XRD. XRD analysis reveals that natural powder

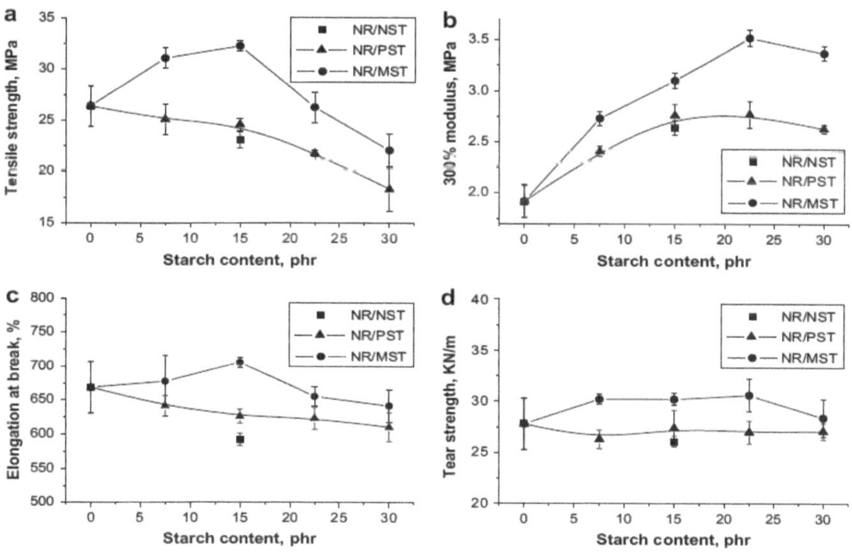

Fig. 4.7 Mechanical properties of NR/starch composites; NR/NST-direct starch powder filled, NR/PST-starch in unmodified gelatine phase form and NR/PST-starch in modified gelatine phase form. It shows **a** Tensile strength, **b** Modulus, **c** Elongation at break, and **d** Tear strength as a function of starch content in par per hundred rubber (Liu et al. 2008)

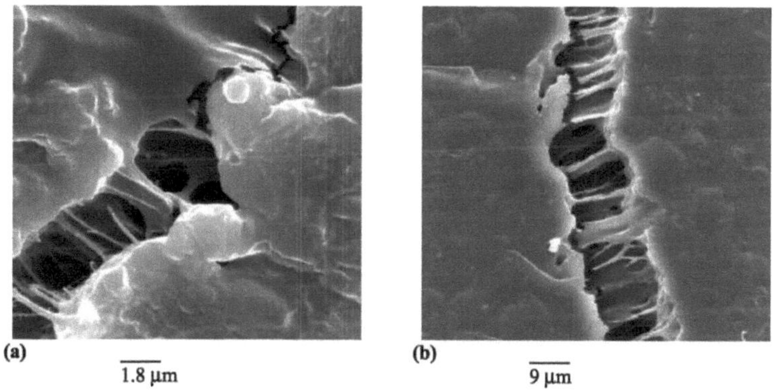

Fig. 4.8 Small fractures in NR/starch composites at magnification **a** 1.8 μm and **b** 9 μm (Carvalho et al. 2003)

starch has a semi-crystalline structure, while modification disrupts this structure. The original powder starch shows a high crystalline peak, whereas modified starch prepared using the co-coagulation method does not exhibit any crystalline peaks. This suggests that the modified starch disperses well at the molecular level within the rubber matrix (Liu et al. 2008). The XRD pattern indicates that the shearing force from the rolling mill is insufficient to break the starch particles, leading to reduced

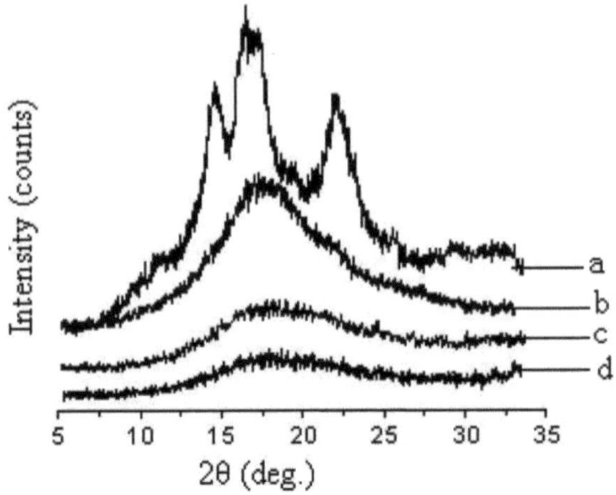

Fig. 4.9 XRD patterns of starch (a) powder starch, (b) unmodified NR/starch (gelatine), (c) modified starch, (d) NR composites (Wu et al. 2006)

mechanical strength in NR/starch composites. Overall, modified starch composites perform better than unmodified starch composites.

Wu et al. (2006) reported that the co-coagulant method can reduce the particle size of starch in NR to less than 1000 nm, resulting in significantly better composites compared to those made with direct loading. This modification lowers the glass transition and melting temperatures of the NR/starch composites. DSC analysis indicates that the glass transition temperature of NR decreases from −70 to −60 °C, while the melting temperature of starch drops from 156 to 84 °C (Feng et al. 2022). Consequently, modified NR/starch composites are more easily broken, softened, and dispersed into the NR phase during compounding and curing. The improved dispersion of fine, grafted starch enhances the mechanical properties of the composites, leading to higher tensile strength. Modified starch, with its lower polarity, provides better dispersion in the NR matrix compared to unmodified starch, which has higher polarity (Liu et al. 2008).

4.2.4 Natural Rubber/PLA-Based Composites

Poly(lactic acid) (PLA) is widely used as a biodegradable plastic, particularly in medical and food packaging applications, due to its transparency, mechanical properties, and suitability for dry goods. PLA is derived from biomass sources such as sugarcane, corn starch, and beet (Sin 2012). While PLA has excellent clarity, high mechanical strength, and modulus, its use is limited by its brittleness and slow crystallization rate (Saeidlou et al. 2012).

Previous studies have explored the use of natural rubber (NR) to enhance the mechanical properties of PLA (Tessanan et al. 2024; Pongpakdee et al. 2024; Ma et al. 2024). NR is chosen for its high elongation at break, which helps mitigate the brittleness of PLA. In NR/PLA composites, NR typically serves as a secondary polymer

Table 4.3 Degree of crystallization, Xc of PLA and PLA/NR composites (Injorhor et al. 2024)

Sample	Degree of crystallization, Xc (%)
Neat PLA	9.91
PLA/$CaCO_3$	61.76
PLA/talc	47.34
PLA/a-CD	34.66
PLA/NR10	19.70
PLA/NR10/$CaCO_3$	29.19
PLA/NR10/talc	27.24
PLA/NR10/α-CD	17.14

to address the weaknesses of PLA. NR acts as a stress concentrator, improving the energy absorption of the brittle PLA and thus enhancing the material's toughness (Pongpakdee et al. 2024). These composites are produced by blending NR into the PLA matrix and incorporating inorganic fillers as nucleating agents. These nucleating agents can accelerate the crystallization rate of PLA and improve the overall mechanical properties of the NR/PLA composites.

Nucleating agents are effective in enhancing the crystallization of PLA when used directly with the PLA matrix, but their impact is limited in PLA/NR blends and composites. In these blends, nucleating agents provide only a modest increase in the degree of crystallization. Table 4.3 presents the degree of crystallization (Xc) for PLA and PLA/NR composites. PLA spherulites are generally not highly sensitive to the presence of natural rubber (Injorhor et al. 2024). However, the addition of fillers, such as nucleating agents, results in smaller spherulites in PLA/NR composites.

Figure 4.10 illustrates the spherulite structure of PLA/NR composites. Figure 4.10a shows that the spherulite size in PLA/NR composites is similar to that of pure PLA (40 μm), indicating that NR has minimal impact on the spherulite size in the composites. With the addition of nucleating agents, as depicted in Fig. 4.10b, c, the spherulite size decreases, with $CaCO_3$ producing a size of 20 μm and talc reducing it further to 10 μm. When α-CD is used as a nucleating agent, the spherulite size in PLA becomes comparable, resulting in better dispersion within the composite phase, as shown in Fig. 4.10d.

Figure 4.11 illustrates the tensile strength and elongation at the break of PLA/NR composites. Blending PLA with NR decreases the tensile strength of the PLA but increases the elongation at the break of the PLA/NR blend, owing to the excellent elongation properties of natural rubber. However, the addition of nucleating agents does not significantly impact the tensile strength and elongation of the composites. Blending with NR makes PLA more ductile (Suksut and Deeprasertkul 2011). Additionally, PLA/NR composites exhibit improved impact toughness. The high degree of crystallinity and smaller spherulite size is advantageous for high strain rate applications, as the smaller spherulites increase the number of tie molecules, leading to improved impact toughness. The elastic nature of NR further contributes to the high toughness of the material.

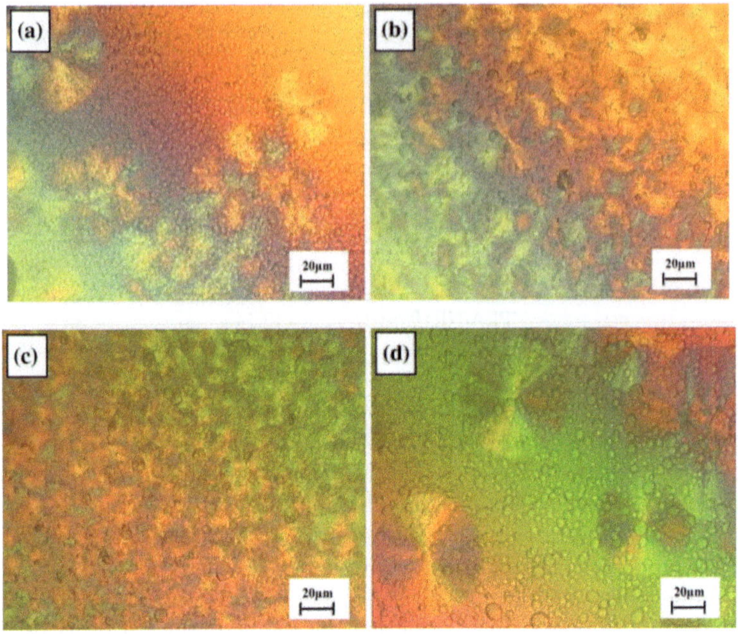

Fig. 4.10 Optical micrographs of **a** PLA/NR blend, **b** PLA/NR with $CaCO_3$, **c** PLA/NR with talc, **d** PLA/NR with α-CD (Suksut and Deeprasertkul 2011)

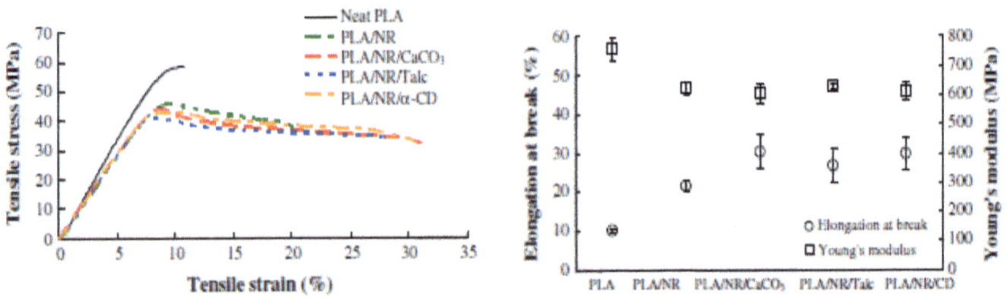

Fig. 4.11 Tensile strength and elongation at break of PLA/NR composites (Suksut and Deeprasertkul 2011)

4.2.5 *Natural Rubber/Lignin-Based Composites*

Lignin is a three-dimensional amorphous natural phenolic polymer derived from various plants and contains hydroxyl, methoxyl, carbonyl, and carboxyl groups (Agustiany et al. 2022). As a stable cell-wall polymer, lignin offers potential as a reinforcing material in natural rubber compounds due to its stabilizing properties. It is abundant, relatively inexpensive, and has garnered significant attention in plastic engineering applications, such as phenol–formaldehyde resins, polyurethane foams,

Table 4.4 Cure characteristics of NR/lignin composites (Yu et al. 2016)

Sample		Minimum torque, ML	Maximum torque, MH	Scorch time, t_2 (min)	Cure time, t_{90} (min)
NR/lignin/silica composites	NR/50 silica	3.3	20.0	2.4	8.5
	NR/40 silica/10 lignin	3.2	19.3	4.3	21.3
	NR/30 silica/20 lignin	2.6	17.9	4.5	26.3
	NR/20 silica/30 lignin	1.9	16.4	5.1	29.1
	NR/10 silica/40 lignin	2.0	14.3	6.4	29.0
	NR/50 lignin	2.0	3.3	6.6	32.545

and other products (Melro et al. 2021). Lignin acts as an active filler, enhancing reinforcement in NR/lignin composites. Their research also found that lignin serves as an excellent binder and stabilizer in cellulose-based materials, suggesting its potential for reinforcement in NR/lignin systems (Barana 2017).

Table 4.4 demonstrates that the use of lignin as a filler slightly affects scorch time, curing time, and torque in NR composites. They observed that lignin alters these parameters in NR/silica/lignin composites (Yu et al. 2016). Specifically, increased lignin content leads to prolonged scorch and curing times, attributed to the large number of hydroxyl groups in lignin, which absorb rubber additives rather than silica, thus slowing down vulcanization. Consequently, higher lignin loading delays the curing process when replacing silica as a filler.

Based on Table 4.5, lignin exhibits active properties that enhance the mechanical performance of NR/lignin composites. Using lignin as a reinforcing filler improves tensile modulus, tensile strength, and elongation, although no significant effect is observed on hardness. This can be attributed to lignin's low molecular weight and polydispersity, which facilitate good miscibility with NR (Li et al. 2022). However, when lignin is used as a replacement for harder fillers like silica, tensile strength, and hardness tend to decrease, as lignin is a softer filler that cannot provide the same reinforcement as silica. Additionally, crosslink density decreases with higher lignin content because lignin is more reactive than silica. Composites with increased lignin content exhibit lower crosslink density (Mohamad Aini et al. 2020). Elongation at break increases with lignin incorporation because the soft filler distributes well within the NR matrix, creating smooth surface distribution and small holes in the composite surface. These holes facilitate bonding during tensile testing, resulting in enhanced elongation properties.

Increasing lignin concentration also improves the thermal stability of NR composites. Table 4.6 illustrates the thermal decomposition of NR/lignin composites, with the temperature of 50% decomposition (T_{50}) serving as an indicator of thermal stability. Higher lignin content enhances thermal stability due to the stabilizing effect of hydroxyl groups. However, in the replacement technique, thermal stability decreases

Table 4.5 Mechanical properties of NR/lignin composites

Sample		Modulus 100% (MPa)	Tensile strength (MPa)	Elongation at break, %	Hardness, shore A	References
NR/lignin/silica composites	NR/50 silica	–	26	500	65	Yu et al. (2016)
	NR/40 silica/10 lignin	–	25	540	58	
	NR/30 silica/20 lignin	–	24	550	57.5	
	NR/20 silica/30 lignin	–	21	570	54	
	NR/10 silica/40 lignin	–	20	610	54	
	NR/50 lignin	–	19	625	53.5	
NR/lignin composites	NR/0 lignin	0.76	16.97	698	44.7	Datta and Parcheta (2017)
	NR/10 lignin	0.86	13.84	725	47.0	
	NR/40 lignin	1.01	10.55	645	52.8	

Table 4.6 Characteristic thermal decomposition of NR/lignin composites

	Decomposition temperature, °C					References
	5%	10%	50%	80%	*T* max	
NR/50 silica	–	349.2	399.9	–	376.8	Yu et al. (2016)
NR/40 silica/10 lignin	–	345.2	396.6	–	378.0	
NR/30 silica/20 lignin	–	341.2	393.8	–	377.9	
NR/20 silica/30 lignin	–	337.9	393.2	–	377.4	
NR/10 silica/40 lignin	–	333.3	391.5	–	376.8	
NR/50 lignin	–	329.4	388.1	–	376.3	
NR/0 lignin	321.4	–	393	407.1	388	Datta and Parcheta (2017)
NR/10 lignin	322.6	–	393	426.4	387	
NR/20 lignin	327.0	–	396	433.4	387	

because silica, an inorganic material, offers superior thermal stability compared to lignin, which is an organic material with lower thermal resistance.

4.2.6 *Natural Rubber/Shell-Based Composites*

Shells, derived from natural renewable resources such as eggshells, cuttlebones, and seashells, are rich in calcium carbonate ($CaCO_3$) and can serve as reinforcing fillers in natural rubber composites (Golakiya 2020). Bio-shells are commonly used in various applications, including cosmetics, biomedical devices, ceramics, coating pigments, and as fillers in thermoplastics. The advantages of using these fillers include their low cost, lightweight nature, and biodegradability (Piras et al. 2024).

The addition of bio-shell filler did not hinder peroxide crosslinking in NR composites, as network chain densities increased even with higher shell filler content. The tensile strength, tensile modulus, and elongation improved due to the small particle size and the presence of chitin, which enhances reinforcement in the composites. The chitin is one of the key components of natural bio-shells and is found in small amounts and covalently linked to proteins. In contrast, calcium carbonate ($CaCO_3$) without chitin, such as talc, leads to a reduction in tensile strength and elongation in NR/bio-shell composites. However, an increase in bio-shell content makes the composites more brittle, reducing their elasticity and tensile properties. Bio-shell, being a rigid filler, increases the hardness of the composites, thus leading to greater brittleness.

The thermal properties of natural rubber filled with bio-shell particles were evaluated using dynamic mechanical analysis (DMA). Figure 4.12 shows the storage modulus (E) and Tan Delta (tan δ) versus temperature for the samples. Both storage modulus and tan δ peak trends remained similar with increasing filler content. The storage modulus increased while the tan δ peak decreased as more filler was added to the composites. Additionally, the glass transition temperatures, as indicated by tan δ peaks, remained consistent across samples, indicating that bio-shell filler did not significantly impact the thermal behavior of NR/shell composites.

4.3 Natural Rubber-Based Nanocomposites

Nanocomposites are materials reinforced with fillers that range in size from 1 to 100 nm in diameter (Njuguna et al. 2008). Compared to micro-sized fillers, nanocomposites offer superior properties, including enhanced mechanical strength with a smaller volume of filler. This lower filler loading not only improves processing but also reduces the overall weight of the composite. Nanocomposites also provide added benefits such as high stiffness, reduced permeability, optical clarity, and improved electrical conductivity (Njuguna et al. 2008). These attributes make them particularly useful for applications in food, beverage, and pharmaceutical packaging, where enhanced barrier properties, strength, and conductivity are essential.

Nanoparticles used in nanocomposites typically vary in shape, including spheres, platelets, and rods. Their characteristic small dimension depends on the particle's geometry—such as the diameter for spheres and rods, or the thickness for platelets

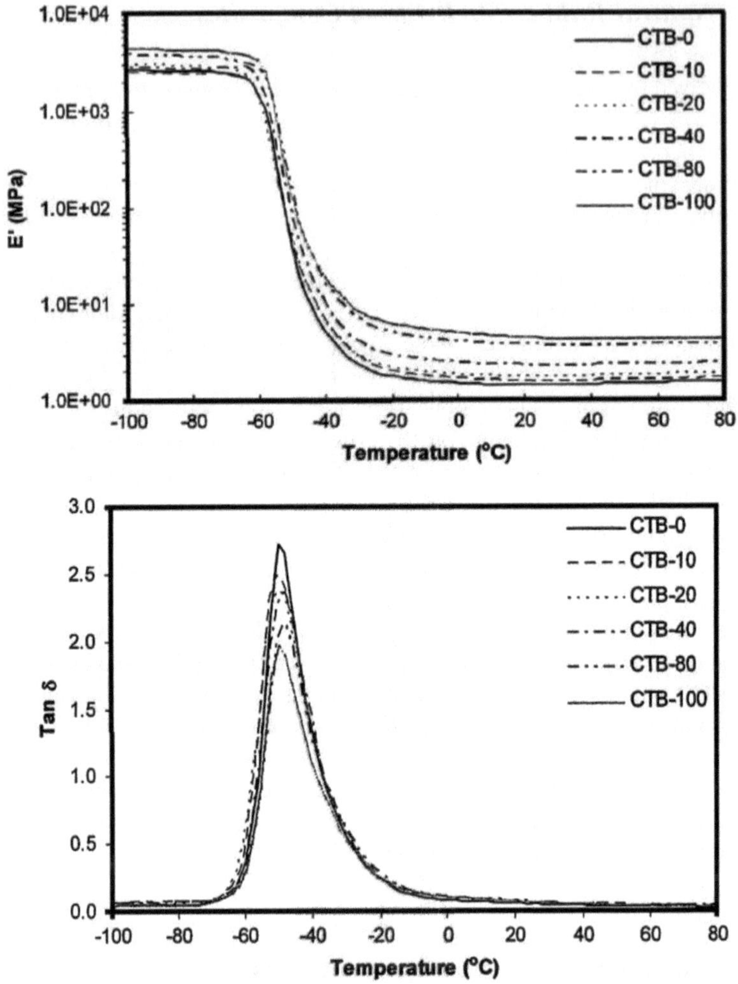

Fig. 4.12 Storage modulus and tan delta of NR/shell composites (Poompradub et al. 2008)

and scales. This unique morphology contributes to the distinct properties of nanocomposites, making them highly versatile in a wide range of applications.

4.3.1 Natural Rubber/Cellulose-Based Nanocomposites

Nanocellulose refers to the nanoscale particles derived from natural cellulose found in plants and agricultural resources (Thomas et al. 2018). Like its macro-sized counterpart, nanocellulose offers advantages such as biodegradability, renewability, non-toxicity, low cost, and high organic content (Zinge and Kandasubramanian 2020). Researchers are increasingly interested in nanocellulose due to its nanoscale dimensions and eco-friendly properties, which make it suitable for applications requiring rapid environmental breakdown (Xue et al. 2017; Thakur et al. 2021). Nanocellulose-based nanocomposites are ideal for use in biodegradable polymer products, as they can decompose quickly, making them environmentally sustainable.

Studies on the morphology of NR/cellulose nanocomposites reveal that nanocellulose fillers can be dispersed homogeneously in the NR matrix without micro-scale aggregation (Visakh et al. 2012). As shown in Fig. 4.13, the fracture surface of neat NR is similar to that of nanocomposites containing nanocellulose fillers. The addition of nanocellulose increases surface roughness but does not create major micro-scale aggregates when using small amounts of filler (Fig. 4.13a–c). However, micro-scale aggregates become apparent with higher filler concentrations, as depicted in Fig. 4.13d.

According to Table 4.7, both tensile strength and Young's modulus increase steadily with higher nanocellulose filler loading in NR/cellulose nanocomposites.

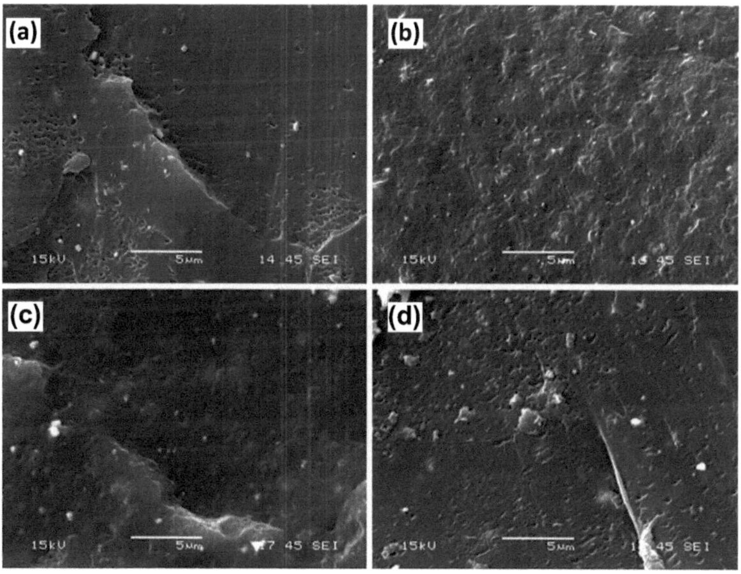

Fig. 4.13 SEM image of NR/cellulose nanocomposites: **a** Neat NR, **b** NR-CNW 2.5%, **c** NR-CNW 5%, **d** NR-CNW 10% (Visakh et al. 2012)

This enhancement is due to the restriction of polymer chain mobility near the cellulosic nanoparticles (Mariano et al. 2014). Conversely, elongation at break decreases with increased filler content, which reduces the elasticity of the nanocomposites. This reduction occurs because the formation of a strong network between the rubber polymer chains and the interaction between rubber and nanocellulose diminishes the material's ability to resist deformation (Kargarzadeh et al. 2017). At higher filler loadings, the elongation at break decreases further due to the increased dominance of filler-filler interactions within the compound.

Figure 4.14 shows that the thermal stability of the nanocomposites improves with increasing nanocellulose filler content. The decomposition temperatures at various stages rise with the addition of nanocellulose filler. This enhancement in thermal stability is attributed to the reduced mobility of the NR phase near the nanocellulose, which slows the diffusion of degradation products from the system.

Table 4.7 Mechanical properties of NR/cellulose nanocomposites

Sample of nanocomposites	Nanocellulose content (%)	Tensile strength (MPa)	Elongation break (%)	Young modulus (MPa)	References
NR/bamboo nanocomposites	0	9.2	554	1.7	Visakh et al. (2012)
	2.5	14	539	2.6	
	5	14.5	477	3.0	
	10	17.3	455	3.8	
NR/palm nanocomposites	0	0.56	575	0.5	Bendahou et al. (2010)
	1	0.86	408	1.7	
	2.5	1.17	358	2.8	
	5	2.71	231	8.4	
	10	8.93	16	11.8	
	15	12.15	14	18.7	
NR/jute fiber nanocomposites	*Non-crosslink*				Abraham et al. (2012)
	0	0.7	816	0.85	
	2.5	3.1	477	2.7	
	5	4.2	325	3.8	
	7.5	6.6	205	5.7	
	10	9.1	101	7.2	
	Crosslink				
	0	1.6	912	1.3	
	2.5	5.2	576	4.2	
	5	6.8	413	6.3	
	7.5	9.8	275	8.1	
	10	12.2	144	9.6	

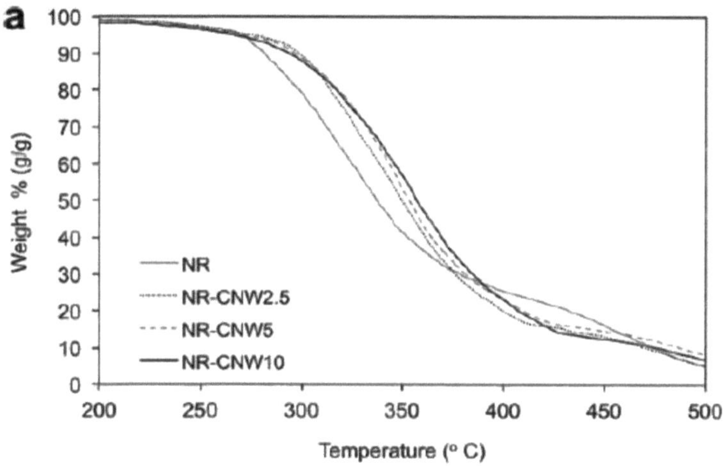

Fig. 4.14 TGA of rubber composites filled with various ratios of nanocellulose bamboo (Visakh et al. 2012)

Figure 4.15 shows that the glass transition temperature, as indicated by the peak in tan *δ*, is around −47 °C for the NR matrix. This peak shifts to higher temperatures with increasing nanocellulose filler loading. The intensity of the relaxation process decreases as the filler content increases, likely due to a reduction in the amount of matrix material and the restricted mobility of the NR polymer chains.

Biodegradation is a process where microorganisms attack and break down natural materials, leading to a reduction in the weight and strength of the composites (Brebu 2020). Cellulose generally degrades faster than natural rubber (NR). As shown in Fig. 4.16a, non-vulcanized composites exhibit higher weight loss and faster biodegradation compared to vulcanized composites. Figure 4.17a demonstrates that microorganisms colonize and attack the non-vulcanized system more rapidly. The rate of degradation depends on void formation, increased porosity, and the loss of integrity in the rubber matrix (Bras et al. 2010). In contrast, vulcanized systems have a robust cross-linked network that limits microbial access and reduces the rate of degradation. The addition of cellulose accelerates the biodegradation process of the composites. As shown in Figs. 4.16b and 4.17b, vulcanized nanocomposites degrade more slowly, with relatively low weight loss. The crosslinking in vulcanized nanocomposites restricts microorganism access, slowing the degradation process (Abraham et al. 2012).

4.3.2 Natural Rubber/Chitin-Based Nanocomposites

Chitosan is derived from chitin, which is found in the exoskeletons of crustaceans like crabs, shrimp, zooplankton, and insects (Nasirov et al. 2023). Chitosan can take

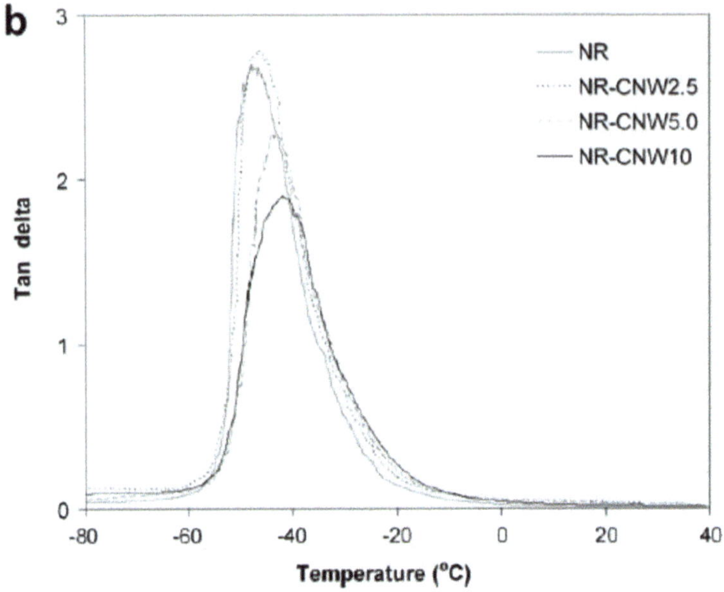

Fig. 4.15 Loss tangent (tan δ) versus temperature for NR/nanocellulose bamboo nanocomposites (Visakh et al. 2012)

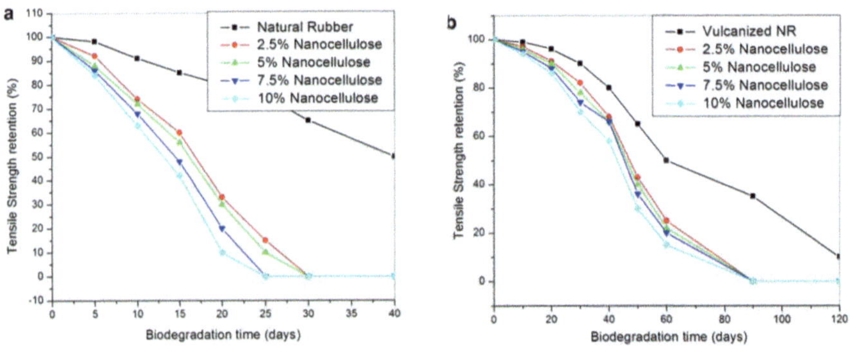

Fig. 4.16 Tensile strength reduction due to biodegradation process: **a** non-vulcanized nanocomposites, **b** vulcanized nanocomposites (Abraham et al. 2012)

various forms, including slender, parallel, and rod-like shapes. Its benefits as a natural filler include low density, renewability, and biodegradability (Bano et al. 2024). In research, NR/chitosan nanocomposites are typically produced by incorporating colloidal chitosan suspension into NR latex (Goosen 1996).

Modification of chitin can be achieved using coupling agents. Figure 4.18 illustrates the morphology of chitin particles and their nanocomposites, both modified and unmodified with a coupling agent. Original chitin particles are characterized by

Fig. 4.17 Degradation process of **a** non-vulcanized and **b** vulcanized NR/cellulose nanocomposites (Abraham et al. 2012)

their sharp rod-like shape, which is easy to identify and measure. In contrast, modified chitin exhibits an entangled structure, making it more challenging to identify and measure (Gopalan Nair et al. 2003).

Tensile strength will increase with the increase in the chitin percentage. In the rubbery state, the elasticity of chitin nanocomposites increases with increase in the

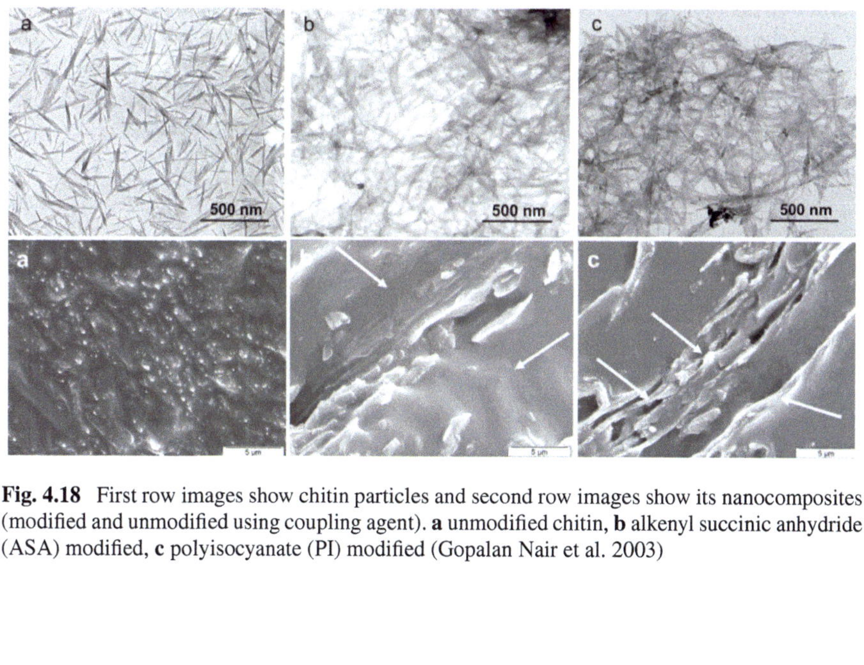

Fig. 4.18 First row images show chitin particles and second row images show its nanocomposites (modified and unmodified using coupling agent). **a** unmodified chitin, **b** alkenyl succinic anhydride (ASA) modified, **c** polyisocyanate (PI) modified (Gopalan Nair et al. 2003)

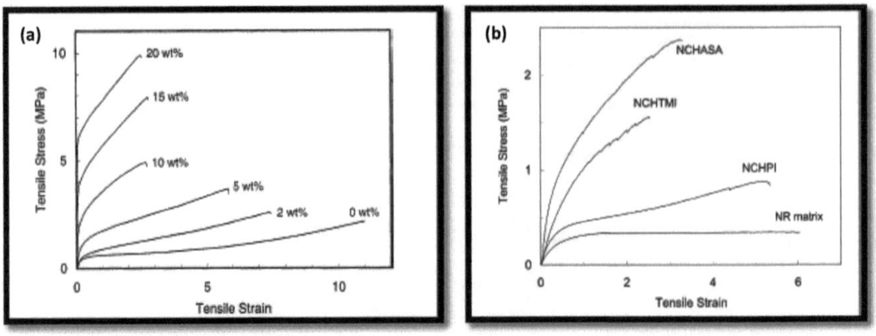

Fig. 4.19 Tensile stress of NR/chitin nanocomposites: **a** unmodified and **b** modified nanocomposites (10 wt%) (Gopalan Nair et al. 2003)

percentage. The soft particle chitin and small particle size make its uniform distribution in nanocomposites. According to Gopalan Nair et al. (2003), unmodified chitin particles cannot be dispersed in toluene because toluene will destroy the chitin properties in nanocomposites. By using a coupling agent for modification, modified chitin can be dispersed in toluene. It also will increase the tensile strength rather than the natural rubber matrix. Performance using coupling agents strongly decreases because the tensile strength of unmodified is higher rather than modified nanocomposites. Based on Fig. 4.19, increasing chitin in NR will increase the tensile strength of the nanocomposites. By using a coupling agent, there are lower increments of tensile strength, and it is less strength than unmodified natural rubber filler.

Figure 4.20 demonstrates that increasing the loading of nanochitin does not significantly affect the glass transition temperature (T_g) of the nanocomposites. However, higher chitin content leads to an increase in the storage tensile modulus of the nanocomposites. In the relaxation process observed at the glass transition, more energy is required to transition the polymer chains into the rubbery plateau region.

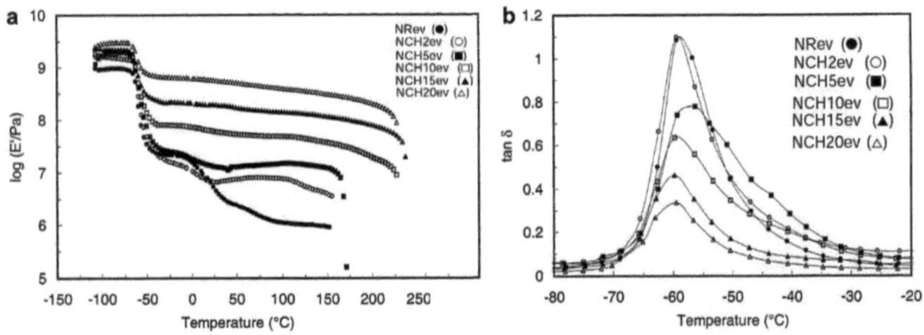

Fig. 4.20 DMA thermographs: **a** log of storage modulus (E') and **b** Tan δ as a function of temperature for NR/chitin nanocomposites (Gopalan Nair and Dufresne 2003)

Figure 4.20 presents dynamic mechanical analysis (DMA) thermographs showing that the T_g remains consistent at approximately −60 °C across all nanocomposite samples. Gopalan Nair et al. (2003) also reported that no distinct melting peak is observed in the DMA analysis of natural rubber/chitin nanocomposites. This lack of a melting peak is attributed to the low degree of crystallinity in chitin, which makes it challenging to detect with DMA.

4.3.3 Natural Rubber/Starch-Based Nanocomposites

Starch is a natural, renewable, and biodegradable resource derived from various crops such as rice, potatoes, oats, sago, cassava bagasse, and wheat. It consists of α-D-glucose units, primarily in the forms of amylose and amylopectin (Höfer 2015). However, the hydrophobic nature of normal starch can lead to poor compatibility with the natural rubber matrix, resulting in suboptimal performance of the final composites (Manimaran et al. 2024). To enhance the compatibility between starch and the natural rubber matrix, starch can be modified through physical or chemical methods. Nanocrystal starch, typically produced via acid hydrolysis (e.g., using sulfuric acid), has a platelet-like structure and offers advantages such as high stiffness, strength, and elastic behavior (Murtaza et al. 2024).

Figure 4.21a shows the smooth and uniform surface of natural rubber (NR). As starch loading increases, the surface begins to exhibit homogeneous white dots, as seen in Fig. 4.21b–d. Figure 4.21e–h demonstrate that modified starch creates a rough surface due to non-uniform dispersion, resulting from the formation of large, compact aggregates during the modification process. Ungrafted nanocrystal starch provides better uniformity compared to modified nanocrystal starch [grafted with Alkenyl Succinic Anhydride (ASA) and Polyisocyanate (PI)], which tends to be less uniform.

Figure 4.22 illustrates that tensile strength and modulus increase sharply with rising nanocrystal starch loading in NR/starch nanocomposites, while elongation at break decreases significantly. The improvement in mechanical performance is attributed to the rigid filler's ability to endure stress and the strong interaction between nanocrystal starch and the polymer matrix (Sowińska-Baranowska et al. 2022). Although modification of nanocrystal starch enhances adhesion between the filler and matrix, it may also lead to a sharp decrease in mechanical performance due to the partial or destruction of the three-dimensional network within the nanocomposites (Ruamcharoen et al. 2023).

Figure 4.23 shows that the XRD patterns of nanocrystal starch particles and NR/starch nanocomposites exhibit similar peak patterns. Since natural rubber is a fully amorphous polymer, no crystallinity peaks are present in the composites. However, the addition of nanocrystal starch strengthens the A-type peak, indicating a higher degree of crystallinity in the NR/starch nanocomposites.

PLA and NR have a weakness of lack of compatibility between these two polymers. Because of that, compatibilizer is required to improve the morphology of

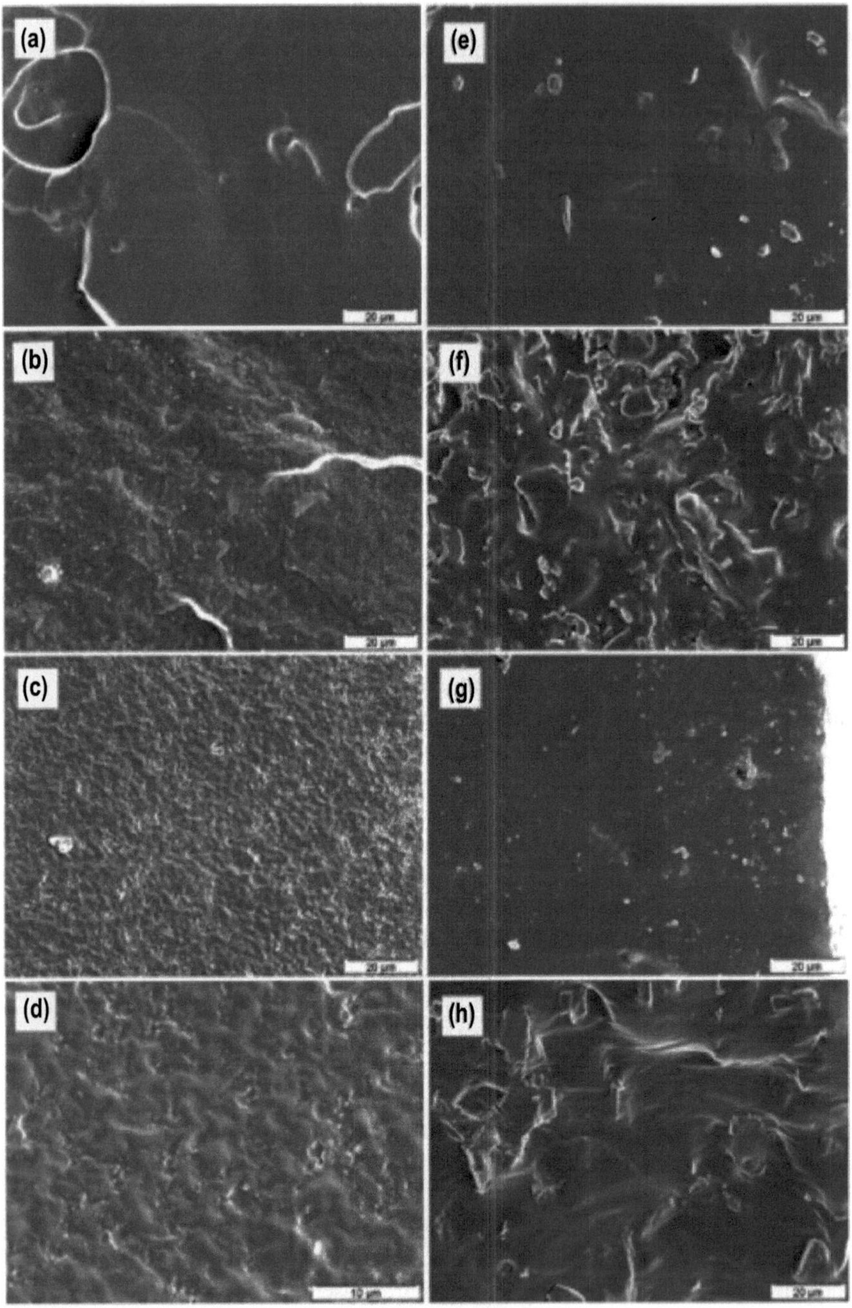

Fig. 4.21 Surface morphology of: **a** natural rubber; **b**, **c**, and **d** natural rubber with ungrafted nanocrystal starch particles; **e** and **f** natural rubber with nanocrystal starch particles grafted with alkenyl succinic anhydride (ASA); **g** and **h** natural rubber with nanocrystal starch particles grafted with polyisocyanate (PI) (Corre and Angellier-Coussy 2014)

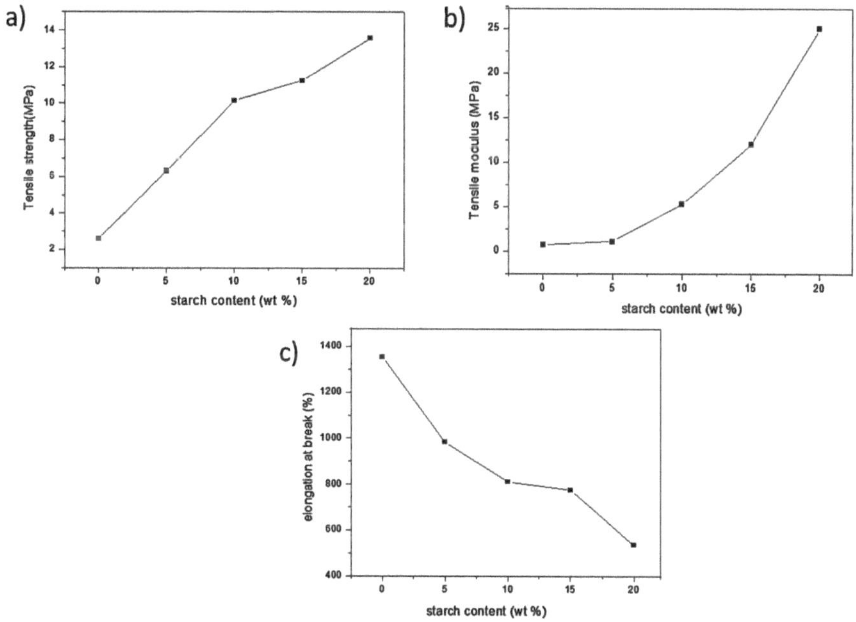

Fig. 4.22 Mechanical properties of NR/starch nanocomposites (Rajisha et al. 2014)

these polymer nanocomposites. Many researchers used organoclay to improve the quality of PLA and NR. Nanofiller (organoclay) acts as a compatibilizer between two polymer phases. The advantages of organoclay are the high surface area and high aspect ratio will improve the physical properties of the nanocomposite systems (Mogy and Lawandy 2024). Two types of organoclay that usually used, it is C15A (modified with dimethyl dehydrogenated tallow alkyl ammonium cation) (Mogy and Lawandy 2024) and C30B (modified with bis (2-hydroxyethyl)-methyl tallow alkyl ammonium cation) (Misman et al. 2018).

A key factor in understanding the compatibilization mechanism in nanocomposites is the location of the nanofiller within the polymer blend. This factor plays a crucial role in stabilizing the system by reducing the interfacial tension between the polymers. The presence of rigid fillers around the dispersed polymer helps inhibit coalescence. Changes in the viscosity ratio between the polymers and the interaction between the filler and polymer can significantly affect the mechanical properties of the nanocomposites.

The addition of organoclay to NR/PLA nanocomposites enhances their properties. As NR serves as the second phase in the matrix, the presence of organoclay markedly decreases the size of the natural rubber particles. This reduction in particle size results in a more homogeneous and uniform distribution, improves compatibilization, and alters the viscosity ratio of the blended phases.

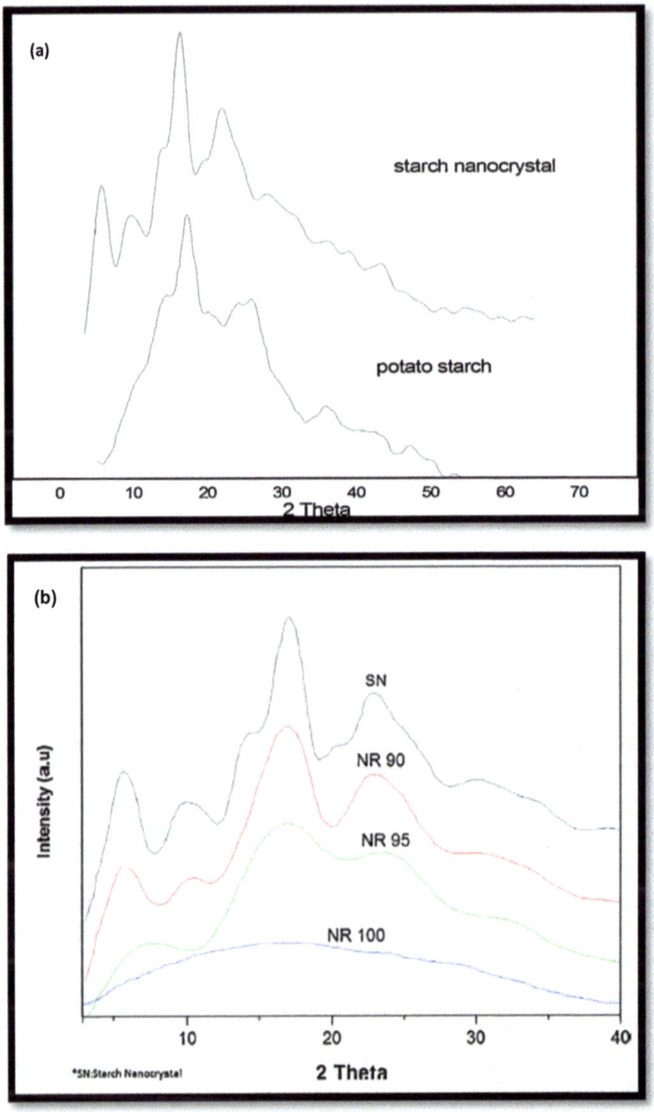

Fig. 4.23 XRD patterns of: **a** nanocrystal starch particles and **b** NR/starch nanocomposites (Rajisha et al. 2014)

4.3.4 Natural Rubber/PLA-Based Nanocomposites

Poly(lactic acid) (PLA) is a renewable biomass polymer known for its excellent mechanical properties, biodegradability, biocompatibility, thermal stability, and ease of production (Li et al. 2020). However, PLA's major drawbacks include brittleness,

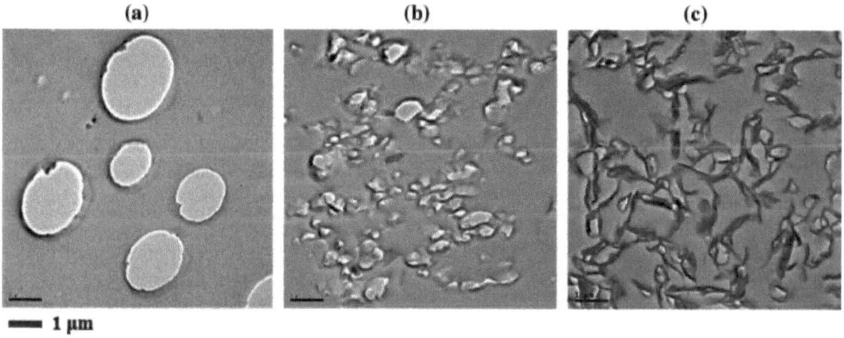

Fig. 4.24 Droplet size with **a** 1 wt%, **b** 3 wt%, **c** 5 wt% of Ock et al. (2016)

a low crystallization rate, and poor gas barrier properties. Most polymers are immiscible with each other, leading to separation during the blending process (Li et al. 2020). To address these weaknesses, strategies such as using plasticizers and incorporating rubber can be employed. Since both PLA and natural rubber (NR) originate from natural resources, they can be used to create eco-friendly polymer blends. Figure 4.24 illustrates that increasing clay concentration enhances the viscosity ratio, dispersion, and modulus of the nanocomposites while reducing the droplet size. The degree of dispersion is crucial in NR/PLA blends. Higher clay concentrations lead to smaller droplet sizes of natural rubber and promote the formation of a branching structure.

Table 4.8 shows that while elongation at break increases with higher clay concentration, tensile stress and Young's modulus decrease in NR/PLA nanocomposites. The high elongation properties of NR help transform the brittle PLA into a more ductile material. Among the tested nanocomposites, a 1 wt% clay concentration is the most effective. At this concentration, the droplet size and distribution remain consistent at the interface. The improved tensile strength, Young's modulus, and elongation at 1 wt% clay concentration are due to the clay's role in facilitating stress alignment and redistribution among polymer chains. Additionally, hydrogen bonding contributes to the increased Young's modulus by enhancing the interaction between the clay and the continuous phases in the nanocomposites.

The degradation process of NR/ PLA nanocomposites happens due to many factors like hole formation, water absorption, the formation of molecular weight compounds, or an increase in PLA crystallinity rate. PLA and NR are natural resources that can be degraded by microorganisms. Figures 4.25 and 4.26 shows the degradation process of the nanocomposites. Nanocomposites without loading organoclay are faster in the degradation process rather than nanocomposites with organoclay loading. The organoclay is homogeneously dispersed in the PLA matrix and its improvement of barrier properties slow the degradation process (Bitinis et al. 2014).

Table 4.8 Mechanical properties of NR/PLA nanocomposites (Bitinis et al. 2012)

	Young modulus (MPa)	Tensile strength (MPa)	Elongation at break (%)
Processed PLA	2710	65	3.9
NR/PLA	2240	42.9	161
NR/PLA/C15A 1 wt%	2200	37.5	196
NR/PLA/C15A 3 wt%	2290	30.1	82
NR/PLA/C15A 5 wt%	2050	27.3	42
NR/PLA/C30B 1 wt%	2210	38.2	118
NR/PLA/C30B 3 wt%	2420	39.6	7.6
NR/PLA/C30B 5 wt%	2630	38.1	2.7

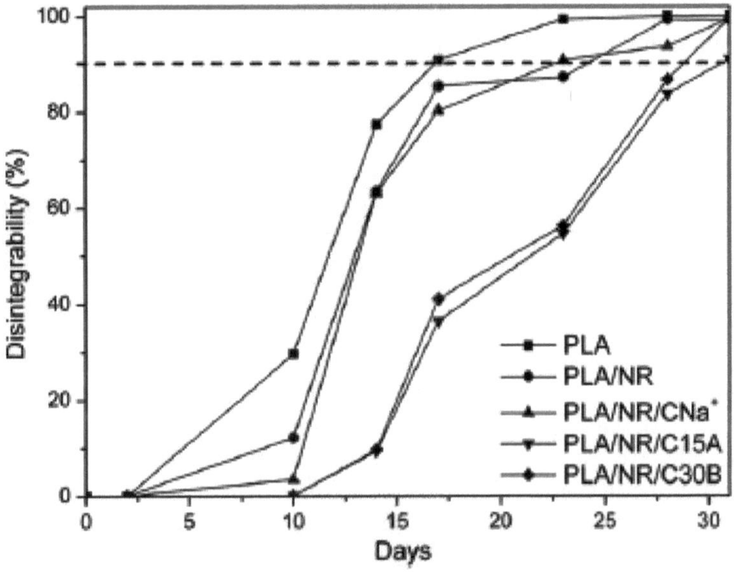

Fig. 4.25 Degradation of PLA/NR nanocomposites (Bitinis et al. 2014)

4.4 Conclusion and Future Perspective

There are various types of natural fillers that can be used as reinforcements in natural rubber composites, including cellulose, chitosan, starch, lignin, and others. These materials can naturally be processed into micro and nano sizes. Given the unsustainable nature of carbon black, natural resources offer a practical alternative for reinforcing fillers. In filled natural rubber composite systems, some fillers require modification to enhance their quality and performance.

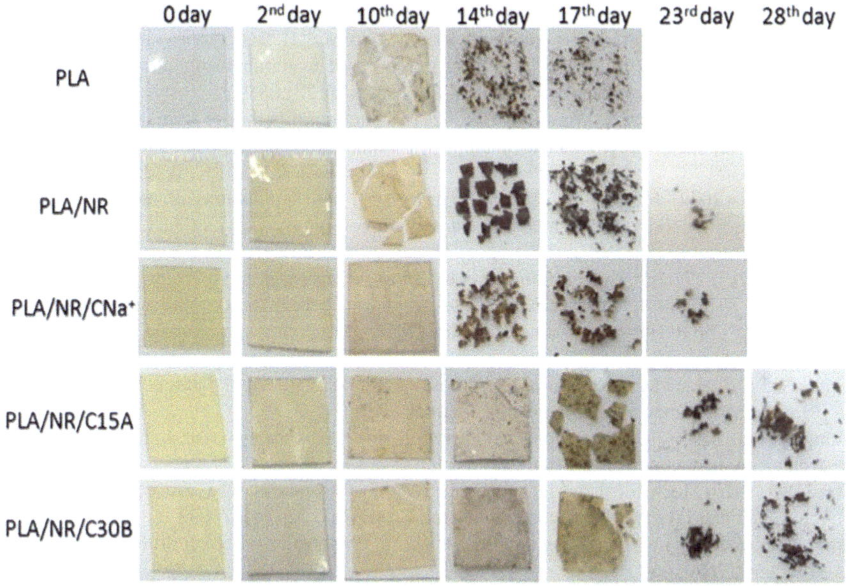

Fig. 4.26 Degradation process of PLA/NR nanocomposites (Bitinis et al. 2014)

The effectiveness of a reinforcing filler generally depends on several factors, including particle size (ideally smaller than 100 nm), particle structure, specific surface area, and surface chemical activity. For optimal reinforcement, both the size of the filler and its interaction with the rubber matrix are crucial. Typically, smaller filler sizes and appropriate interaction between the filler and rubber enhance the rubber's strength and modulus. Cellulose, starch, and nanocrystal PLA can benefit from specific modification techniques to improve their mechanical properties.

Research interest in alternative reinforcement fillers for natural rubber composites and nanocomposites is increasing. While many studies focus on cellulose, starch, and PLA, fewer have explored chitosan, lignin, and other novel fillers. This indicates that there is significant potential for discovering new fillers with reinforcing capabilities. Utilizing natural fillers not only helps protect the environment but also reduces the cost of rubber composites.

References

Abraham E et al (2012) X-ray diffraction and biodegradation analysis of green composites of natural rubber/nanocellulose. Polym Degrad Stab 97(11):2378–2387

Agustiany EA et al (2022) Recent developments in lignin modification and its application in lignin-based green composites: a review. Polym Compos 43(8):4848–4865

Ali G et al (2024) Applications of chitin and chitosan as natural biopolymer: potential sources, pretreatments, and degradation pathways. Biomass Convers Biorefin 14(4):4567–4581
Amaraweera SM et al (2021) Development of starch-based materials using current modification techniques and their applications: a review. Molecules 26(22):6880
Anandkumar R, Rajkumar S, Wondimu W (2024) Extraction of cellulose from soybean hulls for the development of polymer matrix composites with fishtail palm fibers. Mater Res Express 11(3):035309
Bano S et al (2024) Morphology, rheology, properties, and applications of nanocellulose and nanochitin-filled polymer blends. In: Nanofillers for binary polymer blends. Elsevier, pp 303–341
Barana D (2017) Lignin-based elastomeric composites for sustainable tyre technology
Bendahou A, Kaddami H, Dufresne A (2010) Investigation on the effect of cellulosic nanoparticles' morphology on the properties of natural rubber based nanocomposites. Eur Polym J 46(4):609–620
Bitinis N et al (2012) Physicochemical properties of organoclay filled polylactic acid/natural rubber blend bionanocomposites. Compos Sci Technol 72(2):305–313
Bitinis N et al (2014) Thermal and bio-disintegration properties of poly (lactic acid)/natural rubber/organoclay nanocomposites. Appl Clay Sci 93:78–84
Bras J et al (2010) Mechanical, barrier, and biodegradability properties of bagasse cellulose whiskers reinforced natural rubber nanocomposites. Ind Crops Prod 32(3):627–633
Brebu M (2020) Environmental degradation of plastic composites with natural fillers—a review. Polymers 12(1):166
Carvalho AJF et al (2003) Thermoplastic starch/natural rubber blends. Carbohydr Polym 53(1), 95–99
Chand O et al (2025) Extraction of micro fibrous cellulose from coconut husk by using chlorine free process: potential utilization application as a filter aid. J Mol Struct 1319:139325
Chandran AJ et al (2024) Micro/nano fillers for value-added polymer composites: a comprehensive review. J Vinyl Addit Technol
Chang BP et al (2021) Bioresourced fillers for rubber composite sustainability: current development and future opportunities. Green Chem 23(15):5337–5378
Cholilie IA et al (2024) The characteristics of art paper produced from palm oil empty fruit bunches at different cooking temperatures. J Fibers Polym Compos 3(1):74–86
Dáger-López D et al (2024) Advances in the production of sustainable bacterial nanocellulose from banana leaves. Polymers 16(8):1157
Datta J, Parcheta P (2017) A comparative study on selective properties of Kraft lignin–natural rubber composites containing different plasticizers. Iran Polym J 26(6):453–466
Daud S, Ismail H, Bakar AA (2016) The effect of 3-aminopropyltrimethyoxysilane (AMEO) as a coupling agent on curing and mechanical properties of natural rubber/palm kernel shell powder composites. Procedia Chem 19:327–334
Dave PN, Macwan PM (2024) Biocomposites with natural fibers. In: Advances in biocomposites and their applications. Elsevier, pp 81–120
Dutta B, Ray SK (2024) Characterization and transport properties of ceramic filler incorporated natural rubber composites. J Appl Polym Sci 141(7):e54966
El Mogy SA, Lawandy SN (2024) Enhancement of the cure behavior and mechanical properties of nanoclay reinforced NR/SBR vulcanizates based on waste tire rubber. J Thermoplast Compos Mater 37(2):520–543
Feng Y et al (2022) Co-amorphous delivery systems based on curcumin and hydroxycinnamic acids: stabilization, solubilization, and controlled release. LWT 170:114091
George J (2023) Electrical and mechanical characteristics of carbonaceous composites. ResearchSpace@ Auckland
Golakiya G (2020) Extraction of calcium carbonate from waste eggshells as fillers in composites. University of Saskatchewan
Goosen MF (1996) Applications of chitan and chitosan. CRC Press

Gopalan Nair K, Dufresne A (2003) Crab shell chitin whisker reinforced natural rubber nanocomposites. 2. Mechanical behavior. Biomacromolecules 4(3):666–674
Gopalan Nair K et al (2003) Crab shell chitin whiskers reinforced natural rubber nanocomposites. 3. Effect of chemical modification of chitin whiskers. Biomacromolecules 4(6):1835–1842
Hakimi NMF et al (2021) Surface modified nanocellulose and its reinforcement in natural rubber matrix nanocomposites: a review. Polymers 13(19):3241
Höfer R (2015) Sugar-and starch-based biorefineries. In: Industrial biorefineries and white biotechnology, pp 157–235
Injorhor P et al (2024) Effect of modified natural rubber on the mechanical and thermal properties of poly (lactic acid) and its composites with nanoparticles from biowaste. Polymers 16(6):812
Ismail H, Shuhelmy S, Edyham M (2002) The effects of a silane coupling agent on curing characteristics and mechanical properties of bamboo fibre filled natural rubber composites. Eur Polymer J 38(1):39–47
Ismail H, Shaari S, Othman N (2011) The effect of chitosan loading on the curing characteristics, mechanical and morphological properties of chitosan-filled natural rubber (NR), epoxidised natural rubber (ENR) and styrene-butadiene rubber (SBR) compounds. Polym Test 30(7):784–790
Izmar M, Afiq M, Azura A (2012) Effects of different additions of sago starch filler on physical and biodegradation properties of pre-vulcanized NR latex composites. Compos B Eng 43(7):2746–2750
Jantachum P, Khumpaitool B, Utara S (2023) Effect of silane coupling agent and cellulose nanocrystals loading on the properties of acrylonitrile butadiene rubber/natural rubber nanocomposites. Ind Crops Prod 195:116407
Jantachum P, Phinyocheep P (2024) Compatibilization of cellulose nanocrystal-reinforced natural rubber nanocomposite by modified natural rubber. Polymers 16(3):363
Jiang T et al (2020) Starch-based biodegradable materials: challenges and opportunities. Adv Ind Eng Polym Res 3(1):8–18
Johns J, Rao V (2008) Characterization of natural rubber latex/chitosan blends. Int J Polym Anal Charact 13(4):280–291
Kamal MM, Clarke J (2013) Effect of starch loading on the properties of natural rubber compounds. J Rubber Res 16:1
Kargarzadeh H et al (2017) Recent developments on nanocellulose reinforced polymer nanocomposites: a review. Polymer 132:368–393
Kazemi H et al (2022) Effect of cellulose fiber surface treatment to replace carbon black in natural rubber hybrid composites. Rubber Chem Technol 95(1):128–146
Kumar R, Ranjan N, Sidhu RS (2024) Role of fillers and various sustainable innovations in the use of filler material for the preparation of composites. Green Compos Manuf Sustain Approach 19:79
Le Corre D, Angellier-Coussy H (2014) Preparation and application of starch nanoparticles for nanocomposites: a review. React Funct Polym 85:97–120
Leblanc JL (2009) Filled polymers: science and industrial applications. CRC Press
Lee XM, Azura AR (2013) Effect of modified tapioca starch on mechanical properties of styrene-grafted natural-rubber/starch composites. Int J Mater Eng Innov 4(3–4):258–268
Li G et al (2020) Synthesis and biological application of polylactic acid. Molecules 25(21):5023
Li M et al (2022) Design of a lignin-based versatile bioreinforcement for high-performance natural rubber composites. ACS Sustain Chem Eng 10(24):8031–8042
Liu C, Shao Y, Jia D (2008) Chemically modified starch reinforced natural rubber composites. Polymer 49(8):2176–2181
Ma C et al (2024) Green synthesis of bio-based flame retardant/natural rubber inorganic-organic hybrid and its flame retarding and toughening effect for polylactic acid. Int J Biol Macromol 256:128378
Mahadi N et al (2024) Synergistic effects of cellulose nanocrystal on the mechanical and shape memory properties of TPU composites. Int J Biol Macromol 134842

Manimaran M et al (2024) Critical assessment of the thermal stability and degradation of chemically functionalized nanocellulose-based polymer nanocomposites. Nanotechnol Rev 13(1):20240005
Mariano M, El Kissi N, Dufresne A (2014) Cellulose nanocrystals and related nanocomposites: review of some properties and challenges. J Polym Sci Part B Polym Phys 52(12):791–806
Masek A, Cichosz S, Piotrowska M (2021) Biocomposites of epoxidized natural rubber/poly(lactic acid) modified with natural fillers (Part I). Int J Mol Sci 22(6):3150
Mathew S et al (2024) Effect of inorganic fillers on natural rubber latex foam vulcanisates. J Rubber Res 27(1):127–136
Melro E et al (2021) Revisiting lignin: a tour through its structural features, characterization methods and applications. New J Chem 45(16):6986–7013
Misman M, Rashid AA, Yahya SR (2018) Modification and application of starch in natural rubber latex composites. Rubber Chem Technol 91(1):184–204
Mohamad Aini NA et al (2020) Lignin as alternative reinforcing filler in the rubber industry: a review. Front Mater 6:329
Murtaza S et al (2024) Starch-nanostructured-based active packaging for food applications. In: Starch based nanomaterials for food packaging. Elsevier, pp 103–160
Nasirov F et al (2023) Chitin and chitosan-biopolymer materials of the 21st century. Process Petrochem Oil Refin 24(4)
Njuguna J, Pielichowski K, Desai S (2008) Nanofiller-reinforced polymer nanocomposites. Polym Adv Technol 19(8):947–959
Ock HG et al (2016) Effect of organoclay as a compatibilizer in poly (lactic acid) and natural rubber blends. Eur Polymer J 76:216–227
Pandey R, Mathur G (2024) Current trends in chitosan functionalization methods and their applications. Starch-Stärke, p 2300248
Parambath Kanoth B et al (2015) Biocomposites from natural rubber: synergistic effects of functionalized cellulose nanocrystals as both reinforcing and cross-linking agents via free-radical thiol–ene chemistry. ACS Appl Mater Interfaces 7(30):16303–16310
Piras S et al (2024) Biomimetic use of food-waste sources of calcium carbonate and phosphate for sustainable materials—a review. Materials 17(4):843
Pongpakdee C, Akaradechakul K, Pattamaprom C (2024) The influence of processing parameters on the process and product properties of polylactic acid/natural rubber blown films. Mater Des Process Commun 2024(1):5568117
Poompradub S et al (2008) Cuttlebone as reinforcing filler for natural rubber. Eur Polymer J 44(12):4157–4164
Rajisha KR et al (2014) Preparation and characterization of potato starch nanocrystal reinforced natural rubber nanocomposites. Int J Biol Macromol 67:147–153
Rani MSA et al (2024) Development and characterization of crab-based chitosan filler–reinforced polymer composites. In: Polymer composites derived from animal sources. Elsevier, pp 171–187
Rao V, Johns J (2008) Thermal behavior of chitosan/natural rubber latex blends TG and DSC analysis. J Therm Anal Calorim 92(3):801–806
Riseh RS et al (2024) Agricultural wastes: a practical and potential source for the isolation and preparation of cellulose and application in agriculture and different industries. Ind Crops Prod 208:117904
Rødland ES et al (2024) Are eco-friendly "green" tires also chemically green? Comparing metals, rubbers and selected organic compounds in green and conventional tires. J Hazard Mater 476:135042
Roy K et al (2021) Application of cellulose as green filler for the development of sustainable rubber technology. Curr Res Green Sustain Chem 4:100140
Ruamcharoen J, Munlee R, Ruamcharoen P (2023) Eco-friendly bio-based composites of cassava starch and natural rubber compatibilized with nanoclays. Polym Compos 44(2):1071–1082
Saeidlou S et al (2012) Poly (lactic acid) crystallization. Prog Polym Sci 37(12):1657–1677

Sajith M, Sambhudevan S (2023) Natural rubber and Gutta-Percha rubber: applications. Handbook of biopolymers. Springer, pp 1–35
Seidi F et al (2021) Chitosan-based blends for biomedical applications. Int J Biol Macromol 183:1818–1850
Shaari S, Hanafi I, Othman N (2011) The effect of chitosan loading on the properties of chitosan filled epoxidized natural rubber compounds. Key Eng Mater 471:851–856
Sin LT (2012) Polylactic acid: PLA biopolymer technology and applications. William Andrew
Songtipya L et al (2024) Enhancing thermo-mechanical properties of thermoplastic starch/natural rubber blends through the synergistic combination of PEG and modified natural rubber. J Polym Environ 32(4):1868–1878
Sowińska-Baranowska A, Maciejewska M, Duda P (2022) The potential application of starch and walnut shells as biofillers for natural rubber (NR) composites. Int J Mol Sci 23(14):7968
Suksut B, Deeprasertkul C (2011) Effect of nucleating agents on physical properties of poly (lactic acid) and its blend with natural rubber. J Polym Environ 19:288–296
Suzuki A, Nakamura Y, Asada C (2024) Production of polylactic acid biocomposite reinforced with environmentally friendly cellulose nanofiber derived from steam-treated bamboo. Biomass Convers Biorefin 14(15):16951–16959
Tessanan W, Phinyocheep P, Amornsakchai T (2024) Sustainable materials with improved biodegradability and toughness from blends of poly (lactic acid), pineapple stem starch and modified natural rubber. Polymers 16(2):232
Thakur V et al (2021) Recent advances in nanocellulose processing, functionalization and applications: a review. Mater Adv 2(6):1872–1895
Thomas B et al (2018) Nanocellulose, a versatile green platform: from biosources to materials and their applications. Chem Rev 118(24):11575–11625
Thomas SK et al (2021) A comprehensive review on cellulose, chitin, and starch as fillers in natural rubber biocomposites. Carbohydr Polym Technol Appl 2:100095
Tom M et al (2022) Approaching sustainability: nanocellulose reinforced elastomers—a review. Rubber Chem Technol 95(4):515–549
Ul-Islam M et al (2024) Chitosan-based nanostructured biomaterials: synthesis, properties, and biomedical applications. Adv Ind Eng Polym Res 7(1):79–99
Visakh P et al (2012) Crosslinked natural rubber nanocomposites reinforced with cellulose whiskers isolated from bamboo waste: processing and mechanical/thermal properties. Compos A Appl Sci Manuf 43(4):735–741
Wang M-J (1998) Effect of polymer-filler and filler-filler interactions on dynamic properties of filled vulcanizates. Rubber Chem Technol 71(3):520–589
Wang Z-F et al (2009) The impact of esterification on the properties of starch/natural rubber composite. Compos Sci Technol 69(11–12):1797–1803
Wawrzyńczak A, Chudzińska J, Feliczak-Guzik A (2024) Metal and metal oxides nanoparticles as nanofillers for biodegradable polymers. ChemPhysChem e202300823
Whba R et al (2024) Intrinsic challenges and strategic approaches for enhancing the potential of natural rubber and its derivatives: a review. Int J Biol Macromol 133796
Wong D et al (2024) A critical review: recent developments of natural fiber/rubber reinforced polymer composites. Clean Mater 100261
Wu Y-P et al (2006) A strategy to prepare high performance starch/rubber composites: in situ modification during latex compounding process. Carbohydr Polym 65(1):109–113
Xuan YY et al (2024) Influence of multi-walled carbon nanotubes on thermal behaviour and mechanical properties of pineapple leaf fibre-based natural rubber composites. J Market Res 30:8608–8619
Xue Y, Mou Z, Xiao H (2017) Nanocellulose as a sustainable biomass material: structure, properties, present status and future prospects in biomedical applications. Nanoscale 9(39):14758–14781
Yadav H, Malviya R, Kaushik N (2024) Chitosan in biomedicine: a comprehensive review of recent developments. Carbohydr Polym Technol Appl 100551

Yapar Ö et al (2024) Sustainable all-cellulose biocomposites from renewable biomass resources fabricated in a water-based processing system by the vacuum-filtration-assisted impregnation method. Polymers 16(13):1921
Yimmut K, Homchoo K, Hinchiranan N (2018) Poly (butyl acrylate-co-fluorinated acrylate)-graft-natural rubber: Synthesis and application as compatibilizer for natural rubber/poly (butyl acrylate-co-fluorinated acrylate) films. Colloids Surf A 540:11–22
Yu P et al (2016) A comprehensive study on lignin as a green alternative of silica in natural rubber composites. Polym Testing 54:176–185
Zhou Y et al (2015) Lignocellulosic fibre mediated rubber composites: an overview. Compos B Eng 76:180–191
Zhou W et al (2024) Preparation and properties of pH-sensitive cationic starch nanoparticles. Food Chem 437:137916
Zinge C, Kandasubramanian B (2020) Nanocellulose based biodegradable polymers. Eur Polymer J 133:109758

Chapter 5
Compatibilization of Natural Rubber-Based Composites and Nanocomposites

Jayalatha Gopalakrishnan

Abstract Natural rubber-based micro- and nanocomposites hold a unique position in industry and research due to their fascinating properties. The incorporation of hydrophilic or polar micro/nano-fillers can adversely affect the mechanical properties of natural rubber composites owing to the poor rubber-filler interaction. The addition of compatibilizer is an efficient strategy to promote the interaction between the rubber and the filler which in turn can significantly alter the overall performance of the composite. This chapter summarizes the utilization of various compatibilizers suitable for fillers such as layered silicates, silica, nylon fiber, and cellulose fibers in developing high-performance natural rubber-based composites. The influence of compatibilizers on the mechanical, viscoelastic properties, rubber-filler interaction, and morphology are discussed in detail.

Keywords Natural rubber · Compatibilizers · Epoxidized natural rubber · Maleated natural rubber · Nano-fillers · Rubber-filler interaction

5.1 Introduction

Natural rubber (NR) is widely used in many applications owing to its numerous attractive properties such as low hysteresis, excellent dynamic properties, fatigue resistance, and high resilience. By virtue of their characteristic feature i.e. strain-induced crystallization, they exhibit superior mechanical properties as compared to several synthetic rubbers (Brydson 1978). Generally, fillers are incorporated into the rubber mainly to achieve improvement of service properties or to reduce the cost of the product. The performance of fillers is governed by their composition, particle size, shape, surface area, surface reactivity, and degree of interaction with the polymer.

J. Gopalakrishnan (✉)
Department of Polymer Science and Rubber Technology, Inter University Centre for Nanomaterials and Devices (IUCND), Cochin University of Science and Technology (CUSAT), Kochi 682022, Kerala, India
e-mail: jayalatha@cusat.ac.in; gjayalatha@gmail.com

© The Author(s), under exclusive license to Springer Nature Switzerland AG 2025

P. M. Visakh (ed.), *Rubber Based Bionanocomposites*, Advanced Structured Materials 227, https://doi.org/10.1007/978-3-031-78557-3_5

Carbon black is the dominant reinforcing filler used for rubber composites. The other fillers that are being used in the rubber industry include silica, clay, and calcium carbonate ($CaCO_3$). The property of main interest in a composite is the mechanical properties required for specific applications.

Based on the size of the dispersed phase, composites can be classified into macrocomposites, microcomposites, and nanocomposites. Conventional macrocomposites have a dispersed phase in dimensions greater than 100 μm. Macrocomposites can be further classified into particulate composite and fiber-reinforced composite. An example of particulate composite is carbon black which is added to rubber matrix so as to enhance the tensile strength and stiffness. On the other hand, fiber-reinforced composite improves the strength and stiffness without increasing the density of the composite. In the case of microcomposites, the dispersed phase has dimensions in the range of 100 μm–100 nm whereas nanocomposites have at least one dimension of the dispersed phase smaller than 100 nm.

Nanocomposites exhibit greatly improved mechanical and thermal properties because of the nanometer scale dispersion of reinforcements and high surface-to-volume ratio. Reinforcing polymers with nano-sized particles is a promising technique for developing materials with enhanced performance without the involvement in expensive synthetic procedures. A low-weight percent of nanofiller will give significant improvements in the strength and stiffness of rubber materials.

One of the main factors that decide the reinforcement of rubber by fillers is the surface of filler-rubber interaction. Most of the fillers/fibers being hydrophilic, when incorporated into a hydrophobic natural rubber matrix can lead to poor dispersion of the fillers. The rubber-filler interaction can be enhanced by the use of coupling agents or compatibilizers. Compatibilization is the term often used for improving the adhesion between the components in a heterogeneous blend. In rubber-filler composites, compatibilization can enhance the properties of the composite while increasing the adhesion between the filler and the rubber, and reducing the interfacial tension. Compatibilizer consists of two parts; one part can either react and link with the filler or form non-bonded interactions such as dipole–dipole interactions or hydrogen bonding. The second part having a similar structure as that of the rubber can easily intermingle with the rubber matrix.

5.2 Natural Rubber/Layered Silicate Nanocomposites

Layered silicates (LS) or clay are those minerals that assemble regularly with the unit crystalline layer at the nanoscale. Clay silicates that have been used as reinforcing fillers for polymers include montmorillonite, hectorite, bentonite, etc. The crystal structure of clay consists of two-dimensional layers with a central octahedral sheet of aluminum or magnesium sandwiched between two tetrahedral silica sheets such that the oxygen ions of the octahedral sheet also belong to the tetrahedral sheets. The layer thickness is around 1 nm while the lateral dimensions may vary from 300 Å to several microns. The layers are stacked by weak dipole forces and exhibit a net negative charge on the lamellar surface. The gallery between the layers is occupied by cations such as Na^+, Mg^{2+}, and Ca^{2+}. The primary particles themselves consist of 10–20 layers in a coplanar orientation referred to as tactoid formation. Clay particles can associate through face-to face, face-to-edge, and edge-to-edge bonding to form aggregates and agglomerates (Vu et al. 2001). Montmorillonite (MMT) is the most commonly used layered silicate because of its natural occurrence.

The characteristic features of layered silicate include high cation exchange property, high surface area, and large aspect ratio (Reichert et al. 2001). Three types of composites can be generally obtained—conventional composites consisting of clay tactoids having stacked layers, intercalated nanocomposites where polymer chains are intercalated into silicate layers, and exfoliated nanocomposites comprising of delaminated silicate layers uniformly dispersed in the matrix. Usually addition of pristine clay into the polymer results in a conventional composite, where the polymer is not intercalated into the silicate galleries. Layered silicates, when added in small amounts into the polymer matrix can increase tensile strength and stiffness with minimal loss of impact resistance, decrease the gas permeability and swelling in solvents as well as improve abrasion resistance, flame retardance, and thermal stability.

The main challenge for nanocomposite formation is the dispersion of hydrophilic clay particles in the non-polar (hydrophobic) rubber matrix. Clay has the ability to undergo interlayer expansion, exposing a large active surface area, and permitting guest molecules entering into the gallery. The cations existing on the internal surfaces of the interlayers can be easily exchanged with organic quaternary alkyl ammonium salts having long chains termed onium (octadecyl trimethyl ammonium salt). Other organic modifiers can be dimethyl dihydrogenated tallow, quarternary ammonium (2M2HT), methyl, tallow, bis-2-hydroxy ethyl, quarternary ammonium (MT2EtOH). The cation exchange capability of layered silicates is exploited to produce organophilic clays (or organoclays) which have less ionic or polar surfaces. Moreover, it widens the gallery distance between the clay platelets. Such organophilic modification improves the compatibilization between the surface of the clay and the hydrophobic rubber matrix whereby the polymers can penetrate more easily into the galleries. The rubber chain molecules can either increase the distance between the layers resulting in intercalated nanocomposites or randomly disperse the separate sheets entirely giving rise to exfoliated nanocomposites. Different methods have been adopted for the preparation of polymer-layered silicate nanocomposites. It includes in situ polymerization, solution blending, latex compounding, and direct polymer melt intercalation.

For Na^+-bentonite clay, octadecylamine has been used as an effective intercalating agent. The exchange of interlayer cations with octadecyl amine is characterized by X-ray diffraction. Figure 5.1 shows the X-ray diffraction patterns of Na^+-bentonite clay and organoclay.

Na^+-bentonite, which shows a diffraction peak at $2\theta = 7°$ is assigned to the interlayer platelet spacing of pristine bentonite with an interlayer distance of 12.6 Å. On

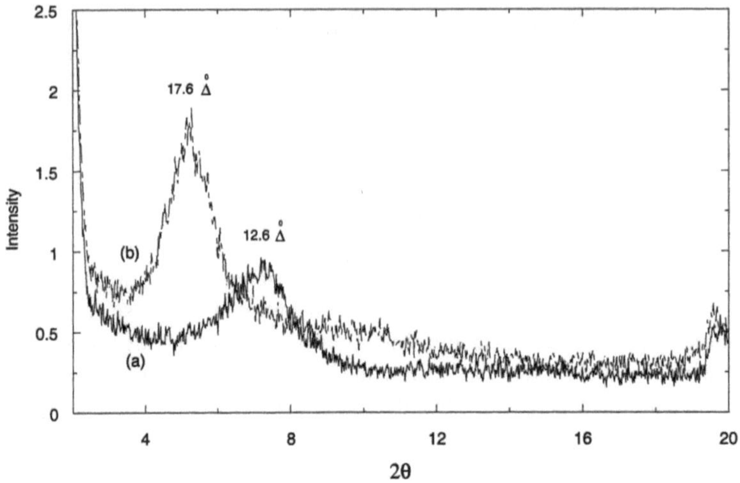

Fig. 5.1 X-ray diffraction patterns of (a) clay, (b) organoclay (Reproduced with permission from *Polymer*, *44*(8), 2447–2453. Copyright (2003) Elsevier.)

modification, the organoclay shows a peak at lower angles of $2\theta = 5°$, corresponding to an interlayer distance of 17.6 Å. The increase in interlayer distance facilitates the incorporation and confinement of NR chains into the silicate galleries. When such organoclays are incorporated in natural rubber, the peak corresponding to the interlayer spacing of clay can disappear which implies that the clay has completely exfoliated or uniformly dispersed in the polymer matrix. It is expected that the intercalation can take place during the compounding stage i.e., on a two-roll mill and exfoliation during vulcanization. Moreover, organoclay also acts as an effective vulcanizing agent for NR due to the formation of a transition complex with amines and sulfur that accelerates the vulcanization process. The thermal stability of NR/organoclay nanocomposite improves as the nanolayers of clay prevents the diffusion of volatile decomposition products generated within the layers (Teh et al. 2004a; Joly et al. 2002; Madhusoodanan and Varghese 2006; Avalos et al. 2008; Lopez-Manchado et al. 2003a; Arroyo et al. 2003).

5.2.1 Epoxidized Natural Rubber (ENR) as Compatibilizer

ENR can be produced via the in situ performic epoxidation reaction of natural rubber latex. The reaction takes place between the C=C bond of the NR molecule and performic acid generated in situ by the reaction between formic acid and hydrogen peroxide (Fig. 5.2). The extent of epoxidation (i.e., mole % of epoxide) in ENR can be varied by varying the reaction time. The ENR latex produced can be subsequently

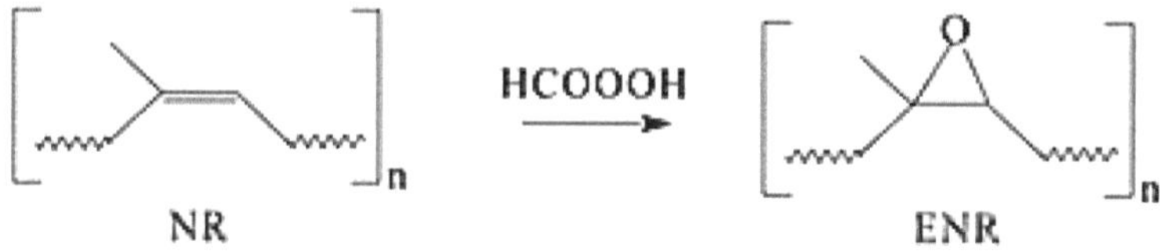

Fig. 5.2 The epoxidation reaction of natural rubber

coagulated with methanol/ethanol, followed by sheeting, washing, and drying. ENR is more polar than NR due to the presence of an epoxide group in its structure.

Teh et al. have reported the use of ENR as a compatibilizer to study the potential of onium-modified MMT (2 phr) in NR vulcanizate and compared the properties with composites containing conventional fillers such as carbon black (50 phr) and silica (30 phr). The organoclay-filled composite exhibits the shortest scorch time and cure time compared to conventional vulcanizates. This is due to the activation of adjacent double bonds by the epoxide group in ENR. The minimum torque (M_L) from the cure studies was lower for ENR compatibilized composites which indicates that ENR improves the processability. The tensile and tear properties of the vulcanizates were highest for organoclay as the filler, followed by silica and then carbon black. The organoclay platelets possess a high aspect ratio and thus act as a reinforcing element (Teh et al. 2004b). Moreover, the amine intercalant octadecyl trimethylamine of the organoclay can react with the epoxide groups of ENR and thereby facilitate the disintegration of the layered clay particles. This can result in an intercalated or exfoliated composite which can improve the interfacial bonding between the rubber and clay layers, and as a consequence enhance the mechanical properties. The tensile strength improved for 2 phr of filler and then deteriorated with increasing filler content. This augmentation is due to the better dispersion of organoclay in the presence of compatibilizer. The tear strength at higher filler content reduces as the agglomerates formed act as stress concentrators which reduce the propagation of tear (Teh et al. 2006). Numerous researchers have reported that usually, the optimum level of organoclay loading in rubber is below 10 phr. However, Nah et al. (2001) have reported that only 15 phr of organoclay is necessary to obtain the optimum tensile and tear properties which is comparable to a compound loaded with 40 phr of carbon black. All these findings indicate that organoclay is not only a filler but can also act as a co-curing agent that facilitates the cure reaction. Higher loadings of organoclay tend to agglomerate and thereby reduce the tensile properties. This could be due to the amount of matrix being less than that required for complete exfoliation.

Figure 5.3a shows that in the presence of ENR, the organoclay is intercalated and exfoliated in the NR matrix whereas Fig. 5.3b indicates poor dispersion of organoclay in the NR matrix. This can be further substantiated by the X-ray diffraction pattern of the organoclay (MMT modified with octadecyl trimethyl ammonium salt) and NR/organoclay composites with and without ENR (Fig. 5.4). For compatibilized composite, the diffraction peak of organoclay shifted to a lower angle ($2\theta =$ 3.7°–2.25°) compared with the organoclay powder. While for the NR/organoclay

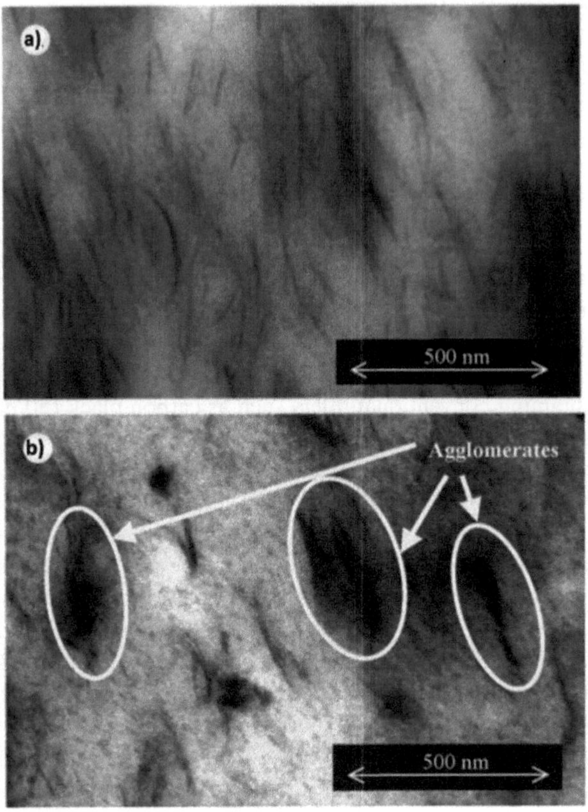

Fig. 5.3 TEM images of **a** NR/ENR/organoclay (2 phr) and **b** NR/organoclay (2 phr) nanocomposites (Reproduced with permission from Eur Polym J 40: 2513–2521, 2004. Copyright Elsevier.)

composites, the 2θ peak shifted to 2.5°. The increase in d-spacing from 2.39 to 3.53 nm indicates that the rubber chains intercalate in between the clay layers by overcoming the Vander Waals forces in between the layers. In the presence of ENR, the tactoids of layered silicate tend to disaggregate into individual sheets (d-spacing 3.93 nm), which are uniformly dispersed in the elastomer matrix. The polar nature of ENR favors the initial insertion into the galleries and due to better interaction with the organoclay improves the dispersion of the layered silicates in the matrix (Teh et al. 2004a; Arroyo et al. 2007). Another interesting observation is that the particle size of the ENR dispersed in NR is reduced in the presence of organoclay. This implies that organoclay may act as an interfacial agent, reducing the interfacial tension with a concomitant break-up of the ENR droplets thereby reducing the particle size.

Even though organic modification of layered silicates can increase the spacing between silicate layers, it is not sufficient for intercalating the non-polar NR chains into the silicate gallery. Polar interactions are vital to produce exfoliated structures.

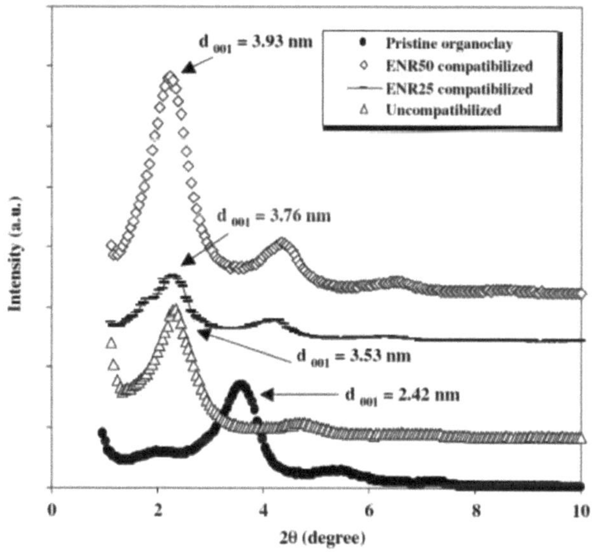

Fig. 5.4 XRD spectra of organoclay, NR/ENR/organoclay, and NR/organoclay nanocomposites (Reproduced with permission from Eur Polym J 40: 2513–2521, 2004. Copyright Elsevier.)

It is therefore necessary to select the compatibilizing agents that are compatible and more polar than the rubber matrix (Doğan et al. 2011).

The improvement in the filler-rubber compatibility by the addition of ENR, not only causes a better dispersion of clay particles in the rubber matrix but also increases the filler-rubber interaction and decreases the agglomeration of filler particles. In the presence of ENR, the reinforcing effect of organoclay is more evident in NR nanocomposites. This behavior may be due to the chemical interaction of the epoxy groups with the hydroxyl groups of the filler surface and the amine layers of the interlayer spacing during the vulcanization process.

Dynamic mechanical studies indicated that the modulus of NR/ENR/organoclay nanocomposites showed a significant increase above the glass transition temperature (T_g) at 10 phr organoclay content compared to NR/ENR gum. The tan δ value at T_g decreased for the organoclay-filled nanocomposite due to the strong interaction between the organoclay and the rubber matrix. The presence of an additional secondary relaxation peak for the tan δ curve represents that part of the rubber that is intercalated.

Influence of Epoxidation Level in ENR

The level of epoxidation can also affect the cure characteristics and mechanical properties of the NR/organoclay nanocomposites. ENR 50 (50 mol % epoxy) containing composites show higher torque values than compared to ENR 25 (25 mol % epoxy)

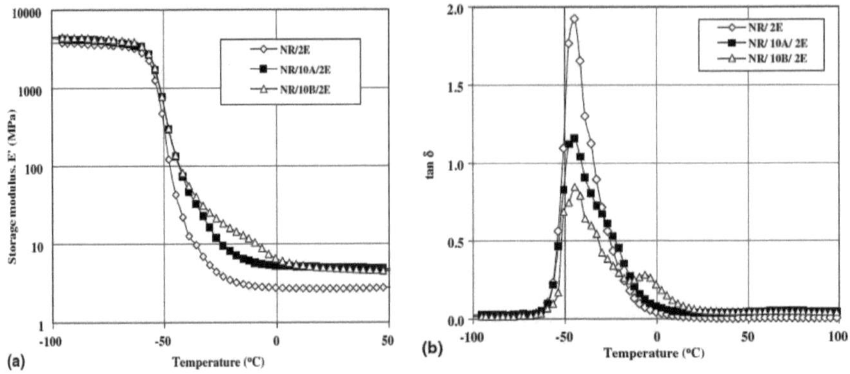

Fig. 5.5 Effect of 10 phr of ENR 25 (10A) and ENR 50 (10B) on the storage modulus (**a**) and loss factor (**b**) versus temperature for NR/organoclay 2 phr (2E) nanocomposites (Reproduced with permission from Eur Polym J 40: 2513–2521, 2004. Copyright Elsevier.)

compatibilized ones. The higher content of epoxy in ENR 50 may affect the activation/cross-linking process which changes the network density and clay dispersion, both of which influence the stiffness. The scorch time and cure time are lower for the ENR 50 mix compared to those with ENR 25. The dynamic storage modulus above glass transition temperature was highest for 10 phr ENR 50 compatibilized composites compared to ENR 25 which was attributed to the good dispersion of the clay layers in the NR matrix (Fig. 5.5a). The height of tan δ peak at T_g is lower for ENR 50 compatibilized composite compared to those of ENR 25 (Fig. 5.5b). The strong interaction between the organoclay and the NR matrix in the presence of a compatibilizer hampers the molecular mobility of the NR matrix. However, a secondary peak observed for ENR 50 compatibilized composites may be attributed to the part of rubber intercalated in between the silicate layers. The improvement in tensile strength, tensile modulus, and tear strength was highest for ENR 50 compatibilized NR/organoclay composites compared to ENR 25 and the un-compatibilized ones (Teh et al. 2004a).

5.2.2 Maleated Natural Rubber as Compatibilizer

Maleated natural rubber (MA-g-NR or MNR) denotes graft copolymers of maleic anhydride and natural rubber. It has been reported that graft copolymerization of maleic anhydride with NR can be carried out both in the molten state (Carone et al. 2000; Nakason et al. 2001, 2002) and solution state (Nakason et al. 2004). Grafting of MA takes place in the molten state in an internal mixer either with or without chemical initiators such as peroxides. In the molten state, during graft copolymerization, the polar maleic anhydride (MA) groups are grafted on the non-polar NR backbone.

Since organoclay can more easily disperse in polar polymers than in non-polar ones, the functionalization of NR via introducing polar groups such as maleic anhydride (MA) can help in the dispersion of organoclay. Comparing both NR/organoclay nanocomposites with and without maleic anhydride-g-natural rubber (MNR) for varying loadings of organoclay (2–10 phr), the scorch time and cure time is lower for composites with MNR. This may be because of the reaction between the accelerator and free maleic anhydride molecules or succinic acid formed by the ring-opening reaction of the grafted maleic anhydride. The tensile strength and modulus of the composites are higher for composites with 8 phr loading in the presence of MNR. This is due to the formation of strong H-bonding between the anhydride group of MNR and the amine group of the clay surface which promotes the interaction between the organoclay and the NR matrix. At 10 phr, the tensile strength and modulus drop due to the agglomeration of the clay at high loadings. In the presence of MNR, the NR/organoclay composites exhibit higher matrix tearing and rougher surfaces compared to those without MNR (Lopez-Manchado et al. 2003b).

Dogan et al. (2011) have compared the effect of two compatibilizers viz. maleic anhydride-g-polybutadiene (MA-g-PB, M_n-5400) and ENR 50 (50 mol % epoxy) in NR-based composites containing Cloisite 30B (2 phr) and carbon black (50 phr). It was found that intercalation of the clay was more with composites containing ENR compared to MA-g-PB. The hydroxyl groups present in the structure of Cloisite are able to react with the epoxy groups of ENR 50, which favors dispersion and intercalation. The higher reactivity of epoxy groups compared to the maleic anhydride group gives rise to the more intercalated structure. The polar nature of the compatibilizer determines the extent of intercalation (Vu et al. 2001; Varghese et al. 2003). MA-g-PB reduces the cross-link density of the organoclay-modified composite which may be due to the retardation effect of maleic anhydride groups on the sulfur curing pathways.

The filler networking effect arises from the tendency of filler aggregates to form agglomerates at high filler loadings. Such networks are usually broken down under dynamically applied strains. The breakdown of the filler network caused by increasing strain amplitude leads to a decrease in modulus termed the Payne effect. The rheological studies of the uncured compound revealed that the modulus at all strain values increased for the composites with compatibilizer. The intercalated polymer chains are immobilized due to polymer–clay adhesion and occlusion which increases the effective volume of filler. The effective increase in modulus of the composites in the presence of compatibilizer is due to strong interactions between carbon black, organoclay, and polar components of the compatibilizer. Tan δ is the ratio of loss modulus to elastic modulus. When dynamic strains are applied to rubber, a certain amount of energy is converted to heat while the remaining energy is used to recover the sample into its initial state. The important factor that governs the tan δ of uncured rubber products is the filler networking effect. The presence of compatibilizer increases the tan δ value of the composite. The interaction between the polar groups on the surface of fillers and the polar groups of the compatibilizer reduces the interaction between the filler particles and thereby increases the rubber-filler interaction which contributes to an increase in tan δ. The tensile strength is higher for the composites containing ENR as a compatibilizer than compared to those containing MA-g-PB. Although MA-g-PB facilitates intercalation, some amount of clay remains unintercalated forming weak spots and inhomogeneities leading to less improvement in tensile strength.

5.2.3 *Silane Coupling Agent*

Silane coupling agent is used to enhance the effectiveness of filler strengthening by increasing the interaction between the filler and rubber through chemical bonding (Sae-Oui et al. 2005). Organoclays can give high dispersion in natural rubber via solution mixing in the presence of coupling agent, bis(triethoxysilylpropyl)tetrasulfide (TESPT) rather than mill mixing. The improvement in properties such as modulus, strength, and hardness is due to better filler-rubber compatibility. The coupling agent can interact with the –OH group of the silicate through H–bonding. However, the presence of a coupling agent can increase the cure time of the natural rubber/organoclay due to the interaction of the silane group with the amine groups in the silicate, thereby hindering the accelerating effect of the amine groups (Lopez-Manchado et al. 2004).

5.3 Natural Rubber/Silica Composites

Silica is widely used as filler in white and colored rubber products because of its reinforcing effect similar to that of carbon black. The addition of silica to a rubber compound offers two advantages: reduction in heat build-up when used to partly replace carbon black improvement in tear strength, and cut and chip resistance. Due to the high surface area, compounds with silica offer advantages to tire users such as better wet grip, reduced rolling resistance, and reduced rolling noise with abrasion resistance similar to that of carbon black (Donnet 2003).

Silica consists of silicon and oxygen arranged in a tetrahedral structure of a three-dimensional lattice. The presence of siloxane and silanol groups on the surface of silica makes the filler acidic and moisture adsorbing (Ansarifar et al. 2007). In addition, the presence of functional groups such as –OH on the surface leads to the formation of strong agglomerates through hydrogen bonding. The highly polar nature and high structure of silica make it very difficult to mix with non-polar rubbers. Another problem that lies in silica is its tendency to adsorb basic accelerators which in turn retards the cure of the rubber compound (Kosmalska et al. 2003). Moreover, when a large amount of silica is added, the processing becomes more difficult because the viscosity increases significantly and more heat is generated during mixing (Ostad-Movahed et al. 2008). This reduces the dispersion of silica particles into the rubber matrix and gives poor adhesion between the silica particles and the rubber matrix.

5.3.1 Silane Coupling Agent

Silane coupling agent improves the performance of silica by increasing the adhesion between the two phases. It modifies the characteristic surface of silica from hydrophilic to hydrophobic. Vinyl- and mercapto-silanes reduce the silica aggregate size without re-agglomeration of silica. However, modification with an amino-silane tends to increase the silica agglomerate size due to its hydrophilicity leading to the formation of hydrogen bonds between particles/aggregates (Jesionowski and Krysztafkiewicz 2001; Lee et al. 2011). The incorporation of silane coupling agent in rubber compounds lowers the rolling resistance of the tire and decreases the tread wear index while maintaining the wet traction (Wolff and Wang 1992). It also plays an important role in improving some shortcomings associated with silica filler such as cure retardation, and difficulties involved in the compounding and dispersion of filler. Silane coupling agents can enhance the cure by improving the dispersion of filler in the rubber matrix. It can reduce the adsorption of accelerators on the silica surface resulting in lower cure time.

The two silane coupling agents of commercial significance are mercaptopropyltrimethoxysilane (A189) and bis(3-triethoxysilylpropyl)tetrasulfide (TESPT). Among the various silanes, bis(triethoxysilyl propyl) tetrasulfide (TESPT or Si69) has been the most widely used. Mercaptosilane generally reduces the cure and scorch time depending on the type of accelerator system and elastomer. Literature indicates that chemical bonds are formed between hydroxyl groups of fillers and alkoxy groups of coupling agents. The rubber chemisorption on the surface of filler particles increases and consequently, rubber exhibits increased hardness and strength. The addition of alkoxysilanes can reduce the viscosity of NR/silica compounds. Figure 5.6a shows the mechanism of interaction between silica and TESPT. During mixing, one of the ethoxysilyl groups reacts with the silanol group of silica. This is followed by a condensation reaction between pairs of neighboring silane molecules. The filler-silane intermediate thus formed can now react with the allyl position of unsaturation on the rubber chain (Joseph et al. 2004). During vulcanization at higher temperatures, the reaction takes place between the rubber molecules and the coupling agent via accelerator and polysulfidic pendant group resulting in cross-link formation with the rubber. On completion of the reaction, the pendant accelerator cleaves off and continues to participate in further cross-linking as the vulcanization proceeds (Fig. 5.6b) (Mark et al. 2005). The hydrophobization of the silica surface reduces the silica-silica network and makes the polar silica more compatible with the non-polar rubber. This can improve the dispersion of silica in the rubber matrix. Silanes increase the cross-link density and vulcanizate properties of NR/silica composites (Ismail 1999).

Comparing the effect of untreated and silanized silica for various loadings in NR/silica composites, it was found that the tensile strength increased up to 20 phr and then decreased (Fig. 5.7a). This could be due to the reinforcing effect of silica in the presence of the coupling agent i.e. the small particle size of silica provided a larger surface to contact with the rubber matrix resulting in better filler-rubber interaction.

Fig. 5.6 **a** Mechanism of silanalization reaction **b** interaction of silica/silane intermediate with rubber during vulcanization

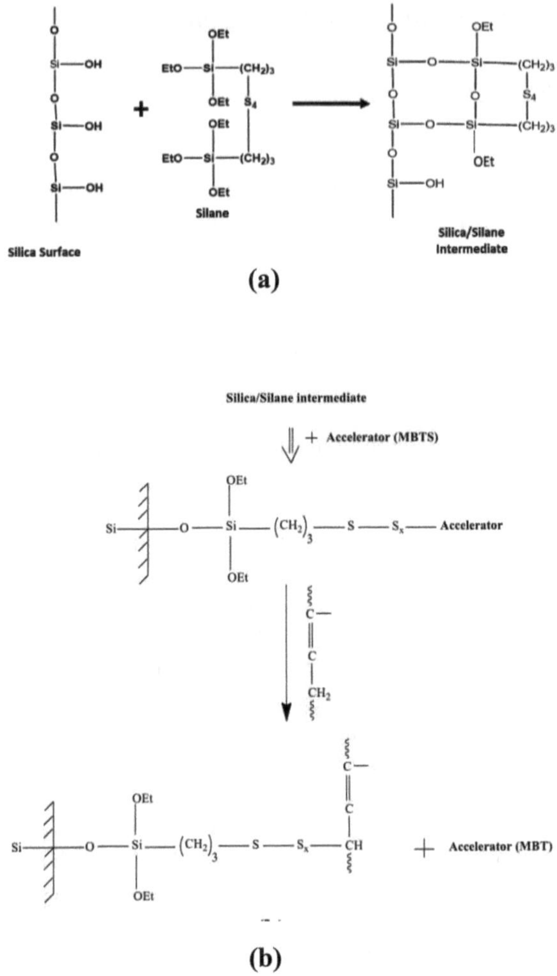

The reduction in tensile strength beyond 20 phr may be due to the poor dispersion and agglomeration of silica due to filler-filler interaction. Also, there would not be enough rubber matrix to hold the silica particles (Sae-oui et al. 2007; Chandra et al. 2017). For composites with untreated silica, the modulus increased up to 20 phr and then decreased, while the silanized silica increased with filler loading (Fig. 5.7b). Silanized silica has a higher modulus since the presence of Si69 improves the cross-link density due to better interaction between the filler and the rubber matrix (Idrus et al. 2011). The scanning electron micrographs (SEM) (Fig. 5.8a, b) reveal that untreated silica formed large agglomerates (40 μm) and has lower wettability with the rubber matrix resulting in more holes. In the presence of silane, silica is more uniformly dispersed in the rubber matrix and is smaller in size (about 250 nm) (Ansarifar et al. 2004).

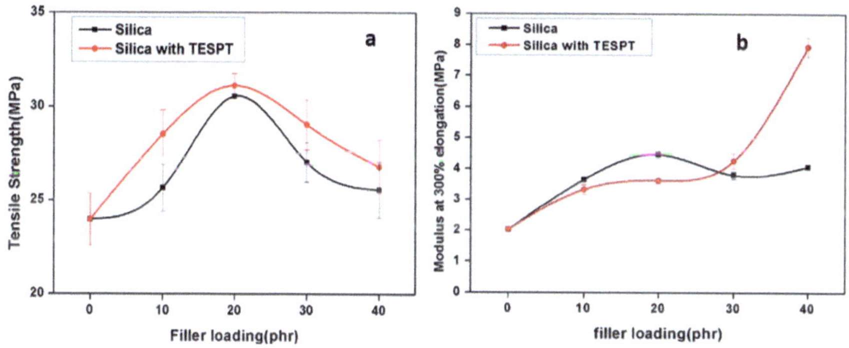

Fig. 5.7 **a** Tensile strength and **b** modulus at 300% elongation of NR/silica composites with untreated and silanized silica (Reproduced with permission from Polym Test 60: 187–197, 2017. Copyright Elsevier.)

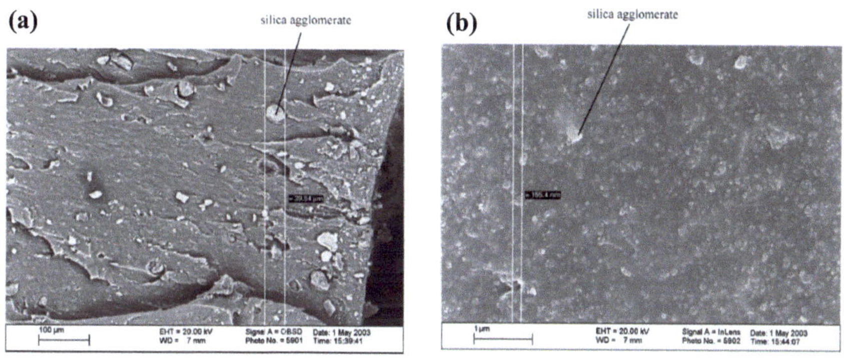

Fig. 5.8 SEM micrograph of the tensile fracture surfaces at NR/silica (30 phr) vulcanisates **a** untreated silica **b** silica with TESPT (Reproduced with permission from Int J Adhesion Adhes, 24: 9–22, 2004. Copyright Elsevier.)

Dynamic mechanical behavior of NR/silica composites containing various filler loading under strain sweep (Fig. 5.9) revealed that the storage modulus is independent of strain at low strain levels. But at higher strain levels, all the moduli tend to converge to a common point. The Payne effect is more prominent for 40 phr silica composites which may be due to filler-filler networking. The decrease in modulus with increasing strain is due to the breakdown of filler-filler networks. The Payne effect is lower for composites containing silanized silica. This is due to the reduction of filler-filler interaction and an increase in filler-matrix interaction (Chandra et al. 2017).

However, there are some drawbacks to the use of silane coupling agents. In silane, one or two out of the three ethoxy groups are split off as ethanol. The substantial ethanol emission occurring during mixing and downstream processing can lead to health and fire safety risks. Furthermore, the ethanol remaining in the compound

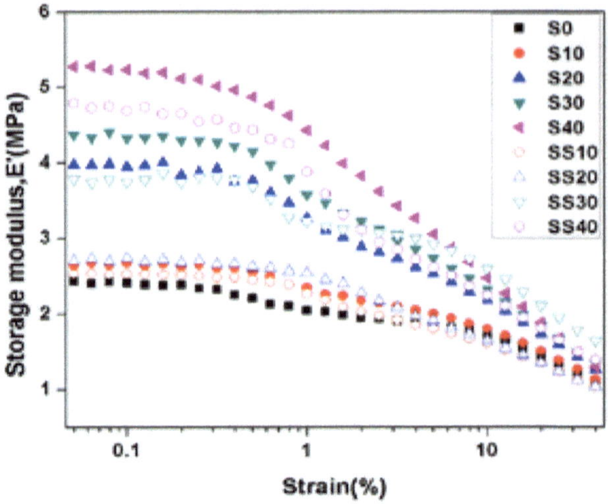

Fig. 5.9 Dynamic strain sweep of NR, NR/silica (S), and NR/silanized silica (SS) composites for various loadings of silica at 100 °C (Reproduced with permission from Polym Test 60: 187–197, 2017. Copyright Elsevier.)

can cause porosity and blisters in the vulcanized rubber. Another disadvantage is the high price of silane. One way of reducing the use of silane coupling agent in silica-reinforced rubber compounds is the incorporation of modified elastomers bearing functional groups that can interact with the silanol groups of silica and so improve the processability of the rubber compound and properties of the vulcanizate.

Silane-modified polymer is another route for the improvement of silica reinforcement in polymeric materials. Sengloyluan et al. (2017) have studied the use of 3-octanoylthio-1-propyltriethoxysilane (NXT)-grafted-NR (Fig. 5.10b) as a compatibilizer for NR/silica compound. NXT (Fig. 5.10a) is grafted onto NR using a peroxide initiator at 140 °C in an internal mixer. The free radicals generated by the thermal decomposition of the initiator can transfer its radicals to the sulfur on the coupling agent. The grafting reaction proceeds through an abstraction reaction at the allylic position of the NR molecule more than through the addition reaction. Upon grafting, a covalent C–S–C bond is formed between rubber and silane with pendant alkoxy groups. The use of NXT-g-NR gives a slightly lower Payne effect than the straight use of NXT silane in NR/silica compounds. In addition, higher tensile strength and modulus are exhibited by NXT-g-NR as a compatibilizer than the use of NXT. However, the value is still lower than the use of TESPT.

The storage modulus of NR/silica compound with TESPT in the glassy region has only marginal changes compared to that with NXT-g-NR (Fig. 5.11). However, in the rubbery region, the TESPT system has the highest modulus compared to other vulcanizate. This is attributed to the improved filler-rubber interactions and highest cross-link density contributed by the sulfur present in the TESPT molecules.

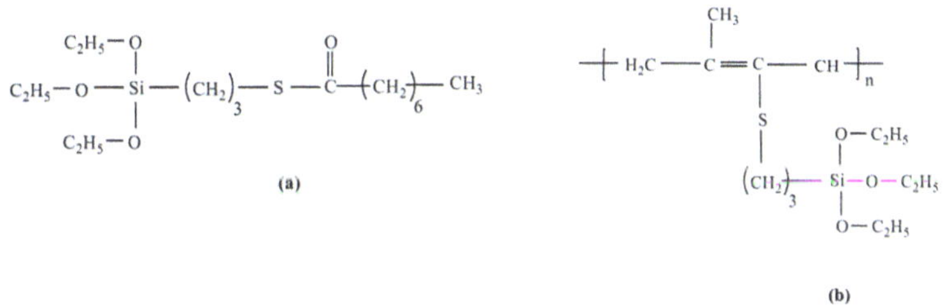

Fig. 5.10 **a** NXT **b** NXT-g-NR

Fig. 5.11 Storage modulus of NR/silica compound, with NXT-g-NR and TESPT (Reproduced with permission from Express Polym Lett 11: 1003–1022, 2017. Copyright Elsevier.)

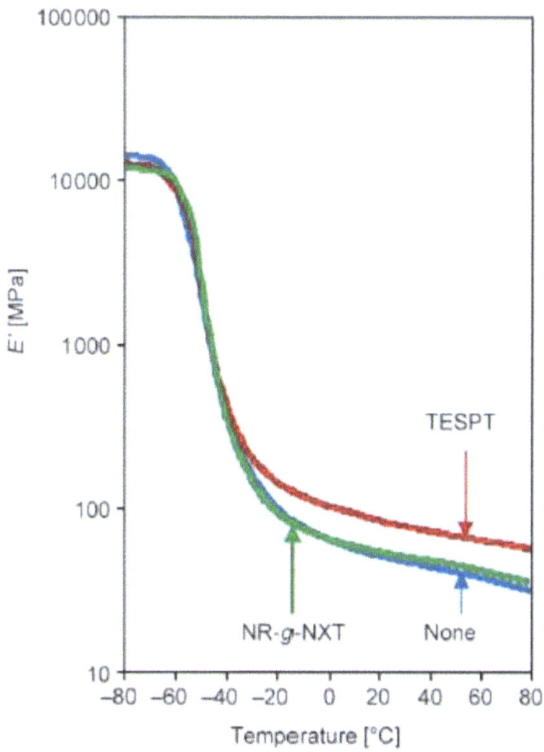

5.3.2 *Epoxidized Natural Rubber as Compatibilizer*

ENR, due to its polar functional epoxide groups interacts with hydroxyl groups on the silica surface while the unmodified part of ENR is compatible with the NR matrix. It has been reported that ENR in tire compounds can improve silica dispersion leading to improved processability, better storage stability, and tire performance (Segatta et al. 1995; Schaal et al. 2002; Lin and Hergenrother 2005).

Payne effect which is an indication of filler-filler interaction can be determined using a Rubber Process Analyser by varying the strains. The difference in storage moduli at low strains and high strains is indicated as the Payne effect. In NR/silica compounds, the Payne effect decreases as the ENR content is increased as shown in Fig. 5.12. The compound without any compatibilizer has the highest filler-filler interaction. Compounds with higher mole % of epoxide show a similar value of the Payne effect but lower than that of ENR-10. The lower Payne effect indicates less silica-silica interaction and better dispersion of silica in the matrix. The Payne effect in compounds with TESPT is similar to that of higher contents (7.5–15 phr) of ENR having higher epoxide mole %.

The reduction of filler-filler interaction in the presence of ENR in silica-filled compounds can be attributed to interactions between silanol groups of silica and epoxide groups of ENR through hydrogen bonding and/or chemical bonding that occur after possible ring opening of the epoxy either during mixing or compression molding (Fig. 5.13) (Rocha et al. 2004; Manna et al. 1999).

Rubber-filler interaction parameter (I) is defined as the ratio of the static to dynamic moduli of the filled rubber compound (Ayala et al. 1993).

$$I = \frac{\sigma}{\eta}$$

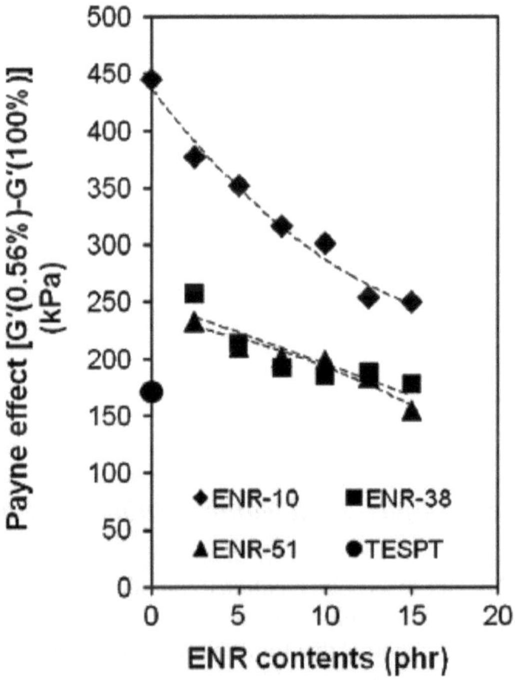

Fig. 5.12 Payne effect of silica-filled NR compound with ENR as compatibilizer (Reproduced with permission from Eur Poly J, 51: 69–79, 2014. Copyright Elsevier.)

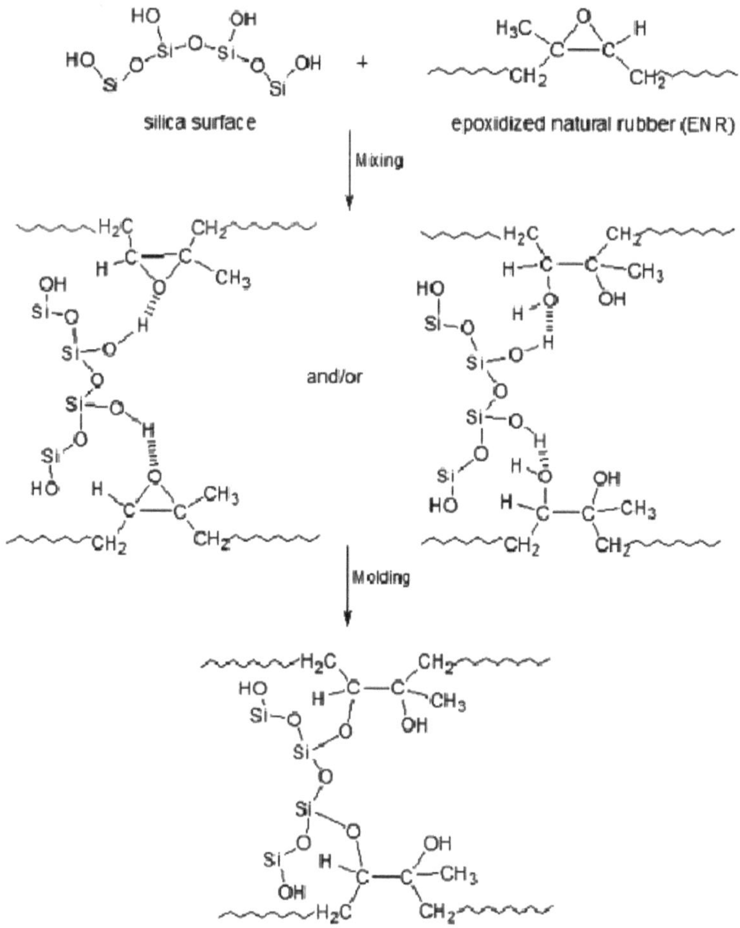

Fig. 5.13 Possible mechanism for interactions and bonding between ENR and silica (Reproduced with permission from Eur Poly J, 51: 69–79, 2014. Copyright Elsevier.)

where σ is the slope of the stress–strain curve of a cured vulcanizate taken at a relatively linear region at low elongation and η is the filler-filler networking parameter calculated from the storage moduli at 0.083 and 5% strain of the compounds without curatives. Increasing the mole % of epoxide groups of ENRs increases the rubber-filler interaction parameter. However, the ultimate effect is lower than that of a silica-filled compound containing TESPT. In the silica-silane system, TESPT can form covalent bonds with silica via silanization reaction and generate cross-links with the rubber phase via sulfur atoms in its structure (Brinke et al. 2003; Kaewsakul et al. 2012). An increase in ENR contents and mole % of epoxide increases the scorch time and cure time of NR/silica compounds. Even though part of the silanol groups on silica are assumed to have interacted with epoxide groups in ENR, the remaining free silanol groups and epoxide groups can possibly form hydrogen bonds with polar accelerators causing accelerator adsorption on the polar surface. In addition, due to the polarity difference between NR and ENR, the more polar curatives can migrate into the ENR phase and so cause a delay of vulcanization reaction in the NR matrix. The silica-filled compound with TESPT displays shorter scorch and cure times. The tensile strength increases with increasing ENR mole % but is lower than that of the vulcanizate containing TESPT. The loss tangent or tan δ at 60 °C of vulcanizates is an indication of the rolling resistance of tires. NR/silica vulcanizates containing TESPT have an outstanding low rolling resistance compared to that containing ENR as a compatibilizer. With increasing mole % of epoxide in ENR, the tan δ value increases. This is associated with the damping properties of ENR as the glass transition temperature of the vulcanizate is shifted to higher values with increasing mole % of epoxide (Gelling and Porter 1988; Gelling 1985). In addition, increasing ENR content having different polarity leads to phase separation with NR. The weak interfacial adhesion results in more energy losses under dynamic deformation and hence increases the loss tangent (Sengloyluan et al. 2014).

Saramolee et al. (2016) have reported the effect of epoxidized low molecular weight natural rubber (ELMWNR) as a compatibilizer in combination with TESPT on the processing and properties of NR/silica compound. ELMWNR can be produced by depolymerization of ENR latex with periodic acid. The latex can be further coagulated, sheeted, and dried. The use of low molecular weight ELMWNR can further reduce the viscosity of the compound due to the lubricating effect caused by the mobility of the smaller chains. The combination of ELMWNR with TESPT can partly weaken the filler-filler interaction and promote filler-rubber interaction. In addition, the low molecular weight rubber molecules can act as internal plasticizers in the rubber compounds, thus improving the mobility of both rubber chains and filler, and facilitating filler dispersion.

The possible interactions that occur in the TESPT/ELMWNR compatibilized NR/silica system can be explained as follows. The addition of ELMWNR can introduce the formation of hydrogen bonds between epoxide groups of ELMWNR and silanol groups on the silica surface. Also, chemical reactions between the epoxide groups with the silanol groups and rubber molecules may occur during the mixing stage. In addition, the carbonyl groups of ketone and aldehyde at the chain ends of ELMWNR may also interact with silanol groups on the silica surface. The presence of polar epoxide groups in ELMWNR can also lead to interaction between themselves, and during vulcanization can further co-cross-link with the NR chains. When TESPT is added in combination with ELMWNR, rubber-filler interaction is further enhanced through the silanization reaction with the remaining silanol groups on the silica surface. Further, the tetrasulfide groups of TESPT may split under heat to react with the rubber chains forming covalent bonds during the vulcanization process.

The vulcanizates compatibilized with ELMWNR and TESPT have higher moduli and tensile strength than those of vulcanizates that contain TESPT only. However, the reinforcement index ie, the ratio of moduli at 300–100% elongations, for vulcanizates with the combination of compatibilizer increases with TESPT content but is still lower than that containing TESPT alone. This is due to the presence of low molecular weight rubber molecules. The addition of ELMWNR can improve the wet grip resistance which is indicated by higher tan δ in the temperature range 0–20 °C, but negatively affects the tire rolling resistance.

5.3.3 Liquid Natural Rubber as Compatibilizer

He et al. have used LNR as a compatibilizer in NR/silica composites prepared by wet compounding technology (He et al. 2018). LNR with an average molecular weight of 8000 contains functional groups such as –OH and –SH. Silica added to emulsified LNR and subsequently diluted, followed by high shearing can produce stable silica slurry. This can be mixed with NR latex and subjected to co-coagulation resulting in NR/silica/LNR masterbatch. The masterbatch can be further melt processed with curing agents and vulcanized. The functional terminal groups in LNR could interact with the silanol groups of silica thus obtaining homogeneous silica dispersion and strong rubber-filler interfacial interactions thereby improving the compatibility between NR and silica. The Payne effect of the compounds with LNR is lower than that without LNR. The CRI of the composites with LNR increased compared to that without LNR which indicates that the vulcanization reaction is accelerated in the presence of LNR. The incorporation of LNR in NR/silica composites contributes to better mechanical properties such as tensile strength, modulus, and tear strength. The dynamic mechanical properties, especially the tan δ maximum peak shifted to higher temperatures for composites with the LNR system. This increase is attributed to the greater number of cross-linked points produced due to the incorporation of LNR which constrained the motion of rubber molecular chains. The system also showed the best combination of high wet-skid resistance and low rolling resistance which is correlated with the tan δ values at 0 °C and 60 °C, respectively. The thermal stability which is reflected by the decomposition temperature also improved for composites with LNR. This is ascribed to the strong interactions between NR and silica which restricted the thermal motion of the polymer chains.

Rubber-filler interaction can be better understood by using the Lorenz-Park equation. The lower the Q_f/Q_g value, the higher the extent of interaction between the filler and the matrix.

$$\frac{Q_f}{Q_g} = ae^{-z} + b$$

where subscripts 'f' and 'g' refer to filled and gum vulcanizates, respectively; 'z' is the ratio by weight of the filler to the rubber vulcanizate, a and b are constants.

Q is the solvent absorbed by the rubber, which can be calculated using the following equation:

$$Q = \frac{(W_s - W_d)}{W_i \times \frac{100}{\text{formula weight}}}$$

where W_s is the weight after equilibrium swelling, W_d is the dried weight after equilibrium swelling and W_i is the initial weight of the sample.

Q_f/Q_g value was lower for LNR-containing systems.

5.3.4 Epoxidized Rubber Seed Oil as Compatibilizer

Epoxidized rubber seed oil (ERSO) has been used as a coupling agent and plasticizer in silica-filled NR compounds. ERSO can act as a coupling agent for silica as the epoxy group can react with the silanol group of silica, while the double bond of the higher fatty acid can link with the rubber hydrocarbon by reaction with sulfur added in the compound during vulcanization. A comparative evaluation of TESPT and ERSO as coupling agents in NR/silica vulcanizate has been reported by Joseph et al. The reduced cure time for TESPT-containing compounds compared to that with ERSO reveals that TESPT can also act as an accelerator in addition to its role as a coupling agent. The utilization of part of sulfur by the unsaturated fatty acids present in ERSO results in a higher cure time for the compound. TESPT enhances the reinforcement activity of silica in non-polar rubbers resulting in improved tensile strength, modulus, and abrasion resistance. This is because TESPT can also act as a sulfur donor and increase the cross-link density of the matrix (Joseph et al. 2004).

5.3.5 Miscellaneous

Ismail et al. have compared the effect of the loading of ethylene diamine dilaurate (EDD) as a coupling agent with the commercial coupling agent (Si69) on the cure behavior and mechanical properties of NR composites containing 30 phr silica. At a similar content of coupling agent, the scorch time and cure time of EDD containing NR/silica composites were shorter compared to those containing Si69. The diamine group present in EDD has the ability to promote the vulcanization of olefinic rubber comprising sulfur and primary accelerators through the formation of complexes which are responsible for the fission of sulfur molecules to form cross-links between rubber chains. The tensile strength and modulus increased for the silica-filled NR composites with both EDD and Si69. When more than 2 phr of coupling agent is used, the tensile strength and modulus of EDD are reduced. This might be due to the presence of lauric acid which reduces the interaction between silica and the rubber matrix. However, the elongation at break increased for composites containing EDD which is due to the plasticizing effect of lauric acid in EDD (Ismail et al. 2005b).

5.4 Natural Rubber/Ash Composites

Rice husk via burning process can yield rice husk ash. The two grades of rice husk ash are white rice husk ash (WRHA) and brown rice husk ash (BRHA). WRHA consists of nearly 96% silica content whereas BRHA has lower silica content, typically about 54%, and a substantial amount of carbon content (44%). As the ash contains a large quantity of silica, it can be a potential filler for reinforcing rubber. As with silica, the rice husk ash being hydrophilic has poor compatibility with a hydrophobic polymer matrix.

Ismail et al. have reported the effect of resorcinol formaldehyde as a bonding agent on NR/WRHA composites. The bonding agent enhances the tensile strength and tear strength of the composites up to 10 phr of WHRA. The improved rubber-filler interaction increases the effective stress transfer from the rubber matrix to the filler. The bond formed between the rubber and the matrix results in a higher modulus, cross-link density, and hardness of the rubber vulcanizates (Ismail and Chung 1999).

Ismail et al. have also studied the effect of maleated natural rubber (MNR) as a compatibilizer for palm ash-filled NR vulcanizates. Palm ash contains nearly only 10% silica with the major part being carbon. It also contains K_2O, CaO, MgO and traces of other oxides. The incorporation of MNR in NR/palm ash composite reduces the value of Q_f/Q_g value, which indicates better rubber-filler interaction. MNR enhances the tensile strength and modulus of the composite. In addition, the fatigue life increases as the improved interfacial adhesion helps to reduce the heat build-up during deformation and consequently reduces the chain scission (Ismail and Haw 2008).

5.5 Natural Rubber/Calcium Carbonate Composite

Waste egg-shell derived nanocalcium carbonate has been used as filler in NR composite in conjunction with maleated NR as compatibilizer. This could resolve the problem associated with the disposal of egg-shell waste which would otherwise result in environmental pollution. The egg-shell waste is cleaned by immersing it in boiling water followed by grinding using mortar and pestle. Further, the size is reduced to nanoparticle by subjecting to a planetary ball mill while maintaining the ratio of zirconia ball to powder as 4:1. The egg-shell waste-derived calcium carbonate accelerates the vulcanization of NR due to the presence of organic protein on the surface of $CaCO_3$ which can function as activator. The addition of maleated NR as a compatibilizer slightly delays the curing process as the acidic maleic anhydride groups can adsorb the accelerators. The uniform dispersion of calcium carbonate in the presence of compatibilizer improves the tensile strength to 20.7 MPa in comparison to unreinforced NR (12.7 MPa). The interfacial interaction between NR chains and nano$CaCO_3$ in the presence of MNR is the result of hydrogen bond formation between the succinic acid groups of MNR and amine groups on the $CaCO_3$ surface.

Further, the NR part of MNR can intermingle with the NR chain of the composite. The improved interfacial adhesion also enhances the cross-link density and thermal stability of the composite (Roy et al. 2019a).

5.6 Natural Rubber/Short Fiber Composites

Fiber-reinforced composites are of tremendous importance in end-use applications as they exhibit the combined behavior of soft elastic rubber matrix and strong fibrous reinforcement. When a load is applied to the NR/fiber composite, the load distributes throughout the NR matrix until it reaches the fiber-matrix interface. If the interface is weak, the fibers may debond from the matrix resulting in the fracture of the composites. If the interface is well-bonded, the stress can be transferred from the matrix to the fiber and the fiber then acts as the main stress carrier resulting in high tensile strength of the composite.

The level of reinforcement that can be attained with short fibers is comparatively less compared to long fibers. Nevertheless, it is a viable alternative for long fiber-reinforced and particle-filled rubber composite. The main advantage of using short fibers in rubber composites is the ease of processing. The primary effects of the addition of short fibers in NR are the improvement in modulus, tear strength, hardness, fatigue resistance, hysteresis, and creep resistance while the elongation at break decreases (Jacob et al. 2009). The design of short fiber-reinforced rubber composite depends on several factors such as the aspect ratio of fiber, control of fiber orientation, dispersion, and the existence of a strong interface between fiber and rubber. The main problem when short fibers are compounded with rubber is the poor interfacial adhesion between the hydrophobic rubber and hydrophilic fiber. The issue can be overcome by modifying the fiber surface or by the use of compatibilizers/coupling agents, bonding agents, and pretreatment of fibers using various chemicals (Mathew et al. 2006; Geethamma and Thomas 2004; Jacob et al. 2004; Kondo et al. 2008).

5.6.1 Natural Rubber/Nylon Fiber Composites

Nylon fibers are widely used as the reinforcing material in rubber industries such as tires, conveyor belts, V-belts, etc. However, in order to impart the necessary reinforcement, there should be good adhesion between the fiber and the matrix.

Resorcinol Formaldehyde Bonding Agent

Resorcinol formaldehyde latex (RFL) dip is the most commonly used system to give adhesion between the rubber and fiber.

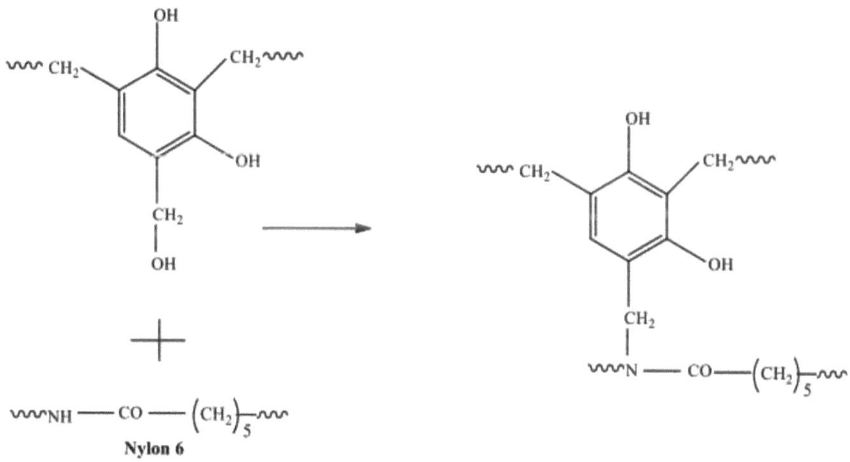

Fig. 5.14 Probable reaction between RF resin and nylon fiber (Adapted from Wootton 2001)

The methylol groups of the RF resin undergo a condensation reaction with the active NH group in the nylon fiber resulting in a covalent bond (Fig. 5.14). The latex part of RFL is a terpolymer containing vinyl pyridine (VP), styrene, and butadiene units. The RF resin may possibly react with the active hydrogen of the butadiene rubber component (Fig. 5.15). Moreover, adhesion can arise from the vulcanization of the rubber component in RFL with the matrix rubber network (Wootton 2001).

The RFL dip is applied to the fiber/fabric before being incorporated into the rubber matrix. The key issue with the RFL dip treatment is the multi-step processing involved, which is expensive, requires time, and is associated with health hazards. An alternative system that has been developed is referred to as a dry bonding agent or in situ bonding agent. Here, adhesion-promoting additives are incorporated into the rubber compound, which later adheres to the untreated fabric. The major additives are resorcinol and formaldehyde donors such as hexamethylenetetramine (HMT). The level of adhesion can be significantly increased by the inclusion of fine particles of hydrated silica. The system works with the production of resorcinol formaldehyde resin in the rubber compound during cure and migrates to the rubber/fabric interface. The silica may act by retarding both the cure of the rubber and also polymerization of RF resin. This ensures that cross-linking occurs only at the fabric interface.

Maleated Natural Rubber

Dong et al. (2013) have used MNR as a compatibilizer for natural rubber/short nylon fiber composites. It was found that the tensile strength of the composites decreased with nylon fiber loading. This is because of the poor dispersion as well as the orientation of the hydrophilic nylon fiber in the NR matrix. The agglomeration of fibers

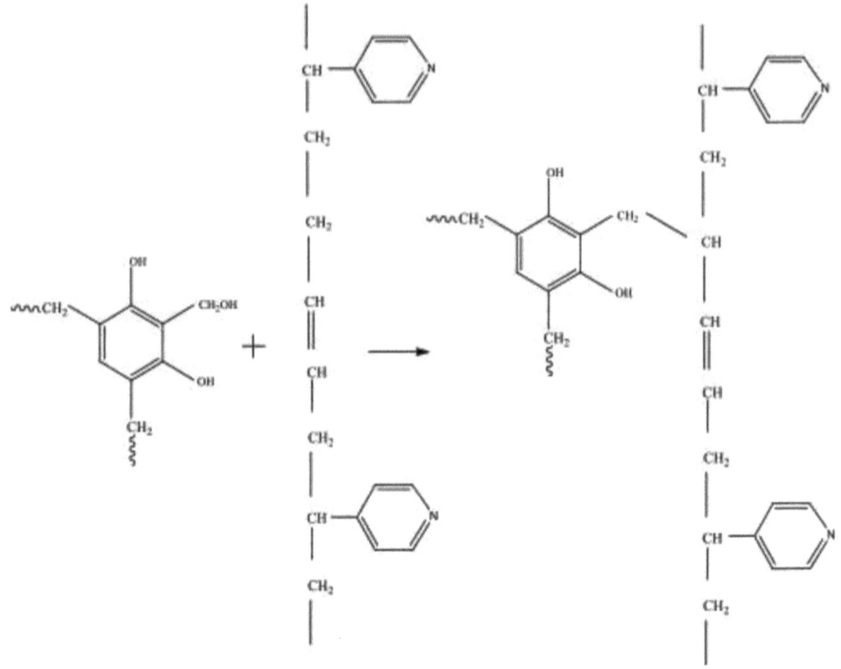

Fig. 5.15 Possible reaction between RF resin and the rubber (Adapted from Wootton 2001)

increases and the de-wetting of the NR matrix at the interface creates stress concentration points. The incorporation of MNR helps to improve the dispersion of nylon fibers and increase the fiber-matrix interfacial adhesion leading to improvement in tensile properties. The interaction between MNR and nylon fiber is shown in Fig. 5.16. The maleic anhydride groups of MNR can react with the free amine groups on the surface of nylon fiber to form covalent bonds. This will reduce the interfacial tension and prevent fiber aggregation. The long NR segments of MNR at the same time can form strong entanglements with the NR matrix thus resulting in compatibilizer efficiency. The modulus of the compatibilized composites increases by about 1.5 times that of un-compatibilized ones for the same fiber loading. The movement of the rubber chain is restricted in the presence of MNR due to the strong adhesion of the fiber with the matrix. The highly bonded system exhibits high resistance to swelling compared to the non-bonded ones.

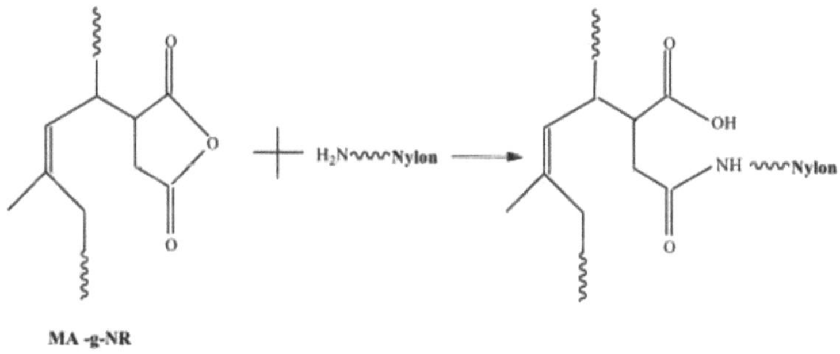

Fig. 5.16 Probable reaction between MNR with Nylon fibers

5.6.2 Natural Rubber/Natural Fiber Composites

Plant-based natural fibers are the major renewable resource material in the world. They have gained immense interest as reinforcing material for the development of environment-friendly green composites. Utilization of natural fibers not only contributes to sustainable development but can reduce environmental issues concerning global-scale pollution. Natural fibers provide many advantages compared to conventional inorganic fibers (glass fibers, carbon fibers) such as abundance and therefore low cost, low density, biodegradability, less machine wear during processing, and tunable fiber aspect ratio. However, these fibers exhibit large variations in quality and are sensitive to moisture. The plant fibers are composed of cellulose, hemicellulose, lignin, and pectin with small quantities of extractives. Of these, cellulose is the main component of all plant fibers. Cellulose can be derived from a variety of sources such as wood, pulp; cotton, coir (seed fibers); flax, jute, sisal (bast fibers); bagasse, bamboo (grass fibers), etc. Cellulose-based plant fiber is a suitable and effective filler for rubber composites in place of commonly used fillers in place of carbon black and silica.

Among the various forms of natural fibers that can be used as filler, short fiber-reinforced NR composites have gained much attention due to their low specific weight, being a renewable resource, ease of processing, and economic advantages. The main problem that is encountered when hydrophilic natural fibers are used as reinforcement in hydrophobic polymers is that they exhibit inferior properties because of poor adhesion at the fiber-matrix interface.

Cellulose is a linear condensation polymer consisting of D-anhydroglucopyranose units joined together by β-1,4-glucosidic bonds. The –CH_2OH and –OH groups as well as the glycosidic bonds are equatorial with respect to the mean planes of the pyranose rings. The abundance of hydroxyl groups and the concomitant tendency to form intra- and intermolecular hydrogen bonds results in the formation of linear aggregates. Hence compatibilizers/coupling agents are required to improve the fiber-matrix adhesion.

Resorcinol Formaldehyde Bonding Agent

Coir is a versatile lignocellulosic fiber obtained from coconut trees and is inexpensive among the various natural fibers. Unfortunately, the performance of coir as a reinforcement of polymer composites is unsatisfactory due to various factors such as low cellulose (36–43%) content, high lignin (4–45%) and hemicellulose (0.15–0.25%) content, high microfibrillar angle and large as well as variable diameter (100–400 μm). The efficiency of coir as reinforcement in rubber composites can be improved by enhancing the interfacial adhesion between coir and rubber. Geethamma et al. (1998) have reported that a resorcinol-hexa dry bonding system can enhance the tensile strength, tear strength, and moduli of sodium carbonate-treated coir fiber/natural rubber composites compared to the untreated coir fiber composites.

The effect of resorcinol formaldehyde-hydrated silica-hexa (5:2:5) bonding system on the mechanical properties of oil palm wood flour-filled NR composites has been studied by Ismail and Hasliza (1999) It is reported that the presence of bonding agents increases the tensile strength, tear strength, and hardness. The Q_f/Q_g value obtained from equilibrium swelling studies also decreased compared to composites without bonding agent which indicates a higher extent of interaction between the filler and the matrix.

Utilization of hybrid fibers such as sisal and oil palm fibers in NR composites has been reported by Jacob et al. (2004) Sisal fibers and oil palm fibers can be used after separation from foreign matter and pith. The treatment of these fibers with sodium hydroxide solution helps to remove lignin which is indicated by the decrease in the diameter and weight of the fibers. Resorcinol and hexa (bonding agent) were incorporated in the rubber mix for preparing the composite. It was found that the presence of bonding agents resulted in a longer cure time which was attributed to the direct involvement of bonding agents in the vulcanization. The swelling index is a measure of the swelling resistance of the rubber compound and is calculated using the equation

$$\text{Swelling Index} = \frac{A_s}{W \times 100}$$

where A_s is the amount of solvent absorbed by the sample and W is the initial weight of the sample before swelling. A low swelling index value indicates the restriction of the composite to swelling. The bonding agent is capable of binding the alkali-treated fiber with the rubber and hence exhibits a low swelling index.

Equilibrium swelling method can be used for assessing the rubber-fiber interfacial adhesion using the Kraus equation:

$$\frac{V_{ro}}{V_{rf}} = 1 - n\left[\frac{f}{1-f}\right]$$

where

f volume fraction of fiber
n rubber-fiber interaction parameter
V_{ro} volume fraction of rubber in the swollen gum vulcanizate
V_{rf} volume fraction of rubber in the swollen fiber-filled vulcanizate

The ratio of V_{ro}/V_{rf} represents the degree of restriction of the swelling of the rubber matrix due to the presence of fiber. The volume fraction of rubber in the fiber-filled swollen sample V_{rf} can be calculated according to the equation

$$V_{rf} = \frac{\frac{(w_d - Fw_o)}{\rho_r}}{\frac{(w_d - Fw_o)}{\rho_r} + \frac{(w_s - w_d)}{\rho_s}}$$

where

w_d deswollen weight (or dried weight)
w_o initial weight of the sample
w_s weight at equilibrium swelling (or swollen weight)
F weight fraction of the fiber
ρ_r density of rubber
ρ_s density of solvent

Since the equation is in the form of a straight line, a plot of V_{ro}/V_{rf} versus $f/1 - f$ should give a straight line whose negative slope 'n' indicates the reinforcing effect of fibers. As the fiber loading increases, the solvent uptake of the vulcanizate decreases which increases the V_{rf} value. This decreases the V_{ro}/V_{rf} ratio with fiber loading as V_{ro} is a constant (Fig. 5.17). Further, the V_{ro}/V_{rf} value was lowest for composites containing alkali-treated fibers in the presence of a bonding agent. This revealed that the strong adhesion at the fiber-rubber interface restricted the entry of solvent.

Grass fiber (Cyperus Tegetum Rox b) has been used as filler in NR composites. The main constituents of grass are lignin (40%), cellulose (15%), hemicellulose (21%), water-soluble matter (22%) and ash (2%). The fibers after soaking in RFL (5 pbw of rubber) followed by drying result in coated fiber. The effect of resorcinol formaldehyde latex (RFL) as a bonding agent was compared for alkali-treated and water-leached grass fiber/NR composites (De et al. 2006). The alkali-treated RFL-coated fibers were cured to a greater extent than composites with water-leached RFL-coated grass fibers. This is because the interaction between the RFL and alkali-treated grass fibers is better than with water-leached fibers. However, the presence of RFL prolongs the cure time. RFL-coated fibers showed higher tensile strength and modulus at all levels of fiber loading. Strong adhesion between the fiber and matrix results in higher shear strength at the interface and therefore strong forces are required to overcome the shear strength at the interface, resulting in higher tensile strength. The presence of a bonding agent also improves the swelling value of the composites which is an indirect measure of cross-link density.

The reaction between RF resin and cellulose fiber is shown in Fig. 5.18. The methylol groups of RF resin react with the primary hydroxyl group of cellulose to

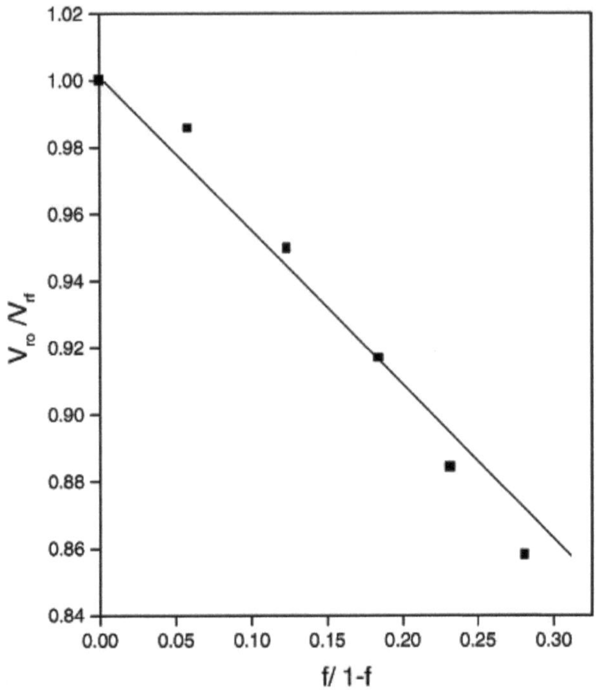

Fig. 5.17 Variation of V_{ro}/V_{rf} with $f/1-f$ of NR/sisal fiber/oil palm fiber composites (Reproduced with permission from Compos. Sci. Technol. 64: 955–965, 2004. Copyright Elsevier.)

form an ether linkage. The RF resin bonds with the fiber and the rubbery part of the latex component bonds with the rubber matrix by co-vulcanization.

Maleated Natural Rubber as Compatibilizer

Ismail et al. have investigated the effect of maleated natural rubber (MNR) as a compatibilizer in paper sludge-filled NR. Paper sludge is a by-product from the pulp and paper industry which is composed of short cellulose fiber and inorganic materials. It was found that the cure time was longer for composites containing MNR which was attributed to the better interaction between the fiber and rubber. Tan δ value derived by dividing S'' by S' is lower for NR composites containing MNR compared to that without MNR. This indicates that MNR reduces the composite hysteresis and improves its resilience. In addition, the tensile strength increases for the MNR system due to better stress transfer from the matrix to the fiber as a result of strong adhesion between the fiber-matrix interface (Ismail et al. 2005a).

The incorporation of MNR in NR/cotton composites improved the tensile modulus and hardness compared to the system without MNR. The Kraus plot of NR/cotton fiber composites with and without compatibilizer indicated that V_{ro}/V_{rf}

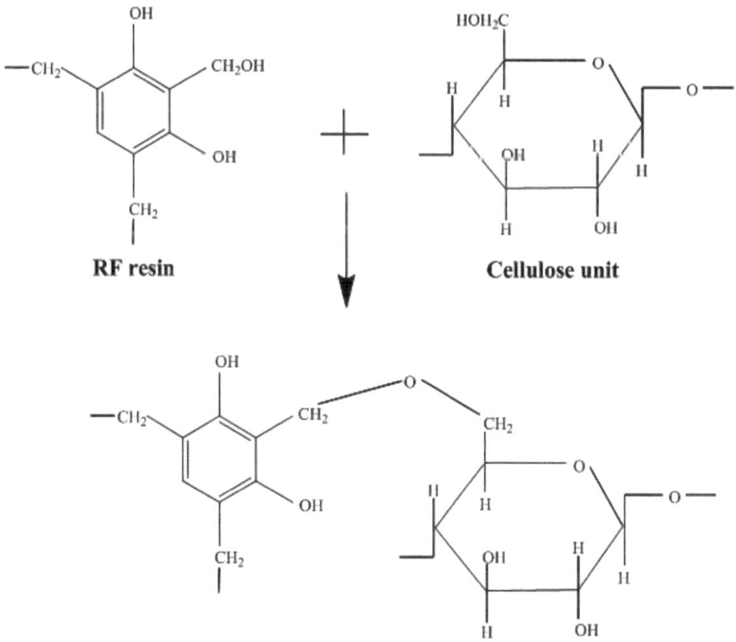

Fig. 5.18 The possible reaction between RF resin and cellulose fiber (Adapted from Wootton 2001)

ratio decreased with fiber loading. In the presence of MNR, the value of V_{ro}/V_{rf} further decreased compared to the composites without MNR. This is attributed to the enhanced fiber-rubber adhesion in the composite. The incorporation of fibers enhanced the storage modulus of the composite both at low temperatures and in the transition region. The presence of MNR further improved the storage modulus than the composites without MNR (Zeng et al. 2010).

Cellulosic fiber comprises both amorphous and crystalline regions. In general, the amorphous regions can be efficiently removed from cellulose using acid hydrolysis to produce microcrystalline cellulose (MCC). Roy et al. have investigated the effectiveness of maleated natural rubber as a compatibilizer in NR/microcrystalline composites. The cure time of MNR compatibilized NR/MCC composite is longer compared to that without MNR. This might be due to the interaction between acidic maleic anhydride groups and accelerators (Wongsorat et al. 2014; Pasbakhsh et al. 2009). Further, the tensile strength and modulus increased for composites containing 10 phr of MNR. The improvement in properties can be due to the interaction between the maleic anhydride groups in MNR and the hydroxyl groups of cellulose which forms covalent bonds and ester linkages thus enhancing the fiber-matrix bonding (Fig. 5.19). The long NR chain of MNR becomes compatible with the NR matrix thus lowering the surface tension of the fiber and thereby increasing the wettability with the matrix. Beyond 10 phr, MNR was not so effective which might be due to the formation of two different rubber phases.

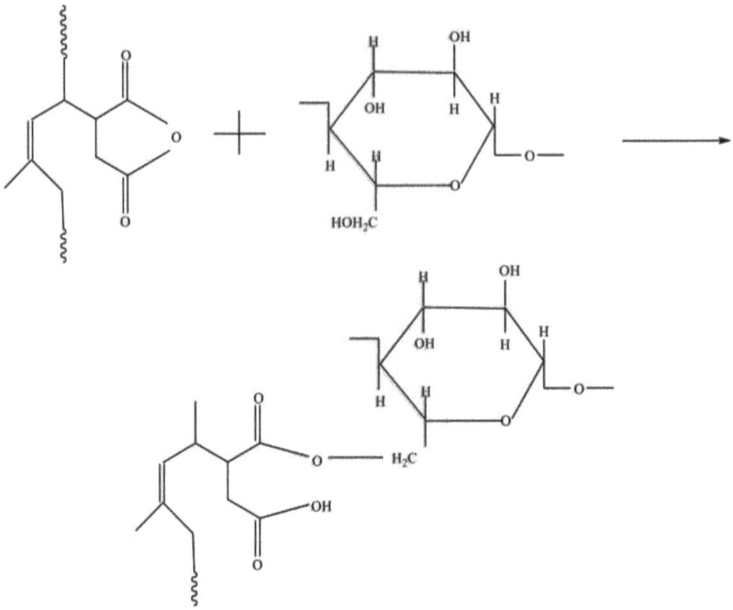

Fig. 5.19 Probable interaction between maleated natural rubber and cellulose

The incorporation of MNR also resulted in high cross-link density. The equilibrium solvent uptake of vulcanizates can be calculated using the following equation.

$$\text{Equilibrium solvent uptake (wt.)} = \frac{W_e - W_o}{W_o} \times 100$$

where

W_o initial weight of the sample
W_e equilibrium weight of the sample.

The solvent uptake behavior of MNR-based NR composites was lower than the un-compatibilized and gum vulcanizate. This is because of the improved interaction between NR and MCC. Furthermore, the uniform dispersion of MCC and the improved interaction between NR and MCC in the presence of MNR restricted the thermal motion of the rubber chain within the network structure. This shifted the degradation region of the composite to a higher temperature and hence enhanced the thermal stability (Roy and Potiyaraj 2018).

Silane Coupling Agent

Bamboo fiber contains typically 48% of cellulose. The silane coupling agents can effectively improve the tensile modulus and hardness of NR/bamboo fiber composites. This is due to the chemical bond formation between the hydroxyl groups of the cellulose fibers and the alkoxy groups of the coupling agent. It also permits shorter vulcanization time with an increase in rubber-bound sulfur (Ismail et al. 2002). Kenaf short fibers have been studied as reinforcing elements in NR-based composites. The incorporation of Si69 reduced the scorch and cure time of the composites compared to that without a silane coupling agent. The addition of a coupling agent improves the dispersion of fiber in the matrix. The decrease in scorch and time is also influenced by the increase of sulfur content in the composites with the addition of silane (Park 2000). This in fact increases the cross-link density (Choi and Choi 2006). The tensile modulus is higher for composites with Si69 as the coupling agent (Ismail et al. 2011).

Epoxidized Natural Rubber

The influence of ENR 50 as a compatibilizer in NR/MCC composites has been explored by Roy et al. The cure rate lowered with the addition of ENR (5 phr) in the composite as the polar accelerators have a tendency to migrate toward the polar ENR in contrast to the non-polar NR leading to accelerator adsorption. The tensile strength enhanced from a value of 12.7–17.8 MPa in the presence of ENR indicating the homogeneous dispersion of MCC in NR. The improved interfacial interaction significantly reduced the solvent uptake of the compatibilized composite (Roy et al. 2019b). The plausible interactions between ENR and cellulose are presented in Fig. 5.20. During melt-mixing, the epoxy group of ENR can form a hydrogen bond with the –OH groups of cellulose fiber. On vulcanization, the epoxy ring may open and form covalent ether linkages with the –CH_2OH groups of cellulose (Balachandrakurup and Gopalakrishnan 2022).

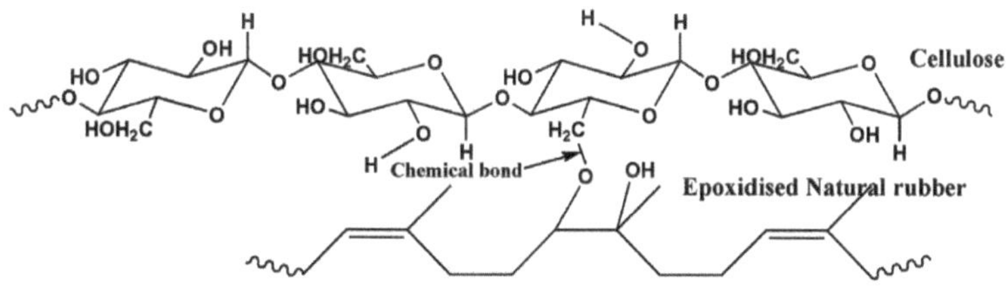

Fig. 5.20 Chemical interactions between cellulose fiber and ENR (Reproduced with permission from *Industrial Crops and Products*, *183*, 114,935 (2022), Copyright Elsevier.)

5.7 Natural Rubber/Nanocellulose Composite

Nanocellulose refers to the form of tiny cellulose that has at least one dimension in the nanometer range. They can be isolated from various lignocellulosic sources via pretreatments, and chemo-mechanical treatments such as cryocrushing, grinding, high-pressure homogenization, ultrasonication, acid hydrolysis, etc. They can occur in the form of rod-like cellulose nanocrystals (CNC) or longer cellulose nanofibrils (CNF). The unique features of nanocellulose include high surface area, low density, high specific strength and modulus, and low coefficient of thermal expansion (Jiang and Hsieh 2013). The methods that have been employed for the preparation of natural rubber/nanocellulose composites are melt compounding, latex stage compounding, etc. (Khalil et al. 2014). Though the aggregation of nanocellulose could be overcome by latex stage compounding, the use of coupling agents/compatibilizers can further enhance the interfacial interaction.

Cellulose nanofiber (CNF) can be easily incorporated in NR latex and co-coagulated followed by melt compounding using traditional compounding ingredients. The effect of MNR as a compatibilizer in NR/CNF composite has been reported by Balachandrakurup et al. The Payne effect value was reduced for the compatibilized composite owing to improvement in interfacial interaction in contrast to the NR/CNF composite. With 5 phr of CNF, the tensile strength increased to 36.8 MPa for the compatibilized composite facilitating better stress transfer from fiber to matrix and vice-versa, while the un-compatibilized one exhibited a value of 30.9 MPa. The reinforcing index and tear strength enhanced substantially for the NR/CNF/MNR composite (Balachandrakurup et al. 2021).

Rather than incorporating silane coupling agents separately, CNF has been surface-treated with silane coupling agent (TESPT) to improve the adhesion between the nanofiber and NR. The hydroxyl group of CNF can easily interact with the hydrolyzable alkoxy group of silane leading to high bound rubber content in NR/CNF composites. Further, treated CNF enhances the cross-link density of the composite due to the sulfur donor ability of TESPT during vulcanization. This further results in higher tensile strength and modulus of the composite in comparison to the untreated one (Somseemee et al. 2021).

Natural rubber that has been modified via oxidative degradation using hydrogen peroxide has been used as a compatibilizer. The reaction with H_2O_2 leads to chain scission and forms polar hydroxyl and carbonyl functionalities in NR. These functional groups can easily interact with the hydroxyl groups of nanocellulose thereby inhibiting the possibility of accelerator adsorption and accelerating the vulcanization process. This could reduce the energy consumption during processing. Such interfacial interaction can also enhance the cross-link density and thermal stability of NR/CNC composite (Jantachum and Phinyocheep 2024).

5.8 Natural Rubber/Halloysite Nanocomposites

Halloysite nanotube (HNT), a clay mineral of the kaolin group, is the predominant form of natural halloysite with the formula $Al_2(OH)_4Si_2O_5.nH_2O$. The multilayer tubular structure is formed by the wrapping of the 1:1 clay mineral with a mismatch in the oxygen-sharing tetrahedral and octahedral layers under favorable geological conditions. A monolayer of water molecules separates the individual layers of the halloysite. The outer surface of HNT is composed of siloxane groups while the inner surface consists of alumina groups (Yuan et al. 2015). Jehsoh et al. have utilized a biobased compatibilizer, viz., palm stearin which is derived from the refinery processing of crude palm oil. Further, palm stearin comprising triglyceride can easily react with primary or secondary amines forming modified palm stearin (MPS). The MPS consisting of fatty acid eliminates the use of stearic acid in NR/HNT compounding. It being, alkaline in nature accelerates the cure rate of the composite. MPS not only enhances the tensile and tear strength in NR/HNT composite but also promotes strain-induced crystallization. The amide group in MPS can form H-bonds with the silanol groups on the HNT surface while the alkyl part can easily interact with the non-polar chains of NR. This facilitates the dispersion of HNT in the NR matrix (Jehsoh et al. 2021).

The above studies indicate that compatibilizers play a major role in enhancing the technical properties of the nanocomposite thereby increasing the realm of applications.

5.9 Conclusions

This chapter presents an overview of NR-based composite with various fillers such as silica, layered silicate, nylon fibers, and cellulosic fibers. Of these fillers, silica and nylon fibers are being used in the rubber industries. The polar nature of these fillers/fibers leads to problems such as agglomeration and poor dispersion affecting the ultimate properties of the composite. Most researchers have focused on the effect of various compatibilizers on the mechanical, dynamic mechanical, rheological properties, and swelling behavior of natural rubber/filler composites. The type and loading of compatibilizers also play a major role in the improvement of properties. This review forms a foundation for the future use of these fillers with suitable compatibilizers. The studies pertaining to the methods of preparation of compatibilizers need to be addressed in detail.

References

Ansarifar A, Lim HP, Nijhawan R (2004) Assessment of the effect of a bifunctional organosilane on the bound rubber and properties of some natural rubber compounds. Int J Adhes Adhes 24(1):9–22

Ansarifar A, Wang L, Ellis RJ, Kirtley SP, Riyazuddin N (2007) Enhancing the mechanical properties of styrene–butadiene rubber by optimizing the chemical bonding between silanized silica nanofiller and the rubber. J Appl Polym Sci 105(2):322–332

Arroyo M, Lopez-Manchado MA, Herrero B (2003) Organo-montmorillonite as substitute of carbon black in natural rubber compounds. Polymer 44(8):2447–2453

Arroyo M, Lopez-Manchado MA, Valentin JL, Carretero J (2007) Morphology/behaviour relationship of nanocomposites based on natural rubber/epoxidized natural rubber blends. Compos Sci Technol 67(7–8):1330–1339

Avalos F, Ortiz JC, Zitzumbo R, López-Manchado MA, Verdejo R, Arroyo M (2008) Effect of montmorillonite intercalant structure on the cure parameters of natural rubber. Eur Polym J 44(10):3108–3115

Ayala JA, Hess WM, Joyce GA, Kistler FD (1993) Carbon-black-elastomer interaction II: effects of carbon black surface activity and loading. Rubber Chem Technol 66(5):772–805

Balachandrakurup V, Gopalakrishnan J (2022) Enhanced performance of cellulose nanofibre reinforced styrene butadiene rubber nanocomposites modified with epoxidised natural rubber. Ind Crops Prod 183:114935

Balachandrakurup V, George N, Gopalakrishnan J (2021) Effect of compatibiliser on the mechanical, rheological and thermal properties of natural rubber/cellulose nanofibre composites. Mater Today: Proc 47:5345–5350

Brydson JA (1978) Rubber chemistry. Springer, London

Carone E Jr, Kopcak U, Gonçalves MC, Nunes SP (2000) In situ compatibilization of polyamide 6/natural rubber blends with maleic anhydride. Polymer 41(15):5929–5935

Chandra CJ, Bipinbal PK, Sunil KN (2017) Viscoelastic behaviour of silica filled natural rubber composites–correlation of shear with elongational testing. Polym Test 60:187–197

Choi SS, Choi SJ (2006) Influence of silane coupling agent content on crosslink type and density of silica-filled natural rubber vulcanizates. Bull Korean Chem Soc 27(9):1473–1476

De D, De D, Adhikari B (2006) Curing characteristics and mechanical properties of alkali-treated grass-fiber-filled natural rubber composites and effects of bonding agent. J Appl Polym Sci 101(5):3151–3160

Doğan M, Oral DD, Yılmaz B, Savu M, Karahan S, Bayramlı E (2011) Physical properties and cure characteristics of natural rubber/nanoclay composites with two different compatibilizers. J Appl Polym Sci 121(3):1530–1535

Dong ZX, Liu MX, Jia DM, Zhou YH (2013) Synthesis of natural rubber-g-maleic anhydride and its use as a compatibilizer in natural rubber/short nylon fiber composites. Chin J Polym Sci 31(8):1127–1138

Donnet JB (2003) Nano and microcomposites of polymers elastomers and their reinforcement. Compos Sci Technol 63(8):1085–1088

Geethamma VG, Thomas S (2004) Transport of organic solvents through coir-fiber-reinforced natural rubber composites: a method for evaluating interfacial interaction. J Adhes Sci Technol 18(8):951–966

Geethamma VG, Mathew KT, Lakshminarayanan R, Thomas S (1998) Composite of short coir fibres and natural rubber: effect of chemical modification, loading and orientation of fibre. Polymer 39(6–7):1483–1491

Gelling IR (1985) Modification of natural rubber latex with peracetic acid. Rubber Chem Technol 58(1):86–96

Gelling IR, Porter M (1988) In: Robert AD (ed) Natural rubber science and technology, Chapter 10. Oxford University Press

He F, Yuan T, Li C, Sun L, Liao S (2018) Interfacial interactions and properties of natural rubber–silica composites with liquid natural rubber as a compatibilizer and prepared by a wet-compounding method. J Appl Polym Sci 135(30):46457

Idrus SS, Ismail H, Palaniandy S (2011) The effects of silanized ultrafine silica on the curing characteristics, tensile properties, and morphological study of natural rubber compounds. Polym Plast Technol Eng 50(1):1 7

Ismail H (1999) The effects of a "palm oil" faiiy acid additive and a silane coupling agent on the curing characteristics and vulcanizate properties of silica-filled natural rubber compounds. Polym Plast Technol Eng 38(4):805–820

Ismail H, Chung FL (1999) White rice husk ash-filled natural rubber compounds: the effect of bonding agents. Polym Plast Technol Eng 38(5):1059–1079

Ismail H, Hasliza IN (1999) Effects of bonding agents on the mechanical properties of the composites made of natural rubber and oil palm empty fruit bunch. Polym Plast Technol Eng 38(1):137–148

Ismail H, Haw FS (2008) Effects of palm ash loading and maleated natural rubber as a coupling agent on the properties of palm-ash-filled natural rubber composites. J Appl Polym Sci 110(5):2867–2876

Ismail H, Shuhelmy S, Edyham MR (2002) The effects of a silane coupling agent on curing characteristics and mechanical properties of bamboo fibre filled natural rubber composites. Eur Polymer J 38(1):39–47

Ismail H, Rusli A, Rashid AA (2005a) Maleated natural rubber as a coupling agent for paper sludge filled natural rubber composites. Polym Test 24(7):856–862

Ismail H, Suryadiansyah, Azhari B (2005b) The effect of ethylene diamine dilaurate and silane coupling agent on cure characteristics, mechanical, and thermal properties of silica-filled natural rubber composites. Polym Plast Technol Eng 44(8–9):1657–1669

Ismail H, Mahir NA, Ahmad Z (2011) The effect of bis-(3-triethoxysilylpropyl) tetrasulphide (Si-69) as a coupling agent on properties of natural rubber/kenaf fibre composites. Polym-Plast Technol Eng 50(9):893–897

Jacob M, Thomas S, Varughese KT (2004) Mechanical properties of sisal/oil palm hybrid fiber reinforced natural rubber composites. Compos Sci Technol 64(7–8):955–965

Jacob M, Anandjiwala RD, Thomas S (2009) Lignocellulosic fiber reinforced rubber composites. In: Thomas S, Pothan LA (eds) Natural fiber reinforced polymer composites: from macro to nanoscale. Old City Publishing, Inc., Philadelphia, p 252

Jantachum P, Phinyocheep P (2024) Compatibilization of cellulose nanocrystal-reinforced natural rubber nanocomposite by modified natural rubber. Polymers 16(3):363

Jehsoh N, Surya I, Sahakaro K, Ismail H, Hayeemasae N (2021) Modified palm stearin compatibilized natural rubber/halloysite nanotubes composites: reinforcement versus strain-induced crystallization. J Elastomers Plast 53(3):210–227

Jesionowski T, Krysztafkiewicz A (2001) Influence of silane coupling agents on surface properties of precipitated silicas. Appl Surf Sci 172(1–2):18–32

Jiang F, Hsieh YL (2013) Chemically and mechanically isolated nanocellulose and their self-assembled structures. Carbohydr Polym 95(1):32–40

Joly S, Garnaud G, Ollitrault R, Bokobza L, Mark JE (2002) Organically modified layered silicates as reinforcing fillers for natural rubber. Chem Mater 14(10):4202–4208

Joseph R, Alex R, Madhusoodanan KN, Premalatha CK, Kuriakose B (2004) Use of epoxidized rubber seed oil as a coupling agent and a plasticizer in silica-filled natural rubber compounds. J Appl Polym Sci 92(6):3531–3536

Kaewsakul W, Sahakaro K, Dierkes WK, Noordermeer JW (2012) Optimization of mixing conditions for silica-reinforced natural rubber tire tread compounds. Rubber Chem Technol 85(2):277–294

Khalil HA, Davoudpour Y, Islam MN, Mustapha A, Sudesh K, Dungani R, Jawaid M (2014) Production and modification of nanofibrillated cellulose using various mechanical processes: a review. Carbohydr Polym 99:649–665

Kondo, Miyazaki K, Takayanagi K, Sakurai K (2008) Surface treatment of PET fiber by EB-irradiation-induced graft polymerization and its effect on adhesion in natural rubber matrix. Eur Polym J 44(5):1567–1576
Kosmalska A, Zaborski M, Ślusarski L (2003) Adsorption of curatives and activity of silica toward elastomers. In: Macromolecular symposia, vol 194, no 1. WILEY-VCH Verlag, Weinheim, pp 269–276
Lee CH, Park SH, Chung W, Kim JY, Kim SH (2011) Preparation and characterization of surface modified silica nanoparticles with organo-silane compounds. Colloids Surf A 384(1–3):318–322
Lin CJ, Hergenrother WL (2005) U.S. Patent No. 6,845,797. U.S. Patent and Trademark Office, Washington, DC
Lopez-Manchado MA, Arroyo M, Herrero B, Biagiotti J (2003a) Vulcanization kinetics of natural rubber–organoclay nanocomposites. J Appl Polym Sci 89(1):1–15
Lopez-Manchado MA, Herrero B, Arroyo MJPI (2003b) Preparation and characterization of organoclay nanocomposites based on natural rubber. Polym Int 52(7):1070–1077
Lopez-Manchado MA, Herrero B, Arroyo M (2004) Organoclay–natural rubber nanocomposites synthesized by mechanical and solution mixing methods. Polym Int 53(11):1766–1772
Madhusoodanan KN, Varghese S (2006) Technological and processing properties of natural rubber layered silicate-nanocomposites by melt intercalation process. J Appl Polym Sci 102(3):2537–2543
Manna AK, De PP, Tripathy DK, De SK, Peiffer DG (1999) Bonding between precipitated silica and epoxidized natural rubber in the presence of silane coupling agent. J Appl Polym Sci 74(2):389–398
Mark JE, Erman B, Eirich FR (eds) (2005) Science and technology of rubber. Elsevier Academic Press
Mathew L, Ulahannan J, Joseph R (2006) Effect of curing temperature, fibre loading and bonding agent on the equilibrium swelling of isora-natural rubber composites. Compos Interfaces 13(4–6):391–401
Nah C, Ryu HJ, Han SH, Rhee JM, Lee MH (2001) Fracture behaviour of acrylonitrile–butadiene rubber/clay nanocomposite. Polym Int 50(11):1265–1268
Nakason C, Kaesaman A, Samoh Z, Homsin S, Kiatkamjornwong S (2002) Rheological properties of maleated natural rubber and natural rubber blends. Polym Test 21(4):449–455
Nakason C, Kaesaman A, Supasanthitikul P (2004) The grafting of maleic anhydride onto natural rubber. Polym Test 23(1):35–41
Nakason C, Kaesman A, Homsin S, Kiatkamjornwong S (2001) Rheological and curing behavior of reactive blending. I. Maleated natural rubber–cassava starch. J Appl Polym Sci 81(11):2803–2813
Ostad-Movahed S, Ansar Yasin K, Ansarifar A, Song M, Hameed S (2008) Comparing effects of silanized silica nanofiller on the crosslinking and mechanical properties of natural rubber and synthetic polyisoprene. J Appl Polym Sci 109(2):869–881
Park SS (2000) Influence of silane coupling agent on properties of filled styrene-butadiene rubber compounds. Macromol Res 8(6):285–291
Pasbakhsh P, Ismail H, Fauzi MA, Bakar AA (2009) Influence of maleic anhydride grafted ethylene propylene diene monomer (MAH-g-EPDM) on the properties of EPDM nanocomposites reinforced by halloysite nanotubes. Polym Test 28(5):548–559
Reichert P, Hoffmann B, Bock T, Thomann R, Mülhaupt R, Friedrich C (2001) Morphological stability of poly (propylene) nano composites. Macromol Rapid Commun 22(7):519–523
Rocha TC, Schuster RH, Jacobi MM, Samios D (2004) Influence of epoxidation on physical properties of SBR and its interaction with precipitated silica. KGK Kautschuk Gummi Kunststoffe 57(12):656–661
Roy K, Potiyaraj P (2018) Development of high performance microcrystalline cellulose based natural rubber composites using maleated natural rubber as compatibilizer. Cellulose 25(2):1077–1087

Roy K, Debnath SC, Raengthon N, Potiyaraj P (2019a) Understanding the reinforcing efficiency of waste eggshell-derived nano calcium carbonate in natural rubber composites with maleated natural rubber as compatibilizer. Polym Eng Sci 59(7):1428–1436

Roy K, Debnath SC, Pongwisuthiruchte A, Potiyaraj P (2019b) Natural rubber/microcrystalline cellulose composites with epoxidized natural rubber as compatibilizer. Rubber Chem Technol 92(2):378–387

Sae-Oui P, Sirisinha C, Hatthapanit K, Thepsuwan U (2005) Comparison of reinforcing efficiency between Si-69 and Si-264 in an efficient vulcanization system. Polym Test 24(4):439–446

Sae-oui P, Sirisinha C, Wantana T, Hatthapanit K (2007) Influence of silica loading on the mechanical properties and resistance to oil and thermal aging of CR/NR blends. J Appl Polym Sci 104(5):3478–3483

Saramolee P, Sahakaro K, Lopattananon N, Dierkes WK, Noordermeer JW (2016) Compatibilization of silica-filled natural rubber compounds by combined effects of functionalized low molecular weight rubber and silane. J Elastomers Plast 48(2):145–163

Schaal S, Coran AY, Mowdood SK (2002) U.S. Patent No. 6,482,884. U.S. Patent and Trademark Office, Washington, DC

Segatta TJ, Sandstrom PH, Verthe JJ (1995) U.S. Patent No. 5,396,940. U.S. Patent and Trademark Office, Washington, DC

Sengloyluan K, Sahakaro K, Dierkes WK, Noordermeer JW (2014) Silica-reinforced tire tread compounds compatibilized by using epoxidized natural rubber. Eur Polym J 51:69–79

Sengloyluan K, Sahakaro K, Dierkes WK, Noordermeer JWM (2017) Silane grafted natural rubber and its compatibilization effect on silica-reinforced rubber tire compounds. Express Polym Lett 11(12):1003–1022

Somseemee O, Sae-Oui P, Siriwong C (2021) Reinforcement of surface-modified cellulose nanofibrils extracted from Napier grass stem in natural rubber composites. Ind Crops Prod 171:113881

Teh PL, Ishak ZM, Hashim AS, Karger-Kocsis J, Ishiaku US (2004a) Effects of epoxidized natural rubber as a compatibilizer in melt compounded natural rubber–organoclay nanocomposites. Eur Polym J 40(11):2513–2521

Teh PL, Mohd Ishak ZA, Hashim AS, Karger-Kocsis J, Ishiaku US (2004b) On the potential of organoclay with respect to conventional fillers (carbon black, silica) for epoxidized natural rubber compatibilized natural rubber vulcanizates. J Appl Polym Sci 94(6):2438–2445

Teh PL, Mohd Ishak ZA, Hashim AS, Karger-Kocsis J, Ishiaku US (2006) Physical properties of natural rubber/organoclay nanocomposites compatibilized with epoxidized natural rubber. J Appl Polym Sci 100(2):1083–1092

Ten Brinke JW, Debnath SC, Reuvekamp LA, Noordermeer JW (2003) Mechanistic aspects of the role of coupling agents in silica–rubber composites. Compos Sci Technol 63(8):1165–1174

Varghese S, Karger-Kocsis J, Gatos KG (2003) Melt compounded epoxidized natural rubber/layered silicate nanocomposites: structure-properties relationships. Polymer 44(14):3977–3983

Vu YT, Mark JE, Pham LH, Engelhardt M (2001) Clay nanolayer reinforcement of cis-1,4-polyisoprene and epoxidized natural rubber. J Appl Polym Sci 82(6):1391–1403

Wolff S, Wang MJ (1992) Filler—elastomer interactions. Part IV. The effect of the surface energies of fillers on elastomer reinforcement. Rubber Chem Technol 65(2):329–342

Wongsorat W, Suppakarn N, Jarukumjorn K (2014) Effects of compatibilizer type and fiber loading on mechanical properties and cure characteristics of sisal fiber/natural rubber composites. J Compos Mater 48(19):2401–2411

Wootton DB (2001) The application of textiles in rubber. Rapra Technology Ltd

Yuan P, Tan D, Annabi-Bergaya F (2015) Properties and applications of halloysite nanotubes: recent research advances and future prospects. Appl Clay Sci 112:75–93

Zeng Z, Ren W, Xu C, Lu W, Zhang Y, Zhang Y (2010) Maleated natural rubber prepared through mechanochemistry and its coupling effects on natural rubber/cotton fiber composites. J Polym Res 17:213–219

Chapter 6
Starch-Based Rubber Bionanocomposites

Deepti Katiyar, Rajnandani Singh, Shreya Dixit, Debaprasad Ghosh, Manish, and Rashmi Saxena Pal

6.1 Introduction

The waste generated from everyday consumption, such as bottles, leftover food, plastic bags, wrappers, and packaging, is typically disposed of as municipal solid waste. Unfortunately, many waste products, particularly those made of synthetic petroleum polymers containing non-renewable elements, do not naturally decompose (Shey et al. 2006). Furthermore, plastics create a menace to oceans, flora and fauna, community well-being, and economic regions that depend on the well-being of marine life (Beaumont et al. 2019; Jambeck et al. 2015; Borrelle et al. 2020). Therefore, it is essential to build up innovative polymeric materials that are ecological and environmentally gracious to reduce the impact of plastics. Polymers resulting from renewable resources such as starch, rubber, cellulose, and chitins provide an effective solution to accomplish these requirements. A variety of factors, such as the dwindling of fossil fuel reserves, rising costs, environmental concerns related to the impact of degradation or incineration on global warming, recyclable processes susceptible to contamination, and potential toxicity of end products, have led to increased interest in using biopolymers as replacements for those derived from petroleum (Nagalakshmaiah et al. 2019). Biopolymers, which originate from natural substances, have become a recent focus of scientific attention due to their ability to be recycled, environmental friendliness, availability, sustainability, and practicality (Mohiuddin et al.

D. Katiyar (✉) · R. Singh · S. Dixit · D. Ghosh
Department of Pharmacognosy, KIET School of Pharmacy, KIET Group of Institutions, Delhi-NCR, Ghaziabad 201206, Uttar Pradesh, India
e-mail: katiyar_deepti@yahoo.co.in

Manish
Department of Electronics and Communication Engineering, ABES Engineering College, 19th KM Stone, NH-09, Ghaziabad 201009, Uttar Pradesh, India

R. S. Pal
Department of Pharmaceutical Sciences, Lovely Professional University, Phagwara, Punjab, India

© The Author(s), under exclusive license to Springer Nature Switzerland AG 2025

P. M. Visakh (ed.), *Rubber Based Bionanocomposites*, Advanced Structured Materials 227, https://doi.org/10.1007/978-3-031-78557-3_6

2017). The advantages of biopolymers are displayed in Fig. 6.1, and the examples of biopolymers are shown in Fig. 6.2.

The latex extracted from the *Hevea brasiliensis* tree is a milky liquid and the source of natural rubber (NR), an environmentally friendly elastomer, and biodegradable polymer. NR is widely used across various industries due to its renewability, non-toxic properties, excellent physical characteristics, and cost-effectiveness, particularly in applications involving products that come into contact with food or potable water

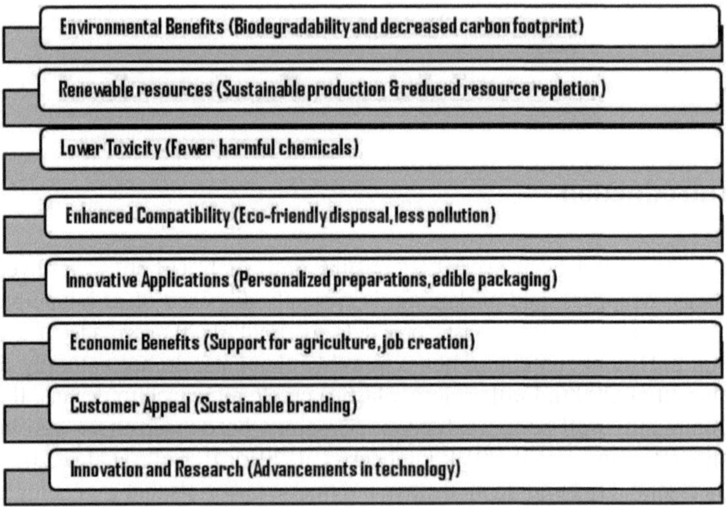

Fig. 6.1 Advantages of bioploymers

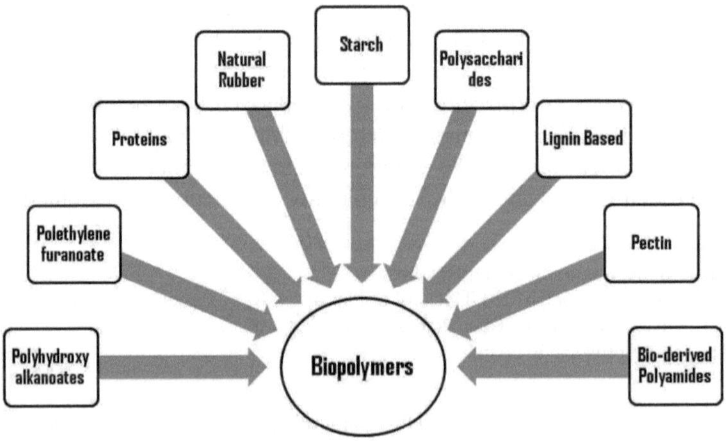

Fig. 6.2 Examples of biopolymers

(Manaila et al. 2018). Despite being categorized as a natural polymer, NR's biodegradation is limited by the high molecular weight of its molecules and the additional chemical crosslinking that occurs during vulcanization (Syuhada and Azura 2021).

Starch, a biodegradable polymer, is composed of two primary constituents, amylose and amylopectin, with their proportions varying based on the plant source, such as corn, cassava, potatoes, sago, rice, and others (Thomas et al. 2021). Rice, the most widely grown cereal grain, is the main staple foodstuff for about half of the world's populace (Wani et al. 2012). Among all types of starch, rice starch (RS) is a popular option for various food and non-food uses owing to its non-allergenic nature, easy digestibility, white tint, tiny grain size (3–10 mm), and an extensive option of amylose/amylopectin ratios (Thakur et al. 2018). Abundant studies have investigated the possible use of starch as a recyclable wrapping material or bio-coating to prolong the shelf life of fresh fruit due to its profusion, cost-effectiveness, biodegradability, and exceptional film-forming properties, either alone or combined with other biopolymers (Thakur et al. 2019). Starch-based films display essential physical uniqueness, counting lucidity, absence of distinctive aroma, and partial permeability to humidity, oxygen, carbon dioxide, and savoring substances. These unique traits can yield outcomes similar to those achieved with a controlled atmosphere. Nevertheless, the limited process ability, constrained tensile potency, and elevated susceptibility to water of starch films restrict their realistic use. Previous studies have revealed that adjustments and fortification of starch can progress the therma, barricade, and mechanical capacity of starch films (Bangar et al. 2022). Within the realm of NR composites, researchers have aimed to beautify biodegradability with the aid of mixing one-of-a-kind styles of starches, consisting of com starch (Wu et al. 2004; Leksawasdi et al. 2021; Senna et al. 2012), waxy corn starch (Valodkar et al. 2010; Angellier et al. 2005), cassava starch (Wang et al. 2009; Riyajan et al. 2012; Vudjung et al. 2014; Riyajan et al. 2015, 2018; Li et al. 2020), potato starch (Bouthegourd et al. 2011; Rajisha et al. 2014), sago starch (Ab Rahman et al. 2019), rice starch (Akbari et al. 2014; Soubam and Gupta 2021), and unspecified starch (Liu et al. 2008; Wu et al. 2018). But, previous studies have predominantly targeted on producing rubber films or adhesives. The apparent immiscibility of hydrophilic starch and hydrophobic NR, however, results in phase-separated blends with bad interfacial characteristics that can be solved by way of compatibilization. Furthermore, gelatinizing starch is a powerful method for enhancing the interfacial interplay between one of a kind phases. The addition of the compatibilizer is likewise typically desired because of its ability to improve the system's properties. The addition of fillers right into a binary polymer which is incompatible has been proven to influence phase morphology as a result of the interaction of the combination's individual components with the solid surface (Ruamcharoen et al. 2023). Composites are crafted from 2 constituent substances, i.e. matrix and reinforcement agent with notably diverse physico-chemical characteristics, template being the incessant section and may be a polymer, steel, and ceramic. The template defends the fibers from the surroundings and scratches maintaining them within the appropriate route. The discrete/discontinuous phase is the strengthening agent that executes numerous features, e.g. increase the physical and mechanical characteristic features of the matrix. Combinations are employed for their structural power, organic, thermal, and ecological purposes.

Nanocomposites are materials composed of more than one phase, with at least one dimension at the nanometer range (10^{-9} m) (Sain and Oksman 2006; Kamel 2007; Kalia et al. 2011; Darder et al. 2007). The 3 specific categories of nanocomposite substances on the premise of matrix substances are (i) polymer matrix nanocomposites, (ii) metal matrix nanocomposites, and (iii) ceramic matrix nanocomposites. Nanocomposites display right mechanical, optical, and barrier properties alongside synergistic flame retardant motion, stepped forward dimensional balance, and better thermal expansion as in comparison to micro- or macro-composites. They provide an opportunity approach to deal with the restrictions of microcomposites and monolithic substances, constituting a swiftly growing subject of look at with an estimated annual boom price of 29.5% from 2017 to 2022 (Anonymous 2022). Nanocomposites are constructed from a matrix material incorporating nano-sized dispersed reinforcement additives, which include fibers, nanotubes, debris, and whiskers. The key advantage of nanofillers over traditional micrometer-sized fillers is that, at equal concentrations, nanofillers usually offer superior residences (Safdari and Al-Haik 2018). Enhancing the properties of nanocomposites, which include optical, mechanical, thermal, and barrier characteristics, stands as certainly one of their primary blessings over alternative composite materials.

Presently, nanocomposites present novel technological and enterprise potentialities throughout diverse industries, with some demonstrating environmentally sustainable attributes (Dong et al. 2015). Figure 6.3 lists the types of nanocomposites while Fig. 6.4 the methods for preparing bionanocomposites.

Biopolymers are polymeric substances received from renewable natural assets. The industrial use of biopolymers is presently restricted because of troubles in

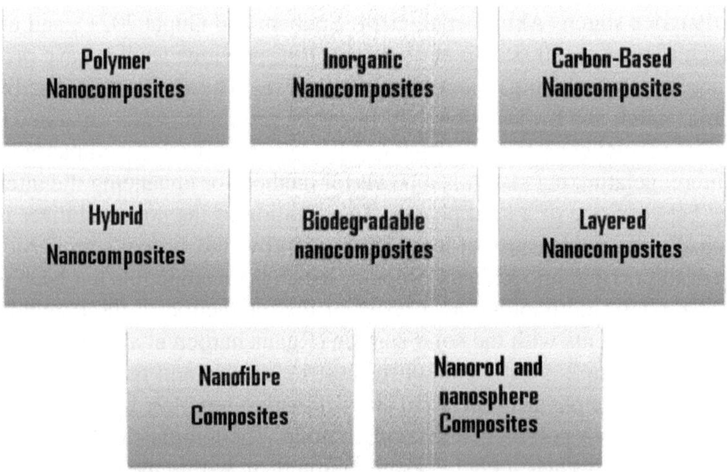

Fig. 6.3 Types of nanocomposites

Fig. 6.4 Methods for preparing bionanocomposites

their overall performance, processing, and value. Bionanocomposites encompass a biopolymer matrix strengthened with natural products. These are occasionally called biocomposites, nanocomposites, nano-biocomposites, inexperienced composites, biohybrids, or bio-based plastics (bioplastics). They can be categorized depending upon their matrix sorts, beginning, form, and length of the supports. Thinking about the figure of particle reinforcement, bionanocomposites may be labeled into particulate, elongated particle, and layered systems. These have accurate mechanical residences, biocompatibility, and biodegradability. These may be applied as biodegradable, value-effective, and renewable film substances for diverse packages along with meals packaging with stepped forward anti-microbial activity (Sharma et al. 2020). The general applications of bionanocomposites are mentioned in Fig. 6.5.

6.2 Applications of Starch-Based Natural Rubber Nanocomposites

Nanocomposites have these days surged in reputation greater than conventional amalgamated materials and find vast use in various fields like food, packaging, biomedicine, electronics, energy storage, optics, automobile, environmental cleanup, textiles, and more (Ates et al. 2020; Saad et al. 2021). Polymer nanocomposites are constituted of polymer matrices infused with nanofillers (Puiggalí and Katsarava 2017). Petroleum-derived polymers are manufactured appreciably globally. They may be non-biodegradable, derived from finite assets, and generate harmful materials that create menace to both human healthiness and the surroundings (Madhumitha et al. 2018). Biopolymer-based nanocomposites offer a tenable substitute for

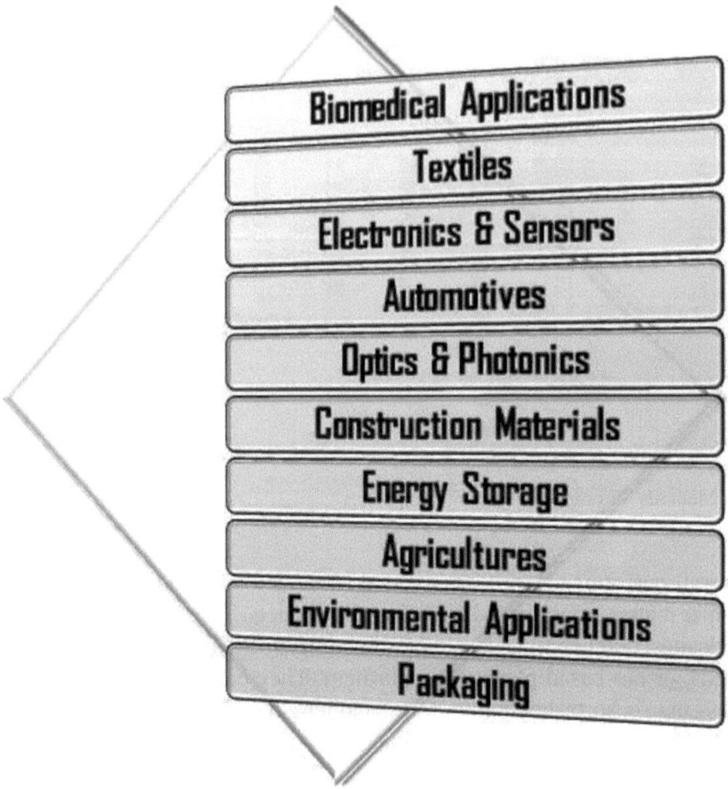

Fig. 6.5 Applications of bionanocomposites

petroleum-based counterparts across diverse packages, way to their biodegradability, eco-friendliness, renewability, cost-effectiveness, low toxicity, abundance, and greater thermal, mechanical, physical, barrier, and useful traits (Puiggalí and Katsarava 2017; Madhumitha et al. 2018; Mohiuddin et al. 2017).

Natural rubber (NR), a biodegradable elastomer derived from the *Hevea Brasiliensis* tree as latex, possessing exceptional physical properties, is renewable, non-toxic, and value-effective. Because of these attributes, natural rubber is the most extensively utilized elastomer worldwide, specifically in the industries and applications involving contact with meals or potable water (Manaila et al. 2018).

A ramification of natural biopolymers, which includes starch, pectin, cellulose, lignin, rubber, alginates, chitin/chitosan, hyaluronic acid, terpenes, gelatin, gluten, and polyhydroxyalkanoates (PHAs), sourced from flora and fauna, algae, and microorganisms, can be more advantageous with starch nanoparticles/starch nanocrystals and various nano polymers which include nanoclay (e.g. montmorillonites [MMTs], halloysite nanotubes [HNTs]), carbon nanotubes (CNTs), nanofibers, and nanowhiskers (e.g. cellulose, chitin), in addition to metal and metal oxides (e.g. TiO_2 NPs, ZnO NPs), to gain preferred characteristics and create doubtlessly environment friendly sustainable nanocomposite materials (Madhumitha et al. 2018; García et al. 2015; Le Corre and Angellier-Coussy 2014). Lifecycle tests performed on amylum and its combinations confirm their reduced ecological footprint and viability as sustainable options to petrochemical based polymers (Venkatesh et al. 2018; Kakadellis and Harris 2020). Sago starch, derived from the sago palm (*Metroxylon* spp.) located across Southeast Asia, stands out because of its better content of fiber, phenolic compounds, waxes, and lipids compared to different starches.

This particular composition permits sago starch films to showcase a slower swelling charge in water improving their ability of sago starch (SS) as a biopolymer film for packaging. However, numerous drawbacks restrict sago starch-based polymers, ensuing in subpar inherent traits, which include enormously modest mechanical properties, inadequate water resistance, and barrier abilities (Le Corre and Angellier-Coussy 2014).

While categorized as a natural polymer, NR reveals limited biodegradability owing to its excessive molecular weight and the formation of chemical crosslinks at some stage in vulcanization, hindering its decomposition process (Syuhada and Azura 2021). A study appraised the production of NR-starch composites by sulfur pre-congealing and latex co-coagulation strategies. It was seen that amylum could improve the biodegradability of natural rubber. But, hydrophilic property of amylum brought on incompatibility with NR. Consequently, the direct amalgamation of starch decreased the mechanical qualities of the NR amalgamates. The review disclosed a huge variety of implications to enhance the affinity of the NR composites which include (i) the advent of a compatibilizer, (ii) matrix amendment, and (iii) nanostarch modification with the aid of esterification, etherification, acetylation, cross-linking and grafting. Besides, the thermal and mechanical properties could be encouraged by way of numerous elements: the quantity of filler loading, filler distribution, and framework.

Riyajan et al. (2012) prepared gelatinized cassava starch and handled them with potassium persulfate (K2S2O8) to shape the tailored amylum with lesser molecular weight. The thermal in addition to mechanical qualities of NR were stepped forward with grafting. The NR embedded starch was additionally employed to encapsulate urea which could then be released with the aid of engorgement with water, consequently providing for the controlled release of urea. In another study, Riyajan (2015) organized an admixture of NR grafted cassava starch and maleated epoxidized NR. The mixing helped in improving the biodegradability, tensile strength, and thermal steadiness of NR grafted starch. The hydrophilicity of natural rubber (NR) was improved by grafting with changed cassava starch (ST) (NR-g-ST) through the use of potassium persulfate (K2S2O8) as a catalyst. The swelling ratio of NR-g-ST was determined in water and the findings confirmed that the swelling ratio of the changed NR reduced as a function of ST. This product with precise controlled-release and water-retention might be particularly beneficial in agricultural and horticultural programs.

Wang and Zhang suggested a relative observation at the mechanical characteristics of starch-bolstered NR composites, and indicated the use of co-coagulation and direct mixing (roll mill) strategies. The composites manufactured with the aid of co-coagulation approach confirmed stepped forward mechanical residences, for instance, tensile strength, elongation at break, and tear strength than the equivalent process. It became attributed to (i) decrease in starch particle length and (ii) higher dispersion of the debris within the matrix.

Natural rubber (NR) latex gloves are unparalleled when it comes to mechanical and barrier characteristics, consolation, and lesser cost compared to artificial latex gloves. However, due to their single use, it contributes to the growing stable waste disposal trouble. An endeavor to integrate NR latex gadget with bio-substances to support biodegradation of NR latex films upon disposal had been made by several researchers. The starch has numerous traits which includes reasonably priced, evidently plentiful, renewable assets and excessive receptiveness to biodegrade, making it one among promising biomaterials to discover. The starch carries a polysaccharide sequence of amylose and amylopectin that are taken up by microbes inside the soil to supply enzymes including amylases and glucosidases which can be used to interrupt the glucosidic linkages in starch. As the starch granule resides in the rubber debris, the breakage of glucosidic linkages in starch can also disrupt the rubber particles, and as a result boost up the degradation method. The various starch-based substances, sago starch (*Metroxylon sagu*) fulfills the requirements with maximum amylose content material in comparison to different kinds of grain. Excessive amylose content material could be very useful for biodegradation system wherein it is able to serve as a supply of energy to the microbes that could boost up the biodegradation method. Every other vital issue of latex gloves that govern with the aid of adjustments in product system (extra of sago starch in latex system) calls for comprehensive research of the product shelf existence. In conclusion, the shelf existence for SSNRL gloves is expected round 7.99 years; which the tensile electricity retention reduced to 50% at shelf temperature of 30 °C. It showed that the SSNRL gloves will have an extended shelf life.

Biodegradable packaging materials can be crafted from cassava starch and natural rubber using glycerol as a plasticizer. The constituted biofilm was evaluated for the physical parameters such as swelling ratio, moisture content, moisture absorption, and mechanical residences. Further, the film decomposed very well in herbal soil (Riyajan and Patisat 2018a, b). Yet another research amalgamated cassava starch (CS) natural rubber latex (NRL) through a sulfur-free crosslinking process employing radiation pre-vulcanization natural rubber latex (RVNRL) in disparity to sulfur pre-vulcanized herbal rubber latex (PvNRL). Scanning electron microscopy (SEM) was used to study the microscopy and X-ray diffraction to determine the crystallinity of the fabricated films while tensile and tear checks were employed to investigate their mechanical properties. Ultimately, it was found that RVNRL films had less crystalline structure than PvNRL films. Sulfur crosslinking showed a greater tendency of tensile property degradation than radiation crosslinking, and at 25 phr of CS content material, both systems exhibited comparable tensile behavior. This is explained by the improved CS dispersion in RVNRL, which a SEM micrograph was able to validate

(Lay et al. 2019). Biocomposite foams made from natural rubber and rice starch have shown to be effective in keeping the quality of banana fruit during of storage and hold promise for use in sustainable packaging answers. NR foam is frequently employed in manufacturing diverse objects such as mattresses, pillows, dolls, cushions, and floors because of its favorable mechanical properties and distinct morphology. It gives the best stability among electricity and weight, making it suitable for such applications (Suethao et al. 2021). This kind of foam is just one instance of the numerous merchandise that can be crafted from both NR latex and dry NR. Research have documented advancements in foams crafted from herbal rubber and starch sourced from agricultural sources (Kiatkamjornwong et al. 2001; Tanrattanakul and Chumeka 2010; Riyajan and Keawittarit 2016). The only studies documenting the manufacturing of NR foam with NR latex as the primary polymer matrix in biocomposites were carried out by Prasopdee et al. (2021). On this take a look at, cassava starch turned into utilized as a light-weight filler to garnish the recovery properties of natural rubber foam (NRF) fabricated via the Dunlop system.

6.3 Biomedical Application

Biodegradable polymers, e.g. starch-based rubber bionanocomposites, are drastically employed as frameworks for tissue engineering, drug delivery systems, wound dressings, surgical sutures, and implants because of their mechanical strength, compatibility with living tissues, ability to break down naturally, and the production of harmless, degradable by-products (Gomes et al. 2002; Lu et al. 2009; Taherimehr et al. 2021). Much like several carbohydrates, starch, a polysaccharide, stands proud as a distinct building block and awesome material for biomedical and pharmaceutical uses due to its affordability, large availability, capability to break down naturally, compatibility with dwelling organisms, lack of dangerous results, and capacity to swell, dissolve, and degrade enzymatically. The unique physical and chemical characteristics, in addition to the functional attributes, of herbal starches, including their brilliant compatibility with residing tissues, ability to break down evidently, loss of harmful effects, and degradation, render them treasured for numerous biomedical applications. Several biodegradable starch polymers, particularly in bone tissue engineering, drug delivery structures, and hydrogels, have gone through a good number of studies in the latest years (Ulery et al. 2011). Pores and skin tissue engineering includes the creation of artificial skin to defend the body from outside stimuli, because the body's largest organ, skin, regulates heat and water retention, serves a crucial protective trait (Fang et al. 2021; Srivastava et al. 2021; Hondzo et al. 2022). There are two primary methods for repairing broken pores and skin: autografts and allografts. While these methods are effective for treating seriously injured skin, along with good sized burns, they are able to provoke immune responses and bring about complications like morbidity (Yamamoto et al. 2018; Roballo and Bushman 2019). Consequently, tissue engineering appears to be the most suitable technique for repairing tissue harm, serving as a feasible opportunity to natural pores and skin. In addition to possessing robust mechanical properties (skin stretching ratio of 70%, a sizeable aspect), suitable biomaterials for building skin scaffolds need to facilitate cellular attachment, growth, proliferation, differentiation, and migration, at the same time as additionally enabling cellular-to-cell conversation among fibroblasts, keratinocytes, and melanocytes—the three primary types of skin cells—and minimizing immunogenic reactions as much as feasible (Shakir et al. 2018; Quaresma 2019; Kaur et al. 2019; Tarrahi et al. 2022). The various array of biomaterials suitable for skin tissue engineering emerge as enormously promising choices. While hydrophilicity is crucial for all biomaterials employed in tissue engineering, it holds specific importance for pores and skin scaffolds. Salgado et al. (2005) proved that biodegradable bone cements derived from starch offer a brief structural aid that gradually dissipates over time. Moreover, biodegradable polymers derived from starch had been investigated for their ability as scaffolds in bone tissue engineering (Salgado et al. 2005). According to Gomes et al. (2002), an optimal scaffold ought to be made from a biomaterial that degrades at appropriate rates, aligning with the formation of new tissue. For this reason, starch emerges as a suitable choice for scaffolding packages (Gomes et al. 2002). Polymers allow drug incorporation and release or protein functionalization; composites may be used to upshot in angiogenesis located by way of formation of blood vessels in ulcer remedy (Homayun et al. 2019).

The major difficulty with current anti-cancer drugs (toxicity and lack of specificity) can be overcome by incorporating them inside a biopolymer. Therefore, it was suggested to manufacture liposomes, nanoparticles, and microspheres to increase the medication's effectiveness. Cisplatin-based chemotherapy is linked to a number of adverse effects, such as anemia, vomiting and nausea, neurological damage, and kidney damage. In one study, iron oxide-impregnated starch nanoparticles encapsulating cisplatin were synthesized and characterized as a potential and effective medication carrier for magnetically mediated targeted drug transport. The emulsion go-linking approach has been employed for organizing iron-oxide loaded nanoparticles that effectively provide the anticancer medication cisplatin (Gamage et al. 2022).

Hemostatic agents derived from natural assets offer several advantages, together with affordability, loss of toxicity, and affordable degradation costs. The local materials investigated in a research were rice starch and chitosan derived from squid pens, both regarded for their biocompatibility and extensive use in biomedical packages (Seyednejad et al. 2008; Hu et al. 2018). Batool et al. (2019) developed PVA/starch/Ag NPs from *Diospyros lotus* fruit extract. The consequences showed the prolonged swelling and moisture retention ability, reduced water vapor transmission, and higher antimicrobial activity in opposition to *Escherichia coli* and *Staphylococcus aureus*. In some other study, thermoplastic starch (TPS)/beta-tricalcium phosphate (β-TCP) NPs were advanced. The findings revealed that addition of β-TCP at 10% advanced the tensile electricity (from 1.67 to 4.8 MPa) and Young's modulus (from 66.54 to 390.5 MPa), and reduced elongation at ruin (78.56–18.03%) of TPS. They exhibited non-cytotoxicity outcomes and first-rate biocompatibility. These consequences indicated their application in bone tissue engineering substances (Taherimehr et al.

2021). Furthermore, a study developed polylactic acid (PLA)/starch (S)/poly-ε-caprolactone (PCL)/nano hydroxyapatite (nHAp) which showed stepped forward hydrophilicity, hydrolytic degradation, antibacterial activity (against *E. coli* and *S. aureus*), and continuous drug release of PLA/S/PCL film. For this reason, the formulated nanocomposite may be used for the managed release of antibacterial triclosan. Additionally, Starch-itaconic acid/Fe_3O_4 NPs (St-IA/Fe_3O_4) is supplied for the controlled launch of Guaifenesin. The addition of magnetic Fe_3O_4 NPs at 0.83% more suitable the drug launch percentage from 54 to 94% within 24 h in pH 7.4. Adding Fe_3O_4 NPs stepped forward the wound recuperation capacity in mice (healed after 10 days) exhibited low cytotoxicity for human umbilical vein endothelial cells.

6.4 Packaging Application

The packaging materials for food and medicines should be completely safe, have an extended shelf life, and should be able to maintain the quality of the packaged product. Researchers are actively investigating environmental issues such as the consequences of petrochemical plastics (their lack of biodegradability and recyclability), chemical contamination of products, sustainability issues, and health concerns (Hong et al. 2021). Consequently, the primary concern in this industry revolves around the development of bio-based packaging plastics. Additionally, enhancing the sustainability and recyclability of food packaging materials can be achieved through the production of bio-based plastics and polymers.

The current approaches of producing bio-based plastics are divided into two pathways. Initially, bio-based precursors are derived from sources like starch, cellulose, and agricultural residues (Zhang et al. 2021). Subsequently, renewable materials are utilized by gram-positive and gram-negative bacteria to synthesize polylactic acid (PLA) and polyhydroxyalkanoates (PHA) (Venkatesh et al. 2021).

In the second stage, bio-based polymers and plastics are produced from non-biodegradable plastics using modification techniques. For quite some time, antibacterial packaging materials have been acknowledged as one of the most promising active packaging systems. These materials aid in enhancing food safety and extending shelf life by eliminating or restraining spoilage and pathogenic microorganisms that can contaminate food (Maruthapandi et al. 2022; Murugesan et al. 2018).

Antibacterial packaging films are usually created by either attaching antibacterial materials to the surface of the polymer or blending them during the polymer treatment process (Ramirez et al. 2019). Bionanocomposites offer the potential to serve as active food packaging, allowing interactions with food to release beneficial compounds like antimicrobial or antioxidant agents, or to remove undesirable elements such as oxygen or water vapor.

The bionanocomposite can additionally function as intelligent food packaging, capable of detecting properties of the packaged food such as microbial contamination or expiration date. It employs mechanisms to record and transmit information about the quality or safety of the food (de Azeredo et al. 2011; Shakeri and Radmanesh 2014; Sajid and Devasena 2012; Akbari et al. 2007).

Enhanced shelf life and reduced packaging expenses are driving the pursuit of nanocomposites for consumer packaging.

Antibacterial agents often include heavy metals like mercury, cadmium, lead, chromium (Parolo et al. 2011), copper (Bruna et al. 2011), as well as ZnO nanoparticles (Hong et al. 2021) and silver nanoparticles (Abreu et al. 2015), all of which are both hazardous and costly. Alternatively, antibacterial substances such as benzoic acid, sodium propionate, potassium sorbate (Chowdhury and Das 2013), essential oils, volatile oils (Nairetti et al. 2014), proteins from gluten (Gómez-Heincke et al. 2015), and peptides (de Azeredo 2009) are available but have limited availability. Some bioplastic materials, such as egg and soy protein (Jones et al. 2015) and chitosan derived from marine animal shells (Nadrajah 2005), possess antibacterial properties. Additionally, carrageenan compounds from seaweed offer antibacterial (Natrah et al. 2015; Gade et al. 2013) qualities and can serve as bioplastics (Gade et al. 2013).

Starch is one of the most promising biopolymers for the production of food packaging trays for several reasons. It is bio-derived and fully biodegradable; its cost is very low and its production is abundant worldwide. Nevertheless, up to now the introduction of starch in the food packaging market has been restricted because it presents several drawbacks: native starch after being transformed becomes brittle and breaks easily, it is hydrophilic and therefore moist products cannot be packed, and its density is higher (1400 kg/m^3) than common synthetic polymers employed in the food packaging market such as polystyrene (1070 kg/m^3). The properties of NR such as its high elasticity and its hydrophobic nature make it a suitable choice to produce blends with starch. The challenge here is the achievement of good compatibility between both the polymers which was achieved in a research by vulcanization of the blends of these two polymers. Not only a high homogeneity but also high adhesions between both polymers were observed from the pictures and SEM micrographs. Secondly, an improvement of the mechanical properties was also one of its main challenges. Elongation at break was improved with the addition of 3% NR, and higher strengths were also achieved with concentrations of 1 and 3%.

Starch-based bionanocomposites can be employed for preparing coating materials for food packaging. For example, Corn starch/Debranched starch NPs (DSNPs) increased the tensile strength (from 0.95 to 1.73 MPa), decreased the water vapor permeability, and oxygen transmission rate could be potentially used for packaging material. Furthermore, Starch NPs/Ag NPs exhibited antibacterial activity against Staphylococcus aureus, Salmonella typhi, and Escherichia coli indicating that they could be employed as coating materials for food packaging. Additionally, numerous research have proven the potential of starch-based bionanocomposites as material for food packaging.

6.5 Agricultural Applications

6.5 billion pounds of polyethylene are utilized for farming purposes globally annually to enhance yields of crops and preserve agricultural goods following harvesting. These substances are employed in the form of greenhouses, tunnels, mulching material forage films, and bale wraps. Because of this, a substantial quantity of plastic is thrown into the surroundings, dumped in the ground, or burned by farmers, which releases hazardous compounds and has an adverse effect on the ecosystem. The use of biodegradable films in agriculture, which can be deposited either directly into the soil or into a composting system at the end of their useful lives, may be likely to solve this issue.

A variety of benefits can arise from combining recyclable raw materials with nanograde ingredients or bionanocomposites or films, including regulated decomposition, improved longevity, and optical characteristics like ultraviolet, visible, and infrared. Biodegradation in the soil can be attributed to the action of microorganisms like bacteria, fungi, and algae as well as degradation by sunlight and water.

According to a study conducted by Merino et al. (2018), a starch-based nanocomposite with corn starch, chitosan (CS), and native (TPS) or oxidized (TPS-ox) enhanced water resistance (reduced water solubility), radiometric, and antibacterial characteristics could find use in mulch film. A further investigation found that urea, natural char nanoparticles (NCNPs), and corn starch-g-poly (AA-co-AAm) improved the soil's ability to break down organic matter, retain water (35.6% and 33.2% at pH 4.5 and 5.5, respectively, after 6 days), and liberate urea slowly (73% in deionized water and 37% in NaCl). Fertilizers with controlled release could also be made possible by urea encapsulated in starch.

6.6 Structural Application

The structural applications involve the study of the composites denoting a matrix as the primary component holding a more concentrated phase, allowing for a continuous phase, while reinforcements are elements that boost the composite's performance, mechanically or thermally. Matrices could be biodegradable polymers, ideally sourced from renewable origins like plants. Reinforcements might encompass plant fibers, by-products from renewable lignocellulosic sources, synthetic inorganic materials, or natural or modified clays (Klemm et al. 2011). Lignocellulosic polymers, especially those derived from cellulose, attract attention for their ability to reinforce and biodegrade. Additionally, cellulose-based materials exhibit outstanding biocompatibility (Entcheva et al. 2004; Martson et al. 1999).

Lignin, an amorphous polymer, primarily functions as structural support in plants. With a phenolic-based chemical structure, its high carbon and low hydrogen content indicate a highly unsaturated and aromatic nature (Adler 1977). Similar to bone, lignocellulosic materials feature pores that facilitate the transport of nutrients. Both bone and wood share a hierarchical structure spanning from the nano- to the macro-scale (Ratner et al. 2004; Cheng and Teoh 2004). Lignocellulosic materials and bone both possess porous networks, albeit differing in size and connectivity. Nevertheless, they offer channels for nutrient transport and exchange within biological systems (Gross and Ezerietis 2003).

Furthermore, plant fibers like bamboo, sisal, and hemp exhibit an anisotropic structure and serve as significant sources of cellulose and hemicellulose derivatives in the biomedical domain. Bamboo has emerged as the latest innovative material, joining numerous other lignocellulosic woody or fibrous materials in their breakthrough status (Liu et al. 2023; Hasan et al. 2022).

While bamboo's utilization as reinforcement for polymeric composites is appealing due to its renewability and remarkable mechanical and physical characteristics, it remains less popular than synthetic fiber reinforced composites (Nirmal Kumar et al. 2022; Norizan et al. 2022).

One such advancement is the creation of bamboo fiber reinforced biocomposites, as documented in references (catering to various structural needs). These advancements enable their (Nirmal Kumar et al. 2022; Norizan et al. 2022) integration into diverse sectors, spanning transportation, aerospace, construction, and consumer goods.

It's intriguing that bamboo is a multicellular composite fibrous material, with bamboo fiber comprising substantial elementary fibers visible through the microstructure of vascular bundles. SEM images illustrate the surface morphology of bamboo, depicting uplifted filaments tightly packed together, susceptible to removal by external forces due to their weaker binding (Zhang et al. 2015).

The FTIR analysis additionally validates the diverse chemical composition of bamboo fibers, evident through distinct peaks such as the –OH group at 3200–3500 cm^{-1} and the 2900 cm^{-1} peak corresponding to –CH stretching vibration (Zhang et al. 2015). Dimethylaminoethyl methacrylate grafted NR latex was prepared by Rigal and Gilbert and utilized as a modifier for the glycerol-containing starch solution. The modified latex utilized had a concentration ranging from 0 to 30%. In contrast to the original NR latex, the modified NR latex exhibited superior starch compatibility, which resulted in good tensile tenacity and elongation at break. It was attributed to (i) a smaller starch particle and (ii) improved particle distribution inside the matrix. Polybutylacrylate (PBA) was grafted onto starch by Liu et al. (2008). The PBA-g-starch and NR latex were combined to create the composites, and then 10% calcium chloride was used to co-coagulate the mixture. The NR matrix and starch were made more compatible by using the PBA. The NR composites were treated with PBA g-starch and their tensile characteristics, tear strength, and hardness. By using acid hydrolysis, Valodkar and Thakore (2010) were able to extract starch nanocrystals. To alter NR, starch nanocrystals, pure starch, and waxy corn starch were utilized. Two roll-mill methods were used to prepare the composites. Because of their small size, the starch nanocrystal-modified samples showed an improvement in their mechanical properties.

Senna et al. (2012) provided examples on how to synthesize a modified starch-NR latex composite. The NR's hardness, tensile strength, and thermal stability were all enhanced by the addition of the modified starch. Employing electron beam radiation, the impact of illumination on the thermo-mechanical characteristics of composite samples was also investigated. Radiation decreased the composites' hardness and tensile strength while increasing their thermal stability. In 2014, Vudjung and colleagues synthesized crosslinked NR latex starch inter-penetrated network composites and evaluated their biological degradation, absorption of water, and tensile properties. The mechanical findings demonstrated that when the NR level boosted, the elongation at break improved and the tensile strength dropped. In addition, as the amount of rubber increased, less water was absorbed and less biological degradation occurred. Potato starch nanocrystals have been separated by Bouthegourd et al. (2011) and Rajisha et al. (2014), and the NR Latex was reinforced with these nanocrystals. It was decided to vary the nanocrystal concentrations between 5 and 20 weight percent. The existence of potato starch nanocrystals in the NR matrix was verified by the FTIR measurements. The compositions' overall modulus, tensile strength, and crystallinity were all enhanced by the addition of starch to NR. The stiff three-dimensional network forming between the NR latex and starch nanocrystals was the cause of the improvement in tensile characteristics. When the starch content in the composites grew, the starch matrix barrier weakened and free starch was created. This led to an increase in the absorption of water in the composites. Li et al. (2020) used 3-glycidyl-oxypropyl-trimethoxy-silane and sodium isobutyl xanthate to create functionalized starch. It was discovered that the functionalized starch acted as a more effective vulcanization accelerator for NR. Compared to the typical accelerator, the composites' tensile characteristics were superior. Dodecenyl succinate anhydride was used by Wu et al. (2018) to modify porous starch in order to lessen self-aggregation in the NR matrix. The NR matrix was modified using the following starch concentrations: 0, 4.76, 9.09, 13.04, and 16.67 wt%. The largest crosslink density and the best mechanical property improvement were found in the 9.09 wt% composites (Thomas et al. 2021).

6.7 Defense Applications

The enhanced structural benefits of starch-based rubber bionanocomposites have led to a multitude of defense applications in the automotive, aerospace, and marine industries. They may be employed as structural composites in aerospace industry, to attain maximum flexural and to manufacture wind turbine blade (Harussani et al. 2022). Rubber nanocomposites based on starch are used in military applications to improve equipment durability, stealth, and protective gear. They may also provide environmentally friendly alternatives. Advanced protective gear, like body armor and helmets, could be developed using starch-based rubber nanocomposites. By adding nanomaterials, the equipment can absorb and disperse energy from impacts and blasts more efficiently due to its improved mechanical qualities, flexibility, and resistance to impact. These materials can be designed to have particular optical characteristics that aid in the creation of effective camouflage. In order to improve concealment in a variety of operational scenarios, nanocomposites can be engineered to adapt to different environments and lighting conditions. By utilizing nanocomposites in the fabrication of stealth technology components, military equipment and vehicles can have a smaller radar cross-section. Moreover, they may be utilized to reduce or control static electricity in avionics system (Harussani et al. 2022; Cruz-Morales et al. 2023). It is possible to create surfaces that are less detectable by radar by incorporating nanoscale materials with specific electromagnetic properties. Vehicle hulls or protective barriers can be made of starch-based rubber nanocomposites (Harussani et al. 2022).

6.8 Tire Industry

Due to their ability to improve a variety of tire characteristics, starch-based rubber nanocomposites have several promising uses in the tire industry. Tire formulations can be made more resilient to wear by adding starch-based nanocomposites. Tire life can be increased and abrasion resistance can be increased by using nanocomposites to reinforce the rubber matrix. By augmenting the mechanical characteristics of tires, like their elasticity and tensile strength, these nanocomposites can raise their overall performance. Improved stability, comfort, and handling may result from this. Rolling resistance can be decreased by making the rubber compound more optimally formulated using starch-based nanocomposites. This decrease may result in cars using less gasoline, which would save operating expenses and have a smaller negative impact on the environment. Tires' traction and grip on a variety of surfaces can be improved by engineering starch-based nanocomposites. More safety and handling, particularly in difficult driving circumstances, may result from this performance boost. Manufacturing tires with starch is in line with sustainability objectives because it is a renewable and biodegradable material. When it comes to their environmental impact, tires made of starch-based rubber nanocomposites may be less harmful than regular rubber (Cruz-Morales et al. 2023).

6.9 Conclusion

Starch-based rubber bionanocomposites have various applications due to their unique properties, such as biodegradability, sustainability, and improved mechanical performance. Overall, the integration of nanomaterials with starch-based rubbers enhances properties like strength, flexibility, and barrier resistance, making these composites suitable for a wide range of innovative and sustainable applications. The traditional plastics can be replaced with starch-based bionanocomposites for eco-friendly packaging solutions. They offer a sustainable alternative to conventional plastic packaging and can be tailored to provide specific mechanical and barrier properties. Such bionanocomposites can be employed to create more sustainable and fuel-efficient tires. Furthermore, they can be utilized for developing medical implants, wound dressings, and drug delivery systems. Due to their biocompatibility, starch-based rubber bionanocomposites can be used in various biomedical applications. These include wound dressings, drug delivery systems, and biodegradable implants. The traditional rubber components can be replaced with starch-based bionanocomposites for lighter, more sustainable automotive parts. These composites can be used in automotive parts to reduce weight and improve fuel efficiency. They can be found in components such as interior trim, seals, and gaskets. Their enhanced mechanical properties make them suitable for demanding applications. These materials can be used in the construction industry for producing eco-friendly panels, insulation materials, and other building components. Their lightweight nature and thermal insulation properties are beneficial in construction applications. The composites can be used for making biodegradable mulch films. These films help in soil conditioning and reduce the environmental impact of plastic waste in agriculture. Starch-based bionanocomposites can be used in everyday consumer products like footwear, sports equipment, and household items. Their versatility allows for customization in terms of appearance and functionality. More sustainable and comfortable footwear can be manufactured using starch-based bionanocomposites. Emerging research explores the use of these composites in electronic applications, such as insulating materials or flexible electronic components. The biocompatibility and environmental friendliness of the materials are advantageous for developing sustainable electronics. Additionally, they can create biodegradable adhesives and sealants for various industrial applications. Starch-based rubber bionanocomposites can be explored as sustainable materials for 3D printing, used for developing coatings for water-resistant applications, and creating sustainable gloves for medical, food, and industrial uses. Above all, they can be incorporated into textiles for improved durability and sustainability. In the textile industry, starch-based composites are used for creating eco-friendly fabrics and coatings. They provide a biodegradable alternative to traditional synthetic fibers and coatings.

References

Ab Rahman MF, Norfaizal NS, Azura AR (2019) The influence of sago starch dispersion on mechanical properties of biodegradable natural rubber latex films. Mater Today: Proc 17:1040–1046. https://doi.org/10.1016/j.matpr.2019.06.507

Abreu AS, Oliveira M, de Sá A, Rodrigues RM, Cerqueira MA, Vicente AA, Machado AV (2015) Antimicrobial nanostructured starch based films for packaging. Carbohydr Polym 129:127–134. https://doi.org/10.1016/j.carbpol.2015.04.021

Adler E (1977) Lignin chemistry—past, present and future. Wood Sci Technol 11(3):169–218

Akbari Z, Ghomashchi T, Moghadam S (2007) Improvement in food packaging industry with biobased nanocomposites. Int J Food Eng 3(4)

Akbari S, Gupta A, Khan TA, Jamari SS, Ani NBC, Poddar P (2014) Synthesis and characterization of medium density fibreboard by using a mixture of natural rubber latex and starch as an adhesive. J Indian Acad Wood Sci 11:109–115. https://doi.org/10.1007/s13196-014-0124-0

Angellier H, Molina-Boisseau S, Lebrun L, Dufresne A (2005) Processing and structural properties of waxy maize starch nanocrystals reinforced natural rubber. Macromolecules 38:3783–3792. https://doi.org/10.1021/ma050054z

Anonymous (2022) Nanocomposites, nanoparticles, nanoclays and nanotubes: global markets to 2022. https://www.bccresearch.com/market-research/nanotechnology/nanocomposites-nanoparticles-nanoclays-and-nanotubes-global-markets.html. Accessed 19 June 2023

Ates B, Koytepe S, Ulu A, Gurses C, Thakur VK (2020) Chemistry, structures, and advanced applications of nanocomposites from biorenewable resources. Chem Rev 120:9304–9362

Bangar SP, Whiteside WS, Dunno KD, Cavender GA, Dawson P, Love R (2022) Starch-based bio-nanocomposites films reinforced with cellulosic nanocrystals extracted from Kudzu (Pueraria montana) vine. Int J Biol Macromol 1(203):350–360

Beaumont NJ, Aanesen M, Austen MC, Börger T, Clark JR, Cole M, Hooper T, Lindeque PK, Pascoe C, Wyles KJ (2019) Global ecological, social and economic impacts of marine plastic. Mar Pollut Bull 142:189–195

Borrelle SB, Ringma J, Law KL, Monnahan CC, Lebreton L, McGivern A, Murphy E, Jambeck J, Leonard GH, Hilleary MA et al (2020) Predicted growth in plastic waste exceeds efforts to mitigate plastic pollution. Science 369:1515–1518

Bouthegourd E, Rajisha KR, Kalarical N, Saiter JM, Thomas S (2011) Natural rubber latex/potato starch nanocrystal nanocomposites: correlation morphology/electrical properties. Mater Lett 65:3615–3617. https://doi.org/10.1016/j.matlet.2011.07.069

Bruna JE, González V, Rodríguez F, Guarda A, Galotto MJ (2011) Econanocomposites films containing copper as potential antimicrobial active packaging. In: 11th Congresso Brasileiro de Polimeros, pp 3770–3775

Cheng ZY, Teoh SH (2004) Surface modification of ultra thin poly (epsilon caprolactone) films using acrylic acid and collagen. Biomaterials 25:1991–2001

Chowdhury T, Das M (2013) Effect of antimicrobials on mechanical, barrier and optical properties of corn starch based self-supporting edible film. Int J Food Stud 2(2):212–223

Cruz-Morales JA, Gutiérrez-Flores C, Zárate-Saldaña D, Burelo M, García-Ortega H, Gutiérrez S (2023) Synthetic polyisoprene rubber as a mimic of natural rubber: recent advances on synthesis, nanocomposites, and applications. Polymers 15(20):4074

Darder M, Aranda P, Ruiz-Hitzky E (2007) Bionanocomposites: a new concept of ecological, bioinspired, and functional hybrid materials. Adv Mater 19:1309–1319

de Azeredo HMC (2009) Nanocomposites for food packaging applications. Food Res Int 42(9):1240–1253

de Azeredo HC, Mattoso LHC, McHugh TH (2011) Advances in diverse industrial applications of nanocomposites, Chapter 4: Nanocomposites in food packaging—a review

Dong P, Prasanth R, Xu F, Wang X, Li B, Shankar R (2015) Eco-friendly polymer nanocomposite, properties and processing. In: Advanced structured materials, vol 75. Springer, Berlin/Heidelberg, pp 1–15

Entcheva E, Bien H, Yin LH, Chung CY, Farrell M, Kostov Y (2004) Functional cardiac cell constructs on cellulose-based scaffolding. Biomaterials 25:5753–5762

Fang M, Huang G-J, Sung W-C (2021) LWT 137:110494

Gade R, Siva Tulasi M, Aruna Bhai V (2013) Seaweeds: a novel biomaterial. Int J Pharm Pharm Sci 5:40–44

Gamage A, Thiviya P, Mani S, Ponnusamy PG, Manamperi A, Evon P, Merah O, Madhujith T (2022) Environmental properties and applications of biodegradable starch-based nanocomposites. Polymers 14(21):4578

García NL, Famá L, D'Accorso NB, Goyanes S (2015) Biodegradable starch nanocomposites. In: Thakur VK, Thakur MK (eds) Eco-friendly polymer nanocomposites: processing and properties. Advanced structured materials. Springer, New Delhi, India, pp 17–77. ISBN 978-81-322-2470-9

Gomes ME, Godinho JS, Tchalamov D, Cunha AM, Reis RL (2002) Alternative tissue engineering scaffolds based on starch: processing methodologies, morphology, degradation and mechanical properties. Mater Sci Eng C 20:19–26

Gómez-Heincke D, Martínez I, Partal P, Guerrero A, Gallegos C (2015) Development of antimicrobial active packaging materials based on gluten proteins. J Sci Food Agric. https://doi.org/10.1002/jsfa.7525
Gross KA, Ezerietis E (2003) Juniper wood as a possible implant material. J Biomed Mater Res Part A 64:672–683
Harussani MM, Sapuan SM, Nadeem G, Rafin T, Kirubaanand W (2022) Recent applications of carbon-based composites in defence industry: a review. Defence Technol 18(8):1281–1300
Hasan K et al (2022) Effects of sisal/cotton interwoven fabric and jute fibers loading on polylactide reinforced biocomposites. Fibers Polym 23:3581–3595
Homayun B, Lin X, Choi H-J (2019) Challenges and recent progress in oral drug delivery systems for biopharmaceuticals. Pharmaceutics 11:129
Hondzo M, You J, Taylor J, Bartlet G, Voller VR (2022) Geophys Res Lett e2021GL093226
Hong LG, Yuhana NY, Zawawi EZE (2021) Review of bioplastics as food packaging materials. AIMS Mater Sci 8(2):166–184. https://doi.org/10.3934/matersci.2021012
Hu Z, Zhang DY, Lu ST, Li PW, Li SD (2018) Chitosan-based composite materials for prospective hemostatic applications. Mar Drugs 16(8):273. https://doi.org/10.3390/md16080273
Jambeck JR, Geyer R, Wilcox C, Siegler TR, Perryman M, Andrady A, Narayan R, Law KL (2015) Plastic waste inputs from land into the ocean. Science 347:768–771
Jones A, Mandal A, Sharma S (2015) Protein-based bioplastics and their antibacterial potential. J Appl Polym Sci 132:1–11. https://doi.org/10.1002/app.41931
Kakadellis S, Harris ZM (2020) Don't scrap the waste: the need for broader system boundaries in bioplastic food packaging life-cycle assessment—a critical review. J Clean Prod 274:122831
Kalia S, Dufresne A, Cherian BM, Kaith BS, Avérous L, Njuguna J, Nassiopoulos E (2011) Cellulose-based bio- and nanocomposites: a review. Int J Polym Sci 2011:1–35
Kamel S (2007) Nanotechnology and its applications in lignocellulosic composites, a mini review. Express Polym Lett 1:546–575
Kaur A, Midha S, Giri S, Mohanty S (2019) Stem Cells Int 2019:1286054
Kiatkamjornwong S, Surunchanajirasakul P, Tasakorn P (2001) Natural rubber–cassava starch foam by compression moulding. Plast Rubber Compos 30:318–327. https://doi.org/10.1179/146580101322913419
Klemm D, Kramer F, Moritz S, Lindstrom T, Ankerfors M, Gray D, Dorris A (2011) Nanocelluloses: a new family of nature-based Materials. Angewandte Chemie Int Ed 50
Lay M, Samat SN, Hwa KT, Rashid AA (2019) Ecofriendly latex movies from cassava starch-crammed radiation pre-vulcanized natural rubber latex. Radiat Outcomes Defects Solids 174(9–10):741–751
Le Corre D, Angellier-Coussy H (2014) Preparation and application of starch nanoparticles for nanocomposites: a review. React Funct Polym 85:97–120
Leksawasdi N, Chaiyaso T, Rachtanapun P, Thanakkasaranee S, Jantrawut P, Ruksiriwanich W, Seesuriyachan P, Phimolsiripol Y, Techapun C, Sommano SR, Ougizawa T, Jantanasakulwong K (2021) Corn starch reactive blending with latex from natural rubber using Na+ ions augmented carboxymethyl cellulose as a crosslinking agent. Sci Rep 11:19250. https://doi.org/10.1038/s41598-021-98807-x
Li K, You J, Liu Y, Zhu K, Xue C, Guo X, Wang Z, Zhang Y (2020) Functionalized starch as a novel eco-friendly vulcanization accelerator enhancing mechanical properties of natural rubber. Carbohydr Polym 231:115705. https://doi.org/10.1016/j.carbpol.2019.115705
Liu C, Shao Y, Jia D (2008) Chemically modified starch reinforced natural rubber composites. Polymer 49:2176–2181. https://doi.org/10.1016/j.polymer.2008.03.005
Liu X et al (2023) Structural properties and anti-inflammatory activity of purified polysaccharides from Hen-of-the-woods mushrooms (Grifola frondosa). Front Nutr 10:1–15. https://doi.org/10.3389/fnut.2023.1078868
Lu DR, Xiao CM, Xu SJ (2009) Starch-based completely biodegradable polymer materials. Express Polym Lett 3:366–375

Madhumitha G, Fowsiya J, Mohana Roopan S, Thakur VK (2018) Recent advances in starch-clay nanocomposites. Int J Polym Anal Charact 23:331–345
Manaila E, Stelescu MD, Craciun G (2018) Degradation studies realized on natural rubber and plasticized potato starch based eco-composites obtained by peroxide crosslinking. Int J Mol Sci 19:2862. https://doi.org/10.3390/ijms19102862
Martson M, Viljanto J, Hurme T, Laippala P, Saukko P (1999) Is cellulose sponge degradable or stable as implantation material? An in vivo subcutaneous study in the rat. Biomaterials 20:1989–1995
Maruthapandi M, Saravanan A, Gupta A, Luong JH, Gedanken A (2022) Antimicrobial activities of conducting polymers and their composites. Macromol 2(1):78–99
Mohiuddin M, Kumer B, Haque S (2017) Biopolymer compositesin photovoltaics and photodetectors. In: Sadasivuni KK, Ponnamma D, Kim J, Cabibihan J-J, AlMaadeed M (eds) A biopolymer composites in electronics. Elsevier, Amsterdam, pp 459–486. https://doi.org/10.1016/B978-0-12-809261-3.00017-6
Murugesan B, Sonamuthu J, Samayanan S, Arumugam S, Mahalingam S (2018) Highly biological active antibiofilm, anticancer and osteoblast adhesion efficacy from MWCNT/PPy/Pd nanocomposite. Appl Surf Sci 434:400–411
Nadrajah K (2005) Development and characterization of nanocomposite materials. Louisiana State University
Nagalakshmaiah M, Afrin S, Malladi RP, Elkoun S, Robert M, Ansari MA, Svedberg A, Karim Z (2019) Biocomposites: present trends and challenges for the future. https://doi.org/10.1016/B978-0-08-102177-4.00009-4
Nairetti D, Mironescu M, Tita O (2014) Antimicrobial activity of active biodegradable starch films on pathogenic microorganisms. Ann Roman Soc Cell Biol 19(1):75
Natrah FMI, Harah ZM, Sidik BJ, Izzatul NMS, Syahidah A (2015) Antibacterial activities of selected seaweed and seagrass from port Dickson coastal water against different aquaculture pathogens, pp 5438–5466
Nirmal Kumar et al (2022) Mechanical and thermal properties of bamboo fiber–reinforced PLA polymer composites: a critical study. Int J Poly Sci 2022. https://doi.org/10.1155/2022/1332157
Norizan MN et al (2022) Mechanical performance evaluation of bamboo fibre reinforced polymer composites and its applications: a review. Funct Compos Struct. https://doi.org/10.1088/2631-6331/ac5b1a
Parolo M, Fernandez L, Zajonkovsky I, Sanchez M, Baschini M (2011) Antibacterial activity of materials synthesized from clay minerals. In: Science against microbial pathogens: communicating current research and technological advances, pp 144–151
Prasopdee T, Shah DU, Smitthipong W (2021) Approaches toward high resilience rubber foams: morphology–mechanics–thermodynamics relationships. Macromol Mater Eng 306:2100337. https://doi.org/10.1002/mame.202100337
Puiggalí J, Katsarava R (2017) Chapter 7—Bionanocomposites. In: Jlassi K, Chehimi MM, Thomas S (eds) Clay-polymer nanocomposites. Elsevier, Amsterdam, pp 239–272. ISBN 978-0-323-46153-5
Quaresma JAS (2019) Clin Microbiol Rev 32:e00034–00018
Rajisha KR, Maria HJ, Pothan LA, Ahmad Z, Thomas S (2014) Preparation and characterization of potato starch nanocrystal reinforced natural rubber nanocomposites. Int J Biol Macromol 67:147–153. https://doi.org/10.1016/j.ijbiomac.2014.03.013
Ramirez DOS, Varesano A, Carletto RA, Vineis C, Perelshtein I et al (2019) Antibacterial properties of polypyrrole-treated fabrics by ultrasound deposition. Mater Sci Eng C 102:164–170
Ratner BD, Hoffman AS, Schoen FJ, Lemons JE (eds) (2004) Biomaterials science: an introduction to materials in medicine. Elsevier Press, San Diego, CA, 851 pp
Riyajan S-A (2015) Robust and biodegradable polymer of cassava starch and modified natural rubber. Carbohydr Polym 134:267–277. https://doi.org/10.1016/j.carbpol.2015.07.038
Riyajan S-A, Keawittarit P (2016) A novel natural rubber graft-cassava starch foam for oil/gasohol absorption. Polym Int 65:491–502. https://doi.org/10.1002/pi.5062

Riyajan SA, Patisat S (2018a) A novel packaging movie from cassava starch and herbal rubber. J Polym Surround 26:2845–2854
Riyajan S-A, Patisat S (2018b) A novel packaging film from cassava starch and natural rubber. J Polym Environ 26:2845–2854. https://doi.org/10.1007/s10924-017-1172-5
Riyajan S-A, Sasithornsonti Y, Phinyocheep P (2012) Green natural rubber-g-modified starch for controlling urea release. Carbohydr Polym 89:251–258. https://doi.org/10.1016/j.carbpol.2012.03.004
Roballo KCS, Bushman J (2019) Trans Immunol 53:61–71
Ruamcharoen J, Munlee R, Vayachuta L, Ruamcharoen P (2023) Bio-based composites of sago starch and natural rubber reinforced with nanoclays. Express Polym Lett 17(11):1096–1109
Saad EM, Elshaarawy RF, Mahmoud SA, El-Moselhy KM (2021) New Ulva lactuca algae based chitosan bio-composites for bioremediation of Cd(II) ions. J Bioresour Bioprod 6:223–242
Safdari M, Al-Haik MS (2018) A review on polymeric nanocomposites: effect of hybridization and synergy on electrical properties. In: Ismail A, Goh SP (eds) Carbon-based polymer nanocomposites for environmental and energy applications. Elsevier, Amsterdam, pp 113–146
Sain M, Oksman K (2006) Introduction to cellulose nanocomposites. In: Oksman KSM (ed) Cellulose nanocomposites: processing, characterization, and properties. American Chemical Society, Washington, DC, pp 2–8
Sajid P, Devasena T (2012) Synthesis and characterization of silica nanocomposites for bone applications. Int Res J Pharm 3(5):173–177
Salgado AJ, Figueiredo JE, Coutinho OP, Reis RL (2005) Biological response to pre-mineralized starch based scaffolds for bone tissue engineering. J Mater Sci Mater Med 16:267–275
Senna MM, Mohamed RM, Shehab-Eldin AN, El-Hamouly S (2012) Characterization of electron beam irradiated natural rubber/modified starch composites. J Ind Eng Chem 18:1654–1661. https://doi.org/10.1016/j.jiec.2012.03.004
Seyednejad H, Imani M, Jamieson T, Seifalian AM (2008) Topical hemostatic agents. Br J Surg 95(10):1197–1225. https://doi.org/10.1002/bjs.6357
Shakeri A, Radmanesh S (2014) Preparation of cellulose nanofibrils by high-pressure homogenizer and zein composite films. Adv Mater Res 829:534–538
Shakir M, Zia I, Rehman A, Ullah R (2018) Int J Biol Macromol 111:903–916
Sharma R, Jafari SM, Sharma S (2020) Antimicrobial bio-nanocomposites and their potential applications in food packaging. Food Control 1(112):107086
Shey J, Imam SH, Glenn GM, Orts WJ (2006) Properties of baked starch foam with natural rubber latex. Ind Crops Prod 24:34–40. https://doi.org/10.1016/j.indcrop.2005.12.001
Soubam T, Gupta A (2021) Eco-friendly natural rubber latex and modified starch-based adhesive for wood-based panels application—a review. Maejo Int J Energy Environ Commun 3:49–53. https://doi.org/10.54279/mijeec.v3i1.245163
Srivastava HM, Irfan M, Shah FA (2021) Energies 14:2254
Suethao S, Phongphanphane S, Wong-ekkabut J, Smitthipong W (2021) The relationship between the morphology and elasticity of natural rubber foam based on the concentration of the chemical blowing agent. Polymers 13:1091. https://doi.org/10.3390/polym13071091
Syuhada DN, Azura AR (2021) Waste natural polymers as potential fillers for biodegradable latex-based composites: a review. Polymers 13:3600. https://doi.org/10.3390/polym13203600
Taherimehr M, Bagheri R, Taherimehr M (2021) In-vitro evaluation of thermoplastic starch/beta-tricalcium phosphate nano biocomposite in bone tissue engineering. Ceram Int 47:15458–15463
Tanrattanakul V, Chumeka W (2010) Effect of potassium persulfate on graft copolymerization and mechanical properties of cassava starch/natural rubber foams. J Appl Polym Sci 116:93–105. https://doi.org/10.1002/app.31514
Tarrahi R, Khataee A, Karimi A, Yoon Y (2022) Chemosphere 288:132529
Thakur R, Pristijono P, Golding JB, Stathopoulos CE, Scarlett C, Bowyer M, Singh SP, Vuong QV (2018) Effect of starch physiology, gelatinization, and retrogradation on the attributes of rice starch-ι-carrageenan film. Starch 70:1700099. https://doi.org/10.1002/star.201700099

Thakur R, Pristijono P, Scarlett CJ, Bowyer M, Singh SP, Vuong QV (2019) Starch-based films: major factors affecting their properties. Int J Biol Macromol 132:1079–1089. https://doi.org/10.1016/j.ijbiomac.2019.03.190

Thomas SK, Parameswaranpillai J, Krishn J, Krishnasamy S, Begum PMS, Nandi D, Siengchin S, George JJ, Hameed N, Salim NV, Sienkiewicz N (2021) A comprehensive review on cellulose, chitin, and starch as fillers in natural rubber biocomposites. Carbohydr Polym Technol Appl 25(2:100095. https://doi.org/10.1016/j.carpta.2021.100095

Ulery BD, Nair LS, Laurencin CT (2011) Biomedical applications of biodegradable polymers. J Polym Sci Part B Polym Phys 49:832–864

Valodkar M, Thakore S (2010) Thermal and mechanical properties of natural rubber and starch nanobiocomposites. Int J Polym Anal Charact 15:387–395. https://doi.org/10.1080/1023666X.2010.500543

Venkatesh G, Nyflött Å, Bonnerup C, Lestelius M (2018) An economic-environmental analysis of selected barrier-coating materials used in packaging food products: a Swedish case study. Env Dev Sustain 20:1483–1497

Venkatesh S, Mahboob S, Govindarajan M, Al-Ghanim KA et al (2021) Microbial degradation of plastics: sustainable approach to tackling environmental threats facing big cities of the future. J King Saud Univ Sci 33:101362. https://doi.org/10.1016/j.jksus.2021.101362

Vudjung C, Chaisuwan U, Pangan U, Chaipugdee N, Boonyod S, Santawitee O, Saengsuwan S (2014) Effect of natural rubber contents on biodegradation and water absorption of interpenetrating polymer network (IPN) hydrogel from natural rubber and cassava starch. Energy Procedia 56:255–263. https://doi.org/10.1016/j.egypro.2014.07.156

Wang Z-F, Peng Z, Li S-D, Lin H, Zhang K-X, She X-D, Fu X (2009) The impact of esterification on the properties of starch/natural rubber composite. Compos Sci Technol 69:1797–1803. https://doi.org/10.1016/j.compscitech.2009.04.018

Wani AA, Singh P, Shah MA, Schweiggert-Weisz U, Gul K, Wani IA (2012) Rice starch diversity: effects on structural, morphological, thermal, and physicochemical properties—a review. Compr Rev Food Sci Food Saf 11:417–436. https://doi.org/10.1111/j.1541-4337.2012.00193.x

Wu Y-P, Ji M-Q, Qi Q, Wang Y-Q, Zhang L-Q (2004) Preparation, structure, and properties of starch/rubber composites prepared by co-coagulating rubber latex and starch paste. Macromol Rapid Commun 25:565–570. https://doi.org/10.1002/marc.200300125

Wu J, Li K, Pan X, Liao S, You J, Zhu K, Wang Z (2018) Preparation and physical properties of porous starch/natural rubber composites. Starch 70:1700296. https://doi.org/10.1002/star.201700296

Yamamoto T, Iwase H, King TW, Hara H, Cooper DK (2018) Burns 44:1738–1749

Zhang X, Wang F, Keer LM (2015) Influence of surface modification on the microstructure and thermo-mechanical properties of bamboo fibers. Materials 8:6597–6608. https://doi.org/10.3390/ma8105327

Zhang B, Qiao D, Zhao S, Lin Q et al (2021) Starch-based food matrices containing protein: recent understanding of morphology, structure, and properties. Trends Food Sci Technol 114:212–231. https://doi.org/10.1016/j.tifs.2021.05.033

Chapter 7
Polylactic Acid Based Rubber Composites and Nanocomposites

K. M. Navaneethan, Shanmugam Dharmalingam, Sivasubramanian Palanisamy, and Anand Krishnamoorthy

Abstract Natural rubber is the bio-degradable matrix material obtained from trees. The use of natural rubber has gained insight for use in composites. This has created a huge impact on the replacement of traditional materials such as steel, alloys and polymer matrix composites. Natural fibres are also used as a reinforcement phase in such type of composites. Poly Latic Acid (PLA) is derived from renewable sources such as corn starch. In order to improve the specific properties of PLA composites, natural rubbers are integrated to form PLA/natural rubber composites. Polylactic acid based natural rubber composites are preferred for use in many engineering applications including aerospace, automobiles, household appliances etc. PLA is a nontoxic and eco-friendly material that is mostly preferred for use as an alternative for petroleum-based matrix such as polyester epoxy etc. It can be combined or used along with natural rubber for composites fabrication which results in good toughened composite. Nano particles, combined with natural rubber also have a great influence on the resulting properties. Their use can be expanded towards food packing applications also. In this work the use of natural rubber/synthetic rubber along with PLA for different applications are presented and elaborated. It can be found that the PLA based composites combined with natural/synthetic/nano-based materials has a wide potential application from low end house hold parts till high end aerospace applications.

K. M. Navaneethan
Department of Mechanical Engineering, Nachimuthu Polytechnic College, Udumalai Road, Annamalai Nagar, Pollachi 642003, Tamil Nadu, India

S. Dharmalingam
Department of Automobile Engineering, Dr. Mahalingam College of Engineering and Technology, Udumalai Road, Annamalai Nagar, Pollachi 642003, Tamil Nadu, India

S. Palanisamy (✉)
Department of Mechanical Engineering, PTR College of Engineering and Technology, Austinpatti, Madurai–Tirumangalam Road, Madurai 625008, Tamil Nadu, India
e-mail: sivaresearch948@gmail.com

A. Krishnamoorthy
Department of Basic Sciences and Humanities, Adi Shankara Institute of Engineering and Technology, Kalady 683574, Kerala, India

© The Author(s), under exclusive license to Springer Nature Switzerland AG 2025

P. M. Visakh (ed.), *Rubber Based Bionanocomposites*, Advanced Structured Materials 227, https://doi.org/10.1007/978-3-031-78557-3_7

Keywords Natural rubber · Synthetic rubber · Poly latic acid · Applications

7.1 Introduction

Composites made of natural rubber are able to withstand mechanical, chemical, and physical stresses better than other type of matrix. The unique combination of characteristics and adaptability of these composites make them widely employed in a wide range of industrial and consumer applications (Das et al. 2019; Oladele et al. 2020). This article provides a synopsis of natural rubber composites, with an emphasis on their characteristics, production methods, and practical uses in industry. Out of all the biodegradable polymers, polylactic acid (PLA) is the most promising from a commercial standpoint. Production of lactic acid polymer (PLA) typically involves ring-opening polymerization of cyclic lactic dimer. PLA can be made by directly poly-condensing lactic acid. Through the process of bacterial fermentation, lactic acid may be extracted from sugar cane or maize starch (Reddy et al. 2008; John et al. 2009; Vignesh Kumar et al. 2022). The presence of water destroys the building polymer chain to a very low degree; hence it is not possible to directly polymerize lactic acid to a useful product. Each polymerization step yields one molecule of water. To create a cyclic lactide ester, two molecules of lactic acid are esterified once and then catalytically cyclized. Although dimerization also produces water, the considerable decrease in polarity makes it possible to separate it before polymerization. A high molecular weight (MW) of PLA may be synthesized from the dilactate ester using stannous octoate and ring-opening polymerization (Ghalia and Dahman 2017; Díaz Velázquez et al. 2023; Martina et al. 2017). In contrast, tin (II) chloride is used as a catalyst in laboratory demonstrations. Since this process does not contribute to the water supply, the MWs might vary greatly.

There are two optical isomers of lactic acid. When L- and D-lactides are polymerized together, the resulting amorphous polymer is called poly-DL-lactide (PDLLA). The ratio of dipole to spin controls the crystallinity. One inexpensive source of L-lactic acid, a metabolic intermediary, is the fermentation of carbohydrate-containing agricultural and food items (Chaturvedi et al. 2022; Thongchul 2013; Kumar et al. 2016). An energy-intensive process known as thermal dehydration polymerization yields poly(L-lactic acid) (PLLA), with a MW of just a few thousand (Bressanin et al. 2022). Lactide ring-opening polymerization with an appropriate catalyst is used to produce high MW PLA. Coordination, anionic, or cationic processes may all play a role in this. Lewis acids are present in the form of salts of rare earth metals, aluminum, zinc, tin, and titanium, and they are among the catalysts and initiators that are effective. In order to polymerize lactide in bulk, coordinated ring-opening polymerization is a process that can be adopted. Producing PLA via various techniques results in a wide range of characteristics (Farah et al. 2016). The glass transition of PLA is around 58–59 °C. The virtually constant melting point (Tm) of PLA produced by the direct technique is around 163 °C. However, owing to the catalyst and impurities, ring-opened PLA is less stable and has a lower Tm of 178 °C (Chin et al. 2021; Xu 2004; Grignard et al. 2019).

Fiber-based materials made from PLA are used to cover new experimental types grown for breeding and seed production, or they are used as growing surfaces for different crops. Filtration systems, nanosensors, biomedical applications, and electrospun fibers (such as PLA/NR fibers) are used due to their distinctive characteristics (Tertyshnaya et al. 2022, 2023).

7.2 Applications of PLA/Natural Rubber Composites

Natural rubber composites are used in the production of medical devices, including seals, gaskets, and components, due to their biocompatibility, flexibility, and resistance to wear (Chang et al. 2021; Urtekin et al. 2023; Magaña et al. 2022). Due to their resilience, natural rubber composites are employed in seismic bearings, providing shock absorption and to provide stability in buildings and bridges, especially in earthquake-prone areas. Natural rubber composites are utilized to manufacture isolation pads and expansion joints in construction to accommodate movement, dampen vibrations, and provide flexibility while maintaining structural integrity (Fan et al. 2024; Stevenson 1992; Teh et al. 2022). The natural phenolic compound curcumin is well-known for its anti-inflammatory, anti-bacterial, anti-carcinogenic, and antioxidant capabilities. In addition to its antibacterial and wound-healing capabilities, this active ingredient has lately shown promise in treating skin damage. Since curcumin is photodegradable when exposed to both natural and artificial sunshine, so its use in wound-healing dressings is restricted (Sharma et al. 2021; Singhal et al. 2024; Yang et al. 2017; Pozo Morales et al. 2017).

Designing a two-layer asymmetric membrane using electrospun fibers to promote wound healing while preserving curcumin's pharmacological and antibacterial effects and avoiding their photodegradation. The base layer is made of microfibers made of a biodegradable combination of poly(lactic acid) (PLA) and natural rubber (NR), the majority of which contain curcumin (Chagas et al. 2021).

Natural fiber (Fig. 7.1) strengthened PLA material could be used instead of glass fiber, since it is negative for the environment and takes a long time to break down. The composites show promise as a replacement for composites made from gasoline (Chang et al. 2021; Jones et al. 2002). They can be used in many fields, such as the aircraft and automobile businesses. PLA composites with natural fibers are very strong, this is why they are popular in the car industry. Among their many potential applications are the production of furniture, floor panels and seatback linings, and the reduction of ambient noise. Kenaf fiber reinforced composite are used in the car industry, because of its lower pollution, lighter and cheaper parts, and the ability to get it recycled. For kenaf biocomposites to remain always available and to meet the rising demand, improved kenaf fiber cultivation methods are improved (Ramesh 2016; Saba et al. 2015; Lips and Dam 2013).

Fig. 7.1 Various types of natural fiber (Yang et al. 2018)

Additive manufacturing by use of 4D printing technology to make 3D printed objects that can change their shape or qualities over time is also found in literatures. Utilizing smart materials and individualized design necessitates a stimulus such as water, light, heat, or a magnetic field in order to induce deformation. People and animals can get implants, smart valves, and monitors for infrastructure lines, as well as medical devices, drug pills, home tools, shoes and clothes (Qader et al. 2019; Zhang et al. 2019; Haleem et al. 2021).

Recently, natural fiber reinforced plastics (NFRPs) have come up as a different option for boat makers because they have better bio-reinforcement. However, their mechanical qualities and performance aren't clear (Yang et al. 2018). This paper reported that using cellulose fiber reinforced polymer composites instead of manufactured materials is very interesting because it could help cut down on the use of non-renewable or oil-based resources. The cellulose fibers reinforced polymer composites were found to have a density of about 0.9 g/cm^3, good sound reduction, an impact strength of 25 kJ/m^2, and a stiffness of about 3000 MPa and also these properties can be improved even more (Kamath et al. 2017; Johny et al. 2023; Mylsamy et al. 2024). Researchers have discovered that sugar palm fiber shows a lot of promise as a fiber reinforcement for polymer matrix composites (Isma'ila et al. 2016; Palanisamy et al. 2024; Almeshaal et al. 2022). The dynamic qualities of a hybrid material made of woven glass and sugar palm fibers are strengthened with

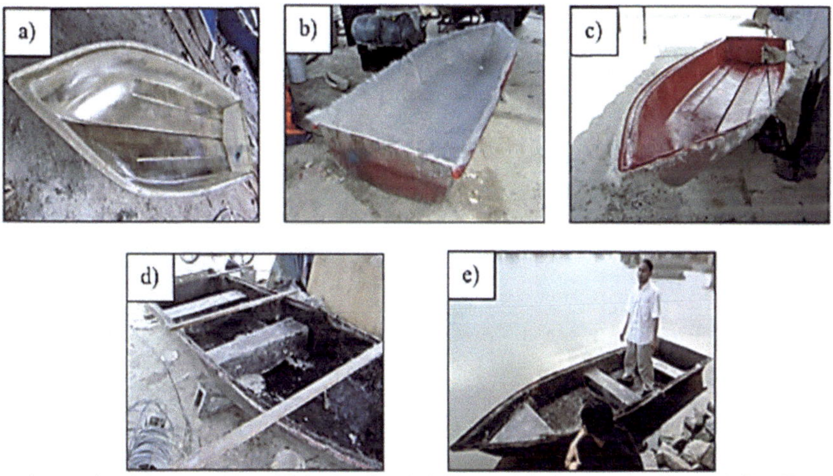

Fig. 7.2 **a–e** Steps of boat fabrication from hybrid material of sugar palm fibre and glass fibre (Misri et al. 2010)

unsaturated polyester. The composites made by compression molding and the lay up method that was used to make the small boats are shown in Fig. 7.2. They came to the conclusion that adding glass fiber had made the tensile qualities of sugar palm fiber composites better.

7.3 Applications of PLA/Natural Rubber Nanocomposites

Polylactic acid (PLA) natural rubber nano composites have a wide range of potential applications due to their unique properties (Bijarimi et al. 2013; Bitinis et al. 2013; Jamshidian et al. 2010). These composites typically exhibit improved mechanical strength, durability, and thermal stability compared to traditional rubber materials. Here are some potential applications for PLA natural rubber nano composites. Cellulose nanofibers have been considered as promising reinforcements. Cellulose nanofibers offer high mechanical properties, sustainability, and abundant availability. The resulting composites hold potential for use in food packaging, construction, and automotive industries (Ilyas et al. 2022; Idumah et al. 2021) as shown in Fig. 7.3.

Automotive Industry: These composites can be used in automotive components such as tires, hoses, seals, and gaskets. Their enhanced mechanical strength and durability make them suitable for withstanding the rigors of automotive applications. Nanocellulose has been studied for usage in lightweight, high-performance

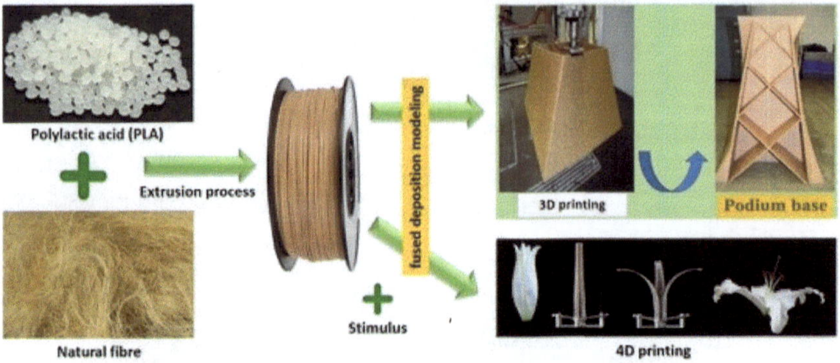

Fig. 7.3 Natural fibre-reinforced PLA 3D printing, as well as in 4D printing for applications. Adapted from Choudhary et al. (2023)

composite materials, showing potential for reducing vehicle weight while maintaining mechanical strength. Its use in foam and composite materials has demonstrated improved thermal and mechanical properties in interior automobile items (Sadiku et al. 2017; Utrera-Barrios et al. 2023; Pradeep et al. 2024).

The progress in biobased plastics for car uses is due to growing worries about the environment and better technology. Nanocellulose is commonly used as a filler in the car business to improve the qualities of plastics. Lightweight, nonabrasive, combustibil, nontoxic, low density, enhanced thermal insulation, less skin irritation, and biodegradability are some of the advantages of natural fiber-reinforced nanocomposites (NFRNs). These materials have been used a lot in the car industry to replace metals (Harun-Ur-Rashid et al. 2023). Researchers all over the world are having trouble making big amounts of polymer–clay nanocomposites. This is especially true when they are trying to turn simple materials into useful technologies. This problem can be turned into a chance by mixing easy materials to make goods with extra value, like nanocomposites for use in cars. Opportunities for the usage of polymer–clay nanocomposite materials in automotive applications may arise as a result of factors such as the rising need for innovative materials with enhanced qualities, which is anticipated to fuel market expansion. Using polymeric nanocomposites to turn polymer and clay into valuable materials is very appealing because the new materials have better qualities, last longer, and are more resistant than the original materials (Bitinis et al. 2011; Alexandre and Dubois 2000; Fu et al. 2019; Ray 2014).

Consumer Goods: PLA natural rubber nano composites can be used in the production of consumer goods such as sports equipment, footwear, and electronic device components. The improved mechanical properties make them ideal for high-performance applications. Numerous food and non-food products have been packed using plastics due to the material's adaptability and widespread use (Jamshidian et al. 2010; Singha et al. 2019; Ranakoti et al. 2022; Yuan et al. 2019). But because they don't break down, their waste poses a health risk and causes a lot of damage to the earth. Biodegradable packaging materials utilizing renewable natural biopolymers like polysaccharides and proteins have garnered more attention due to increasing demand from consumers for high-quality food products and worries about pollution from non-biodegradable plastics. However, natural polymer based packing materials have major drawbacks that make them less useful in industry. For example, they don't have very good mechanical qualities and are more likely to attract water. Nanocomposite films, made by mixing nanocellulose with nano-sized fillers like nanoclays, nanometals, or metal oxides, are one potential solution to the bio-polymeric film challenge. Nanometer-sized diffusion in polymer nanocomposites makes their packing qualities much better. Some of these benefits are higher dynamic strength, lower gas leakage, and higher water resistance. Some bio-nanocomposite films with nanoparticles added also have extra useful features, like being able to kill microbes, protect cells from damage, and block UV light. Because of this, natural bio-polymer-based hybrid packaging materials that are bio-functional have a lot of promise for use in active food packing (Kim et al. 2019).

Medical Devices: Biocompatible and strong nature of PLA rubber-based nano composites made it suitable for use to make medical instruments and equipment, such as medical tubes, seals, and equipment parts. Carbon Nano Tubes (CNTs) are well-suited for a range of biomedical applications due to their distinctive properties (Singh and Agarwal 2016; Sethulekshmi et al. 2022; Liu et al. 2019; Attaran et al. 2017). Because of their hydrophobic properties, CNTs have found several uses in the biological sciences. The CNTs' internalization is facilitated by their interaction with cell surface receptors, which are home to several functional moieties. A lower dosage of a medicine may be possible via these receptor-mediated mechanisms that permit targeted cell loading as shown in Fig. 7.4.

Aerospace and Defence: PLA composites can be used in aerospace and defence applications for manufacturing components such as seals, gaskets, and vibration isolation materials due to their ability to withstand challenging environments and mechanical stress. Manufacturing processes for PLA natural rubber nano composites typically involve melt blending or solution mixing of PLA and natural rubber, followed by the incorporation of nano-sized materials such as nanoparticles or nanofibers to enhance the mechanical properties of the composite (Das et al. 2019; Bhat et al. 2021; Paek et al. 2022; Hawary et al. 2023). In addition, composites used in industrial settings may need specialized processing techniques and formulas to ensure they provide the desired performance. Application of biomaterials such as nanocellulose has also been documented, as it has the use of natural fibers reinforced with nanoparticles to improve ballistic qualities. It was known that shear thinning fluid (STF) worked best for ballistic and protective uses. The nanocellulose's (CNF) function to alter nanosilica STF's characteristics and combine it with Kevlar textiles (Wang et al. 1759; Nurazzi et al. 2021).

The use of carbon-based composites, such as carbon-matrix composites and carbon-reinforced composites, in military technology has recently attracted a lot of

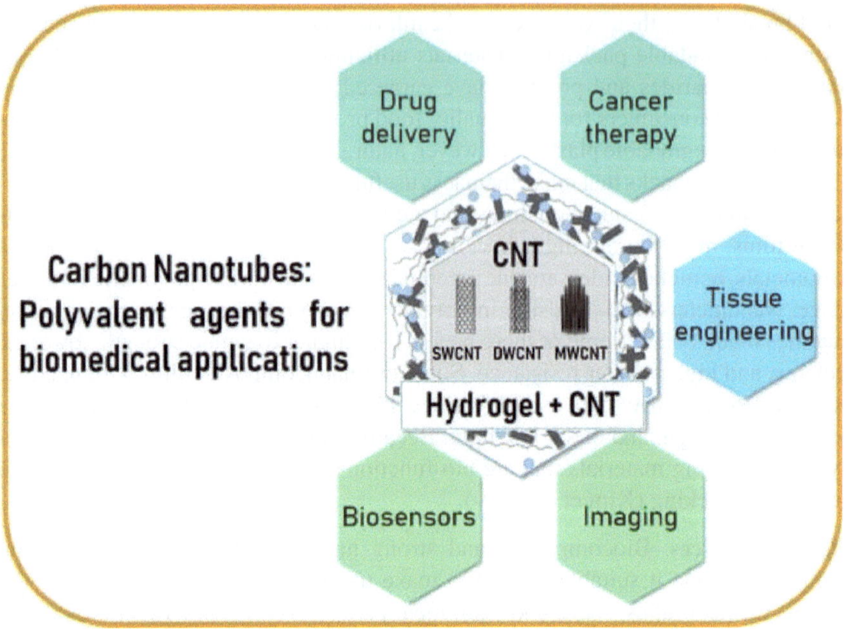

Fig. 7.4 Carbon nanotubes in biomedical applications (Simon et al. 2019)

interest owing to the material's remarkable physical properties, exceptional mechanical and thermal resilience, and low environmental impact (Harussani et al. 2022; Wang et al. 2023; Shirke et al. 2024). The most promising approach to creating cutting-edge defense technology is a carbon-based composite that combines different carbonaceous materials like coke, charcoal, activated carbon, carbon fiber, and various carbon nanomaterials like graphene, carbon nanotubes, and graphite as shown in Fig. 7.5. This paper talks about the properties of carbon-based materials and composites and how they can be used in different areas of defense, such as aerospace, marine, automobiles, electronics, storage of energy, electromagnetic interference (EMI) protection, and buildings. Defense technology holds great potential for carbon-based composites, specifically in the fields of drug delivery agents, chemical sensors, radar technologies, and nanocomposites. This is primarily attributed to the inexpensiveness, accessibility, and processing flexibility of carbon-based composites (Gohardani et al. 2014; Pandey et al. 2022; Kausar et al. 2023).

7.4 Applications of PLA/Synthetic Rubber Composites

Polylactic acid (PLA) synthetic rubber composites offer a wide array of potential applications across various industries due to their unique combination of properties. Some of the key potential applications for PLA synthetic rubber composites include:

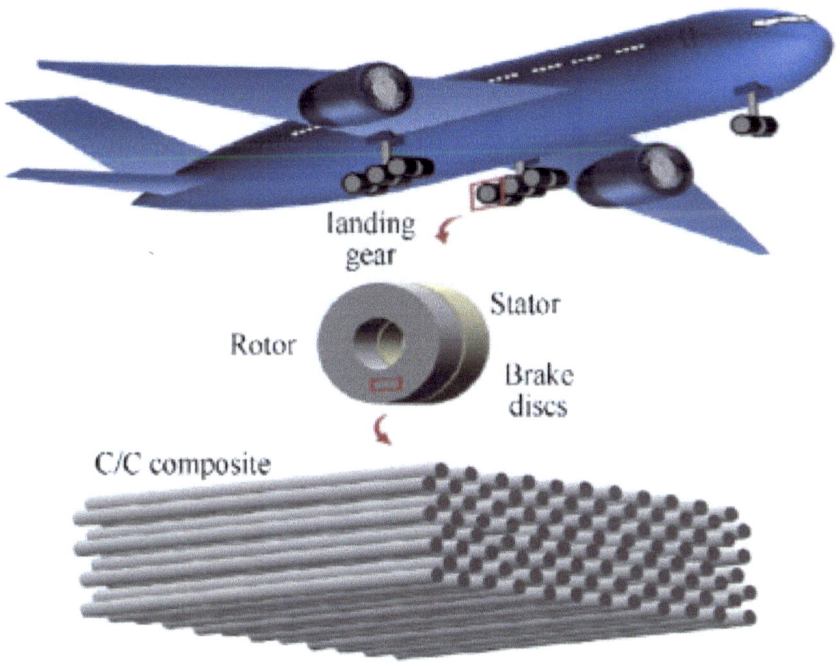

Fig. 7.5 Brake discs of aircraft landing gears made with carbon composite material (Harussani et al. 2022)

Packaging Industry: PLA synthetic rubber composites can be utilized in packaging materials, such as films, trays, and containers. These composites offer good barrier properties, mechanical strength, and biodegradability, making them suitable for sustainable packaging solutions (Attaran et al. 2017).

Textile and Apparel: The composites can be used in the production of textiles and apparel, offering enhanced mechanical properties, such as durability and flexibility, which are desirable for applications in sportswear, outdoor gear, and industrial protective clothing (Kharbanda et al. 2800).

Consumer Goods: PLA synthetic rubber composites can find applications in a wide range of consumer goods, such as electronic casings, toys, and household products, where the combination of mechanical strength, impact resistance, and biodegradability is advantageous (Samsudin and Iñiguez-Franco 2022).

Automotive Components: These composites can be utilized in automotive components, including interior trims, panels, and upholstery which are shown in Fig. 7.6. where their mechanical properties, sustainability, and potential for lightweighting can contribute to improved fuel efficiency and reduced environmental impact (Asim et al. 2018) as shown in Fig. 7.6.

Fig. 7.6 Different types of automotive components (Asim et al. 2018)

Medical Devices: PLA synthetic rubber composites hold potential for use in medical device applications, where their biocompatibility, durability, and resistance to sterilization processes are valuable for surgical instruments, implantable devices, and packaging for medical products. Synthetic rubber composites made of polylactic acid (PLA) are mixed with other types of synthetic rubber, like EPDM or styrene-butadiene rubber (SBR), and sometimes other additives and reinforcing agents are added to make the composites work better for certain uses. The remarkable strength-to-weight ratios, high resistance to heat, and other desirable characteristics of synthetic man-made fibers make them ideal for use in bulletproof and ballistic applications. As a substitute for the traditional metal body armor, these materials allowed for the creation of lightweight and flexible armor that nonetheless offered a high degree of protection (Khan et al. 2014).

7.5 Applications of PLA/Synthetic Natural Rubber Nanocomposites

Polylactic acid (PLA) synthetic natural rubber nanocomposites are advanced materials with a wide range of potential applications due to their unique combination of properties. The incorporation of nanomaterials enhances their mechanical strength, durability, and thermal stability, making them suitable for various industrial and

Fig. 7.7 Composites used in aerospace applications (Asim et al. 2018)

consumer applications. Here are some potential applications of PLA synthetic natural rubber nanocomposites:

Aerospace and Defence: Natural rubber nanocomposites can be employed in aerospace and defense applications for manufacturing components that require high mechanical strength, durability, and lightweight properties. Examples include aircraft seals, gaskets, and vibration isolation materials as shown in Fig. 7.7 (Asim et al. 2018; Aravind et al. 2022).

Automotive Industry: PLA synthetic natural rubber nanocomposites can find applications in the automotive sector for manufacturing tires, engine mounts, and suspension components. The enhanced mechanical properties and thermal stability make them suitable for withstanding the demanding conditions in automotive applications. The incorporation of nanoparticles into natural fibers derived from plants, animals, and geological processes can result in a number of desirable properties, such as reduced skin inflammation, increased thermal insulation, low density, combustibility, nontoxicity, and biodegradability. Natural fiber-reinforced nanocomposites (NFRNs), with their appealing properties, are finding their way into the automotive industry and have been increasingly used as metal substitutes in recent years. There are a number of NFRNs that have been proposed that are either functional or show promise. A better understanding of the factors that influence NFRNs' performance-specific traits has been achieved (Khalid et al. 2017).

Food Packing Applications: Nanotechnology has shown to be a cutting-edge tool that may improve food items in several practical and long-lasting ways. Mixing nanomaterials like Halloysite nano clay, Copper sulfide, and Bentonite nano clay with carrageenan to make new packaging materials with better antimicrobial and mechanical properties, as well as gas and moisture barrier properties that can make food safe and healthy. There has been encouraging evidence that using carrageenan-based bio-nanocomposites as components of food packaging may increase shelf stability and food quality by halting the development of microbes. By incorporating nanomaterials into carrageenan, active packaging methods may be developed to continuously preserve food goods from the farm to the table, regardless of the storage conditions. This will guarantee that foods remain safe and of high quality. Packaging materials made of bio-nanocomposite materials based on carrageenan can assist lessen environmental issues because of its high biodegradability index (Rossi et al. 2017).

Biomedical Devices: PLA synthetic natural rubber nanocomposites hold potential for use in biomedical devices and implants due to their biocompatibility and mechanical properties. Applications include medical tubing, seals, and components for medical equipment. The utilization of materials with a substantial surface area-to-volume ratio in biomedical applications facilitates effective drug loading and transportation to specific cells or tissues. Furthermore, CNTs are desirable for application in tissue engineering due to their unique mechanical characteristics, which allow them to facilitate tissue regeneration and cell development. Because of their tiny size, CNTs may also be utilized in biosensors and bioimaging devices, which allow for the real-time monitoring and identification of biomolecules or cells (Teo et al. 2016).

Electronics and Consumer Goods: The composites can be used in the production of electronic device components, sports equipment, and consumer goods where the combination of mechanical strength, durability, and lightweight properties is beneficial. Biocarbon is achieved by pyrolyzing forestry or agricultural waste in an oxygen-limited environment, and it has gained attention due to its stability, hydrophobicity, and electrical properties. It can be used as a reinforcing agent in biocomposites to impart strength and stiffness to the matrix, with potential applications in electronic casings and automotive interior parts. Synthesized with a pyrolyzed miscanthus biocarbon filler, this biocomposite showed superior flammability resistance and limited frictional wear, making it suitable for durable applications. Nanofillers, including montmorillonite, graphene nanoplatelets, and layered silicate, have been used to develop PLA nanocomposites with potential weight reduction and enhanced properties. Functionalization of nanoparticles controls dispersion and distribution in the matrix, leading to improved interfacial adhesion and mechanical properties. The addition of TPU and GO in PLA enhanced mechanical properties and thermal stability, with potential use in biocompatible materials and 3D printing (Hsissou et al. 2021).

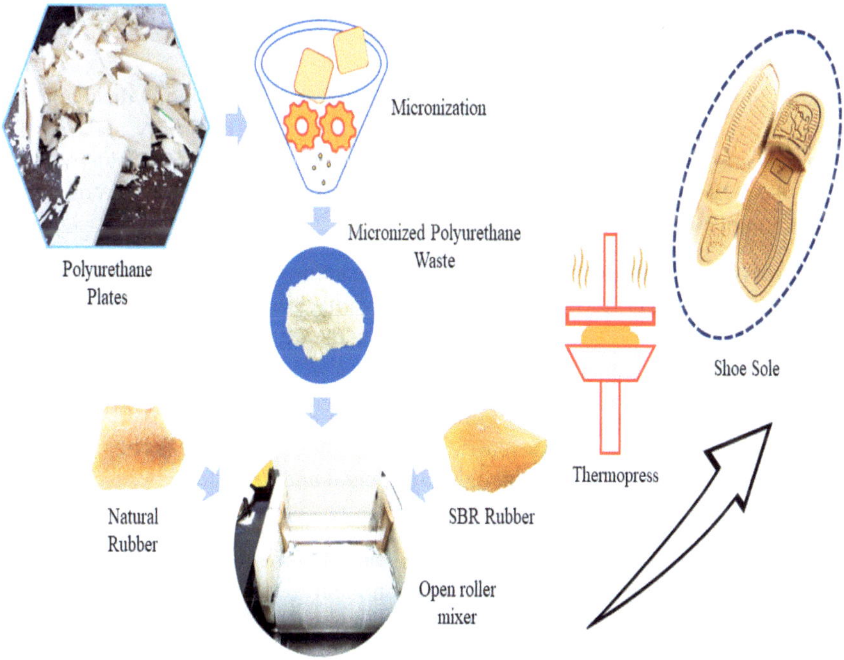

Fig. 7.8 Composites using natural rubber (Carvalho et al. 2024)

7.6 Textile Applications

Textiles launched sunscreen fabric revolution made from Ingeo PLA yarn, which has flame retardant properties. PLA-based yarns are used in textile applications, such as pet collars, harnesses, and straps, as well as in 3D printed clothes and pillows for their moisture wicking properties and fluffiness (Tripathi et al. 2021; Carvalho et al. 2024).

A composite was manufactured by using natural rubber and leather. These composites are used in textiles and foot wear applications. The prototype of shoe sole composite are shown in Fig. 7.8. The viability of the material manufactured is to be tested and classified as a new market trend of the eco-friendly product.

7.7 Electronics Applications

PLA is used in electronic devices to reduce environmental impact, such as in refrigerator and freezer cabinet door liners, organic field-effect transistors, and printed electrodes for electrochemical sensing (Farkas et al. 2024).

7.8 Biomedical Applications

Natural rubber composites have significant potential for various biomedical applications due to their unique properties, such as biocompatibility, flexibility, and mechanical resilience (Guerra et al. 2021). These composites can be used in a wide range of medical devices and components. Some of the key biomedical applications of natural rubber composites include:

Medical Gloves: Natural rubber composites are commonly used in the production of medical gloves due to their flexibility, elasticity, and barrier properties. These gloves provide comfort and protection for healthcare professionals and patients during medical procedures.

Dental Products: Natural rubber composites find applications in dental products such as dental dams, orthodontic elastics, and prosthetic appliances. Their flexibility and biocompatibility make them suitable for dental and orthodontic uses.

Sealing and Tubing: Natural rubber composites can be utilized in the manufacturing of seals, gaskets, and tubing for medical equipment and devices, including infusion sets, catheters, and respiratory equipment. Their sealing properties and biocompatibility are advantageous in medical settings.

Prosthetics and Orthotics: Natural rubber composites are used in prosthetic and orthotic devices due to their flexibility, durability, and shock absorption properties. These composites can be tailored to provide comfortable and functional support for individuals with limb differences or musculoskeletal conditions.

Implantable Devices: In some cases, natural rubber composites with suitable biocompatibility and mechanical properties can be utilized in implantable devices, such as tissue engineering scaffolds, subdermal implants, and drug delivery systems. Nanostructured materials for medicinal uses have been made by electrospinning natural and manufactured polymers, as well as mixes of these polymers. One of the most common manmade plastics used to make wound bandages is poly(lactic acid) (PLA). This is because it can be spun and is biocompatible and biodegradable. When you get latex, which is mostly found on rubber trees (Hevea brasiliensis), you can make natural rubber (NR), which is a polymer made up of isoprene monomers. Additional biomedical uses for nanofibrous materials include dental materials, antibacterial agents, drug delivery vehicles, wound dressings, medical implants, enzyme immobilization scaffolds, along with protective fabrics against biological and chemical hazards.

7.9 Packaging Applications

Particle based nanocomposites were processed using Poly(lactic acid) as a matrix. Fiber Length: up to around 100 nm was fabricated. Process was done by the Melt compounding method with a film thickness of 150 μm. these composites are used for Packaging applications. Particles of Cellulose whisker with a Length: 60–160 nm and thickness of 10–20 nm used for packaging applications. Particle based composites using Montmorillonite as a particle and Poly(lactic acid) at a Thickness: 1–5 nm by the Solution/Casting process at a film thickness of 10–15 μm used for packaging applications. Various reference publications for packaging applications of thickness 1.6 nm by melt processing under a film thickness of 600 μm [Duan et al.], Cellulose nanocrystals as particles using Poly(lactic acid) Length:100–200 nm with a film thickness of 150–250 μm [Fortunati et al.], Montmorillonite as particles based Poly(L-lactic acid) by using the Extrusion/Rolling compression process with a film thickness of 400 μm (Kian et al. 2019).

7.10 Structural Applications

The strips of bamboo fiber and bonded with PLA as the matrix. The research used two types of bamboo, which are Moso (Phyllostachys edulis), culms, and green bamboo (Guadua angustifolia Kunth). Different types of bamboo were used to compare which species can give better strength to the composites for structural application. Although NFRPCs have found extensive usage in the construction and automotive sectors, they might yet benefit from being promoted to more advanced structural uses, such as major structural components for ballistic and bulletproof applications (Pozo Morales et al. 2017; Nurazzi et al. 2021; Brinchi et al. 2013).

7.11 Few Potential Applications

Building and Construction: PLA/natural rubber composites can be utilized in structural elements such as beams, columns, and panels, offering good strength, impact resistance, and durability (Varma et al. 2024).

Automotive Industry: These composites can be used in automotive structural components, such as body panels, interior parts, and under-the-hood applications, benefitting from their lightweight nature and impact resistance. Composites with different materials having variety of applications in automobile sector were shown in Table 7.1. Significant changes are happening in the car industry to deal with important issues involving customers and the environment. Getting cars to use less energy by making them lighter is one of the hardest things to do. Lowering the curb weight by

10% is expected to lower energy use by 6–8% (Liu et al. 2021). Composite materials that are stronger and heavy are one of the best choices for planning, creating, and making lightweight parts. Composite materials could be used in the auto industry to make electric cars lighter and change how well they move through the air. Along with cutting down on toxic fumes and particulate matter, it would also lower the amount of fuel used. Automobile companies and university experts have been studying many changes in these kinds of technologies over the last ten years. The auto industry is interested in fiber-reinforced plastics, especially those made from glass and carbon strands, because they work well and are light (Agarwal et al. 2020; Mallick 2021; Wazeer et al. 2023) (Table 7.2).

Aerospace: The composites can find use in non-critical structural components within aircraft, such as interior panels and non-load-bearing elements, contributing to weight reduction and fuel efficiency (Arora et al. 2024; Ilyas et al. 2019).

In general, an airplane radome is a radar-transparent structure in the shape of a dome that guards radar devices on aircraft from wind loads, bad weather, and bird hits. Figure 7.9 shows that the natural fiber reinforced polymer material might be used in airplane radome. The main material used for RF (radio frequency) transparency is fiberglass-reinforced plastic. Materials that are RF clear allow RF fields to pass through without burning up. Teflon, polypropylene, vinyl chloride, and ultrahigh molecular weight polyethylene are some of the most specialist materials used for high frequency applications. In the last few decades, material scientists, experts, and businesses have become interested in natural fibers because they have certain benefits over manufactured fibers. Natural fibers have low density, are cheap, can be used over and over, and break down naturally. Light airplane radomes often employ composites with a low dielectric constant and a high durability. Potential materials for radome applications include a hybrid of kenaf (processed with NaOH) and glass fibers with epoxy, according to a preliminary evaluation of compilation data of natural fibers (bamboo, banana, kenaf, oil palm, and pineapple leaf fibers) gathered by Haris et al.

Table 7.1 Summary table of fiber reinforced composites for automobiles (Wazeer et al. 2023)

Materials	Results	Application
Steel with foam addition and honeycomb supplement	Superior strength to weight ratio	Automobile bumpers
Integrating steel with carbon fibres	Increase of the strength to weight ratio twice	Roof structure
Composite material made of carbon	Increase in buckling strength and stiffness that is twice as high	Lower arm
Composite material made of carbon	60% reduced weight	Automobile component
Epoxy-carbon based composite	Increase in buckling strength and stiffness that is twice as high	Lower arm
Thermoplastic made of glass and carbon mat (GCMT)	Enhanced impact strength and a 33% reduction in weight	Bump beam

Table 7.2 Current applications of sustainable biocomposites in automotive industry (Agarwal et al. 2020; Mallick 2021; Wazeer et al. 2023)

Company	Model	Biofiber	Matrix	Applications
Audi	A2, A3, A4, A6, A8, Roadster, coupe, Q7	Wood fiber, flax, sisal	PP, epoxy PUR	Seat backs, side and back door panels, boot lining, hat rack, spare tire lining
BMW	3, 5 and 7 series	Kenaf, flax, hemp, wood fiber	PP	Door trim panels, headliner panel, boot lining, seat backs, noise insulation panels, dashboard
Citroen	C5	Wood Fiber, flax	Epoxy	Interior door panelling
Fiat	Punto, brava, marea, alfa romeo 146,156	Flax, sisal, hemp, cotton, coconut fiber	PP	Door cladding, seat back linings, and floor panels, seat bottoms, back cushions, and head restraints
Ford Motors	Ford flex, ford focus BEV, freestar	Wood fiber, wheat straw, coconut coir soy, rice straw	PP, PUR	Interior storage bins, land floor, foam seating, headrest, headliner
General Motors	Cadillac deville, Chevrolet impala, GMC Envoy, trail blazer, terrain, opel vectra	Wood, kenaf, flax, cotton	PP, polyester	Seat belts, trim, rear shelf, cargo floor, door panels, package trays, acoustic insulator, ceiling liner
Honda	Pilot	Wood fiber	–	Floor area parts
Land Rover	2000, others	Kenaf	PP	Insulation, rear storage shelf/panel, door panels, seat backs
Mercedes-Benz	A-class, C-class, E-class, M-class, R-class, S-class,	Abaca/banana, hemp, flax, sisal, jute	PP, PUR, epoxy	Door panels, seat cushion, head restraint, underbody panels, seatbacks, spare tire cover, engine and transmission cover
Mitsubishi	Concept car	Bamboo	PBS	Interior components
Renault	Clio, Twingo	Jute, coir	PP, PUR	Rear parcel shelf

(continued)

Table 7.2 (continued)

Company	Model	Biofiber	Matrix	Applications
Toyota	Prius, Raum	Kenaf, bamboo, corn, starch	PET, sorona EP, PLA, PP	Luggage-compartment, speakers, floor mats, instrument panel, air-conditioning vent, spare tire cover
Volvo	C70, V70	Flax	Polyester	Seat padding, natural foams

PP polypropylene, *PUR* polyurethane, *PLA* polylactic acid, *PBS* polybutylene succinate, *PET* polyethylene terephthalate, *Sorona*® *EP* poly trimethylene terephthalate

(2011) say that the usual mold size for general radome is between 15 and 20 in. The project used hand layup and a vacuum bag with a blend of glass fibers and kenaf fibers to make the object. The overall performance of kenaf fibers is superior than that of bamboo, banana, oil palm, and pineapple leaf (Maiti et al. 2022).

Sporting Goods: The composites can be employed in the manufacturing of sporting equipment like bicycle frames, tennis rackets, and protective gear due to their combination of stiffness, toughness, and impact resistance (Sreejith and Rajeev 2021).

Fig. 7.9 Potential application of natural fibre reinforced polymer composite in airplane radome (Ilyas et al. 2019)

Marine Applications: PLA/natural rubber composites may be used in marine structural components, benefiting from their resistance to moisture and ability to withstand environmental factors (Kuram 2022).

Furniture and Design: These composites can be incorporated into structural elements of furniture and interior design, offering a blend of strength, durability, and aesthetic appeal.

7.12 Military Applications

Good quality fibers were made from knitted fiber that had been treated with silane. So, it was shown that the ramie-woven natural fiber that had been treated with silane could be used as support materials for military bulletproof composite panels. Many different types of businesses use composite parts. They are good in more ways than just being light and strong and stiff compared to other materials (Thiyagu and Narendrakumar 2024). Fundamental ideas include the material and its physical qualities, as well as how they are designed, how to fix tools, check them, and create them. For military use, materials that are both strong and light were used to make helicopters, airplanes, and weapons. Composites can be changed to meet different chemical, mechanical, physical, and structural needs. Synthetic materials are used in many fields, including biology, the military (for things like boats and ships), building (for things like bridges and structures), and the auto business (for things like car bodies) (Bhong et al. 2023). 3D printing with natural fiber reinforcements and its uses in 4D printing for stimuli-responsive polymer applications are discussed in this context (Ilyas et al. 2019, 2021; Mohanty et al. 2023).

People are very interested in using natural fibers in many areas, such as the military, automobile, industrial, furniture, public, and medicinal sectors. In 2018, more than 105 million metric tons of natural fibers were produced around the world. This shows that they are widely used. Because people care about the environment and natural fibers have great qualities, they are still being researched and developed to see if they can replace goods made from petroleum. Natural fibers are becoming more popular as a safe material for industrial uses, such as making parts for cars. They could eventually replace other materials.

Some of the applications include

Protective Gear: These composites can be used in the manufacturing of body armor, helmets, and other protective gear due to their impact resistance and lightweight nature.

Vehicle Components: The composites can be utilized in military vehicle components such as interior panels, non-critical structural elements, and equipment housing, benefiting from their strength, durability, and potential weight reduction.

Temporary Infrastructure: PLA/natural rubber composites can be employed in the construction of temporary shelters, field hospitals, and other infrastructure, offering ease of deployment, durability, and resistance to environmental factors.

Drones and UAVs: The composites may find applications in the manufacturing of drone bodies and components due to their lightweight nature, impact resistance, and potential for aerodynamic design.

Equipment Casings: These composites can be used to manufacture protective casings for sensitive military equipment, providing impact resistance, shock absorption, and potentially electromagnetic shielding properties.

Field Equipment: PLA/natural rubber composites can be utilized in the production of various field equipment such as backpack frames, communication devices, and portable structures, benefitting from their strength, durability, and potentially lightweight nature.

7.13 Tyre Industry

Tyres are an important part of all forms of transportation, including airplanes. They help make life better and the economy grow. Roughly 70 million people possessed personal automobiles and 30 million drove business vehicles as of the end of 2010 (Sienkiewicz et al. 2012). In 2010, to ensure the continued operation of these vehicles, a total of 100 million passenger car tires and 40 million commercial vehicle tires were manufactured. Natural rubber production reached a total of 10.4 million metric tons that year, with the majority of that output coming from Southeast Asia. Not only is natural rubber the main ingredient in most tires, it is also very important for making sure that tires meet safety, performance, and environmental standards. To meet the goals, the tire business tries to improve the design of products, the design of materials, and the production process. From these vantage points, we examine several instances of tire usage in extreme conditions, including airliner tires capable of operating at velocities of up to 380 km/h and off-road tires capable of supporting over 100 per tyre (Hirata et al. 2014) as shown in Fig. 7.10.

At first glance, a tire seems to be nothing more than a piece of black rubber. It is actually made up of layers of "macroscopic" materials with long strengthening fibers or wires wrapped around them. The composition of this rubber material is a result of meticulous engineering. It consists of vulcanized composites of carefully combined raw materials, including polymers (both natural and synthetic), reinforcing fillers (carbon black and silica), and other rubber chemicals. Another layer of rubber covers the outside and is split into four sections: the tread, the shoulder, the sidewall, and the bead.

Fig. 7.10 Types of tyres (Hirata et al. 2014)

7.14 Weight Composition

A tire is mostly made up of things that are used to make rubber mixtures, like NR, synthetic rubber, filler, and rubber chemicals. Different types of synthetic rubber are butadiene rubber (BR), styrene butadiene rubber (SBR), butyl rubber (IIR), along with synthetic polyisoprene (IR).

Natural rubber consists of 30% of weight of a tyre. The synthetic rubber consists of 215 of its weight as shown in Fig. 7.11. By combining both natural rubber and synthetic rubber totally its reaches 60% of polymer components. Comparing both NR and SR tyres, natural rubber has increased shares in the tyres due to the expansion in heavy-duty tyres (Baensch-Baltruschat et al. 2020).

Table 7.3 shows the availability of raw materials in passenger car radial tyres and a truck and bus radial tyres. The ratio of natural rubber is comparatively high in TBR due to the composition of thread in tyre which is the heaviest part of the tyre.

Figure 7.12 shows the stress–strain curve for unvulcanised NR/IR. In this figure, durability, comfort, safety and environmental impact are to be discussed. Rubber components present in the tyres are to be designed to control the changes in the tyres. Rolling resistance is one of the factors for fuel efficiency by reducing it. The elements of natural rubber viscosity will helps in improving the tyre quality.

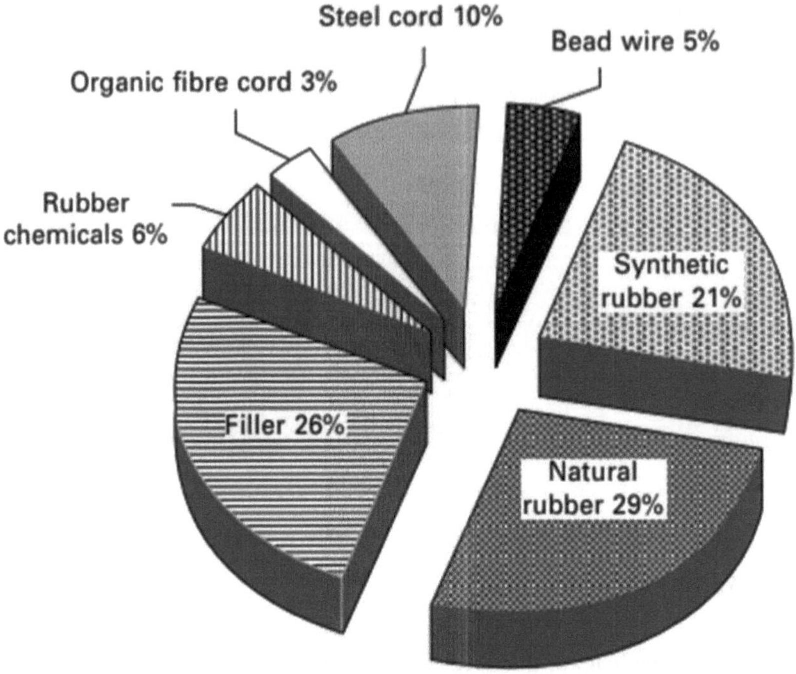

Fig. 7.11 Weight composition of a tyre (Hirata et al. 2014)

Table 7.3 Natural rubber properties required in tyre products (Hirata et al. 2014)

Raw materials	PCR (195/65R15)		TBR (275/80R22.5)	
	General use	Fuel efficient	General use	Fuel efficient
Polymeric components	100.0	100.0	100.0	100.0
Natural rubber	39.0	46.4	77.0	78.8
Synthetic rubber	61.0	53.6	23.0	21.2
Carbon black	50.0	41.3	52.0	47.3
Silica	1.0	16.9	1.0	2.8
Process oil	8.0	9.6	2.0	1.8
Organic rubber chemicals	8.0	13.1	10.0	8.3
Inorganic rubber chemicals	7.0	22.8	9.0	9.9
Zinc oxide	3.0	3.4	5.0	4.4
Sulphur	3.0	2.5	3.0	2.7
Organic fibre cord	10.0	8.0	0.0	0.4
Steel cord	15.0	14.1	33.0	31.5
Beed wire	8.0	9.5	11.0	13.3
Total	206.0	218.4	217.0	212.5

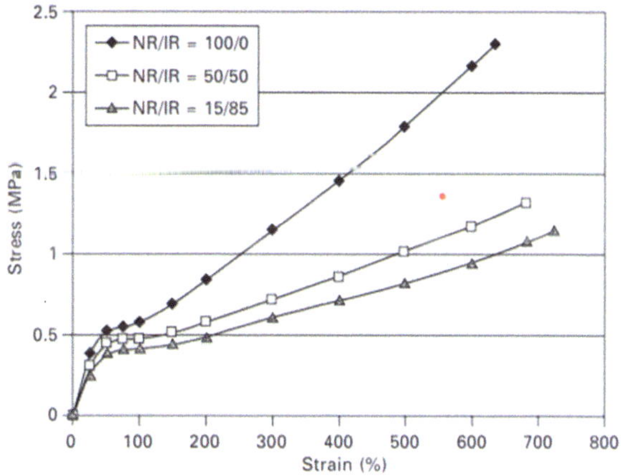

Fig. 7.12 Stress–strain curve for vulcanised NR/IR (Hirata et al. 2014)

7.15 Coating Industry

In heat exchangers, condensing is performed constantly in corrosive environments. The strong acids are formed while condensing flue gases. The materials like carbon steel, aluminium, and copper are not able to survive in the corrosive environments. Perfluoroalkoxy (PFA) based fluoropolymer composites with improved thermal conductivity were developed for coating applications in condensing heat exchangers. The coating developed by these composite materials shows a better surface finish and corrosion resistance as shown in Fig. 7.13.

Fig. 7.13 Metallic substrates coated by graphite-filled PFA: **a** stainless steel coupons, **b** aluminum bare tubes, and **c** finned heat exchangers (aluminum) (He et al. 2015)

7.16 Membrane Technology Applications

Particle based composites Montmorillonite as particles, Poly(lactic acid) as a matrix prepared in a solution/casting process with a film thickness of 25–40 μm used for Membrane application. New biomaterials for intracorneal ophthalmologic implants were designed, manufactured and characterized. A composite material in the form of a membrane was manufactured in a two-stage process. The first stage of the process depended on preparation of a multidimensional (MD-type) fibrous polymer composite. A composite matrix made of a durable terpolymer of polytetraflouroethylene, polyvinylene fluoride, and polypropylene (PTFE-PVDF- PP) served as the basis, while sodium alginate-based biopolymer (NA) was used as the porogenic component, either in the form of short fibers or powder. When it came to mechanical qualities, a membrane made from fibers and particles did better than one made from porogen particles. The membranes are well-suited for use as ophthalmological implants due to their mechanical characteristics and microstructure (Girdthep et al. 2015; Stodolak et al. 2010) as shown in Fig. 7.14.

In the vast array of biomedical applications, composite materials are finding increasing usage in the design and manufacture of implant materials. Eventually, they were useful because they could change the mechanical properties of pure polymers to match the properties of broken tissue. Implants that bear weight after bone surgery may be made from composites that include fiber reinforcement components orientated in various ways (1D, 2D, 3D, and MD) inside the matrix.

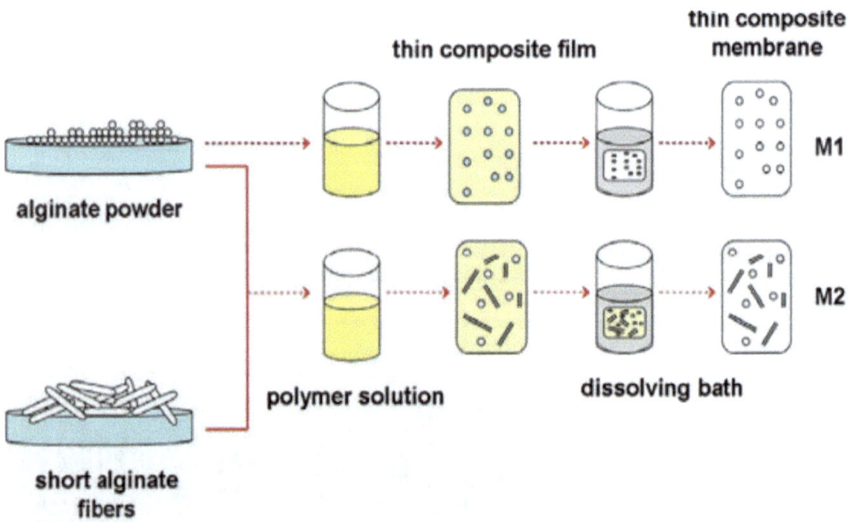

Fig. 7.14 Preparation of composite membranes (Stodolak et al. 2010)

Water Filtration Membranes: PLA/natural rubber composites can be used to develop water filtration membranes for various applications, including wastewater treatment, desalination, and purification of drinking water. The flexibility and biocompatibility of natural rubber, combined with the mechanical strength and biodegradability of PLA, make these composites suitable for producing durable and environmentally friendly water filtration membranes (Bharagava et al. 2023).

Biomedical Membranes: The use of PLA/natural rubber composites in biomedical membranes is promising due to their biocompatibility and mechanical properties. These composites can be employed in applications such as tissue engineering scaffolds, drug delivery systems, and surgical membranes (Liu et al. 2020).

Gas Separation Membranes: PLA/natural rubber composites can also be utilized in gas separation membranes for applications such as carbon dioxide capture, natural gas processing, and air purification. The compatibility of natural rubber with various gases, along with the mechanical strength and process ability of PLA, make these composites suitable for gas separation applications (Zhang et al. 2022).

Food Packaging Membranes: The majority of food packaging is composed of polymers. As fillers, it is usual practice to use combinations of inorganic and organic nanoparticles, often arranged into geometric patterns like flakes, spheres, or threads (Dixit et al. 2022). Filter particles of different sizes and shapes help to strengthen the polymer's structure. Particles with bigger size gaps between them provide the most support. Filler materials like carbon nanotubes, nanoclay, graphene, and silica are widely used as shown in Fig. 7.15.

Various materials used for food packing are proteins and polysaccharides chiefly because of their availability and polymerization into brittle materials. These types of materials show better barrier properties. The reason for using these materials as food packaging films is high sensitivity to moisture. The one material used for packing is Polyvinyl alcohol. It has an extensive property of reusable. Another one example is kraft paper. The main reason for these materials is biodegradability and safety. PLA/natural rubber composites can also be utilized in food packaging membranes due to their biodegradability, flexibility, and barrier properties. These composites can help in developing sustainable and eco-friendly packaging materials with enhanced mechanical strength and barrier performance (Pawar and Purwar 2013; Akelah and Akelah 2013; Silvestre et al. 2011; Grujić et al. 2017).

Electrodialysis Membranes: PLA/natural rubber composites may find use in electrodialysis membranes for applications such as salt removal and ion exchange processes. The electrical insulating properties of natural rubber coupled with the formability and chemical resistance of PLA make these composites suitable for electrodialysis applications (Jasso-Salcedo et al. 2023).

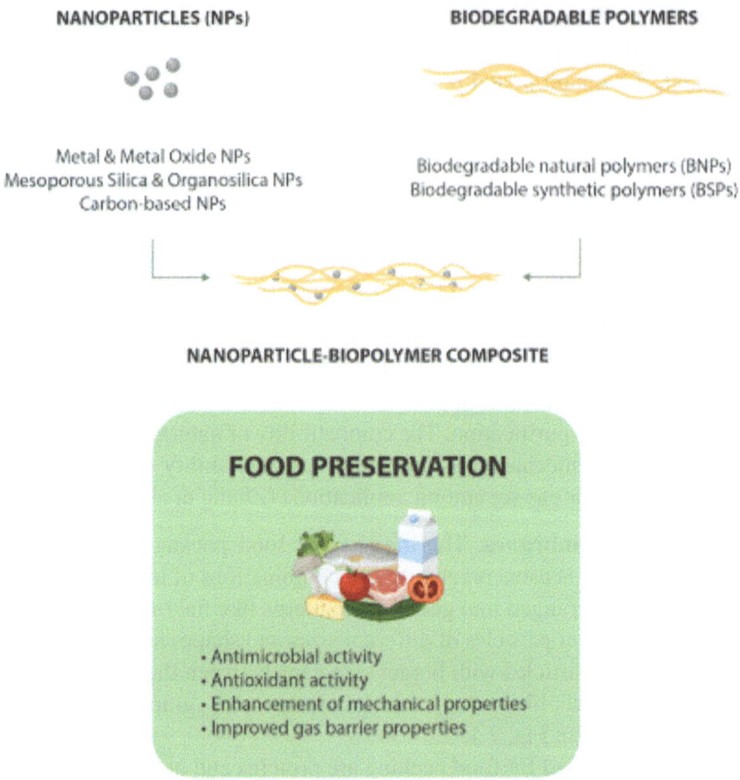

Fig. 7.15 Nanoparticle-biopolymer composite used for food packing applications (Omerović et al. 2021)

7.17 Aerospace Applications

Metal composites, ceramic composites, and polymer composites are the three basic types of composite materials. For many uses, like car interiors, airplanes, spaceships, ships, civil building, packing, and sports goods, polymer composites were popular because they were strong for their weight (Lokesh et al. 2023).

7.18 Aircraft Tyres

The circumstances under which aircraft tires are utilized are peculiar; for example, they are only used for a brief period of time (during landing, taxiing, and taking off), yet under very harsh conditions of speed and weight. A Boeing B747, for instance, has 18 tires, and each one must be able to carry 25 tons. In contrast to OTR tyres, which can be enlarged to accommodate greater loads, AP tyres must be sufficiently tiny and light (approximately 0.5 m in width and 1.3 m in outer diameter for the B747 example) to retract into the landing gear compartment while in flight (Sabat 2015). As the size and inflation pressure of a tire affect how much weight it can hold, AP tires are usually made with a very strong core and high inflation (14.4 kgf/cm^2 for the B747), which is twice as high as for truck and bus (TB) tires. As a simple example, an AP tire needs to be able to carry a lot of weight like an OTR tire while also being able to speed up like a race car tire. You can reach speeds of up to 380 km/h; at cruising altitude, it may be tens of degrees below freezing, but down low, it can be tropical. When you land or take off, the temperature can change very quickly between these two extremes. AP tires can be used for about 200–300 rounds of take-off and landing before they need to be rebuilt for the next cycle of service. However, the conditions are very different for each airplane and operation. Most AP tires can be retreading five to six times, and the skeleton part can be used for about 10,000 km.

AP tires can also be made in bias and radial styles. Bias tires are still widely used in this section because many planes that were made before radial AP tires became popular are still in use and have the bias tires that were originally made for them. Many new planes come with AP radial tires instead of bias tires because they are better in many ways. For example, they use less fuel because they are lighter, last longer because they don't get as hot, and can handle more landings before they need to be replaced. Radial tires are likely to have a bigger share of the AP segment soon than bias tires (Williams and Starke 2003) as shown in Fig. 7.16.

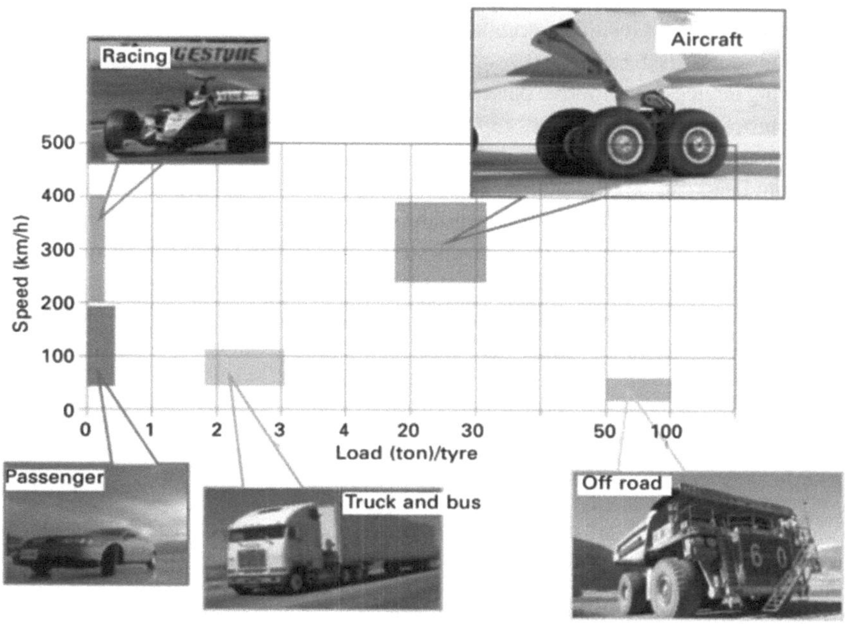

Fig. 7.16 Service conditions of various tyres (Hirata et al. 2014)

The application's extreme demands inform the material selection process for AP tyres, which includes NR in rubber compounds that keep acceptable properties across a wide temperature range and provide durability benefits to support high speed and high load services, as well as very high tensile cords for the skeleton that can withstand extremely high inflation pressure. A lot of rolling resistance (NR) is found in AP tires, just like it was in OTR tires earlier (Greiner et al. 2018). One of the most significant aims is to have a longer and more predictable service life, with safety performance. To do this, microscopic design is used to make tread rubber materials more durable by increasing the number of landing cycles per retread, and macroscopic design is used to make cords stronger so that the skeleton structure lasts longer by adding more retread cycles. Again, planning at the tiny and broad levels work together. To meet the needs of the above-mentioned very strict services, it is once again very important that the NR is of high quality and free of any foreign materials.

7.19 Conclusions

The use of PLA based composites along with natural rubber combined with nano fillers are reported in this chapter. The applications are huge as can be found in literatures. PLA/Natural rubber/natural fiber composites found use in automotive and marine applications such as making of automotive parts and boats. PLA combined with Natural rubber and nano fillers are also found to have a wide variety of applications. The resulting composites are worthy to claim to have less weight reduction, good specific properties etc. Applications include medical devices, food processing devices, automotive parts, consumer goods, defence and aerospace. Also, PLA synthetic based composites represent a range of applications such as medical, textile, automotive, food packaging etc. It can be concluded that the use of PLA is not limited and can be explored further for many applications. The inherent properties such as bio degradability, ease availability, good specific properties and ability to combine and mix with other materials such a natural/synthetic rubber may make this evolve as a most proffered source of material for composites. Any combination possibly resulting in biodegradable natural would be mostly pre-fed and selected to save environmental and related issues.

References

Agarwal J, Sahoo S, Mohanty S, Nayak SK (2020) Progress of novel techniques for lightweight automobile applications through innovative eco-friendly composite materials: a review. J Thermoplast Compos Mater 33:978–1013

Akelah A, Akelah A (2013) Polymers in food packaging and protection. Funct Polym Mater Agric Food Ind 293–347

Alexandre M, Dubois P (2000) Polymer-layered silicate nanocomposites: preparation, properties and uses of a new class of materials. Mater Sci Eng R Rep 28:1–63

Almeshaal M, Palanisamy S, Murugesan TM, Palaniappan M, Santulli C (2022) Physico-chemical characterization of Grewia Monticola Sond (GMS) fibers for prospective application in biocomposites. J Nat Fibers 19:15276–15290

Aravind D, Senthilkumar K, Rajini N, Muthu Kumar TS, Chandrasekar M, Ismail SO et al (2022) Feasibility of elastomeric composites as alternative materials for marine applications: a compendious review on their properties and opportunities. Proc Inst Mech Eng Part M: J Eng Maritime Environ 236:839–855

Arora N, Dua S, Purohit SV, Dash B, Yadav MD, Jena BK et al (2024) Polymer composites: synthesis, application, and basic theoretical aspects. In: Polymer composites: from computational to experimental aspects. Springer, pp 1–41

Asim M, Saba N, Jawaid M, Nasir M (2018) Potential of natural fiber/biomass filler-reinforced polymer composites in aerospace applications. In: Sustainable composites for aerospace applications. Elsevier, pp 253–268

Attaran SA, Hassan A, Wahit MU (2017) Materials for food packaging applications based on bio-based polymer nanocomposites: a review. J Thermoplast Compos Mater 30:143–173

Baensch-Baltruschat B, Kocher B, Stock F, Reifferscheid G (2020) Tyre and road wear particles (TRWP)—a review of generation, properties, emissions, human health risk, ecotoxicity, and fate in the environment. Sci Total Environ 733:137823

Bharagava RN, Goswami L, Kushwaha A, Hussain CM (2023) Bio-based materials and waste for energy generation and resource management: Volume 5 of Advanced zero waste tools: Present and emerging waste management practices. Elsevier

Bhat A, Budholiya S, Aravind Raj S, Sultan MTH, Hui D, Md Shah AU et al (2021) Review on nanocomposites based on aerospace applications. Nanotechnol Rev 10:237–253

Bhong M, Khan TKH, Devade K, Krishna BV, Sura S, Eftikhaar HK et al (2023) Review of composite materials and applications. Mater Today: Proc

Bijarimi M, Ahmad S, Rasid R (2013) Mechanical, thermal and morphological properties of poly(lactic acid)/natural rubber nanocomposites. J Reinf Plast Compos 32:1656–1667

Bitinis N, Hernández M, Verdejo R, Kenny JM, Lopez-Manchado MA (2011) Recent advances in clay/polymer nanocomposites. Adv Mater 23:5229–5236

Bitinis N, Verdejo R, Bras J, Fortunati E, Kenny JM, Torre L et al (2013) Poly(lactic acid)/natural rubber/cellulose nanocrystal bionanocomposites Part I. Processing and morphology. Carbohydr Polym 96:611–620

Bressanin JM, de Mesquita Sampaio IL, Geraldo VC, Klein BC, Chagas MF, Bonomi A et al (2022) Techno-economic and environmental assessment of polylactic acid production integrated with the sugarcane value chain. Sustain Prod Consum 34:244–256

Brinchi L, Cotana F, Fortunati E, Kenny JM (2013) Production of nanocrystalline cellulose from lignocellulosic biomass: technology and applications. Carbohydr Polym 94:154–169

Chagas PAM, Schneider R, dos Santos DM, Otuka AJG, Mendonça CR, Correa DS (2021) Bilayered electrospun membranes composed of poly(lactic-acid)/natural rubber: a strategy against curcumin photodegradation for wound dressing application. React Funct Polym 163:104889

Chang BP, Gupta A, Muthuraj R, Mekonnen TH (2021) Bioresourced fillers for rubber composite sustainability: current development and future opportunities. Green Chem 23:5337–5378

Chaturvedi T, Hulkko LSS, Fredsgaard M, Thomsen MH (2022) Extraction, isolation, and purification of value-added chemicals from lignocellulosic biomass. Processes 10:1752

Chin SY, Shahruddin S, Chua GK, Samsodin N, Setiabudi HD, Karam Chand NS et al (2021) Palm oil-based chemicals for sustainable development of petrochemical industries in Malaysia: progress, prospect, and challenges. ACS Sustain Chem Eng 9:6510–6533

Choudhary S, Haloi J, Sain MK, Saraswat P, Kumar V (2023) Systematic literature review on thermal and acoustic characteristics of natural fibre polymer composites for automobile applications. Mater Today: Proc

Das TK, Ghosh P, Das NC (2019) Preparation, development, outcomes, and application versatility of carbon fiber-based polymer composites: a review. Adv Compos Hybrid Mater 2:214–233

De Carvalho AP, Dos Santos HF, Ribeiro GD, Hiranobe CT, Goveia D, Gennaro EM et al (2024) Sustainable composites: analysis of filler-rubber interaction in natural rubber–styrene–butadiene rubber/polyurethane composites using the Lorenz-Park method and scanning electron microscopy. Polymers 16:471

Díaz Velázquez H, Guzmán-Lucero D, Martínez-Palou R (2023) Microwave-assisted demulsification for oilfield applications: a critical review. J Dispersion Sci Technol 44:1884–1899

Dixit P, Reddy VJ, Parvate S, Balwani A, Singh J, Maiti TK et al (2022) Salt hydrate phase change materials: current state of art and the road ahead. J Energy Storage 51:104360

El Hawary O, Boccarusso L, Ansell MP, Durante M, Pinto F (2023) An overview of natural fiber composites for marine applications. J Mar Sci Eng 11:1076

Fan Q, Duan H, Xing X (2024) A review of composite materials for enhancing support, flexibility and strength in exercise. Alex Eng J 94:90–103

Farah S, Anderson DG, Langer R (2016) Physical and mechanical properties of PLA, and their functions in widespread applications—a comprehensive review. Adv Drug Deliv Rev 107:367–392

Farkas C, Gál L, Csiszár A, Grennerat V, Jeannin P-O, Xavier P et al (2024) Sustainable printed circuit board substrates based on flame-retarded PLA/flax composites to reduce environmental load of electronics: quality, reliability, degradation and application tests. Sustain Mater Technol 40:e00902

Fu S, Sun Z, Huang P, Li Y, Hu N (2019) Some basic aspects of polymer nanocomposites: a critical review. Nano Mater Sci 1:2–30

Ghalia MA, Dahman Y (2017) Biodegradable poly(lactic acid)-based scaffolds: synthesis and biomedical applications. J Polym Res 24:1–22

Girdthep S, Komrapit N, Molloy R, Lumyong S, Punyodom W, Worajittiphon P (2015) Effect of plate-like particles on properties of poly(lactic acid)/poly(butylene adipate-co-terephthalate) blend: a comparative study between modified montmorillonite and graphene nanoplatelets. Compos Sci Technol 119:115–123

Gohardani O, Elola MC, Elizetxea C (2014) Potential and prospective implementation of carbon nanotubes on next generation aircraft and space vehicles: a review of current and expected applications in aerospace sciences. Prog Aerosp Sci 70:42–68

Greiner M, Unrau H-J, Gauterin F (2018) A model for prediction of the transient rolling resistance of tyres based on inner-liner temperatures. Veh Syst Dyn 56:78–94

Grignard B, Gennen S, Jérôme C, Kleij AW, Detrembleur C (2019) Advances in the use of CO_2 as a renewable feedstock for the synthesis of polymers. Chem Soc Rev 48:4466–4514

Grujić R, Vujadinović D, Savanović D (2017) Biopolymers as food packaging materials. Adv Appl Ind Biomater 139–160

Guerra NB, Pegorin GS, Boratto MH, de Barros NR, de Oliveira Graeff CF, Herculano RD (2021) Biomedical applications of natural rubber latex from the rubber tree Hevea brasiliensis. Mater Sci Eng C 126:112126

Haleem A, Javaid M, Singh RP, Suman R (2021) Significant roles of 4D printing using smart materials in the field of manufacturing. Adv Ind Eng Polym Res 4:301–311

Haris MY, Laila D, Zainudin ES, Mustapha F, Zahari R, Halim Z (2011) Preliminary review of biocomposites materials for aircraft radome application. Key Eng Mater 471:563–567

Harun-Ur-Rashid M, Imran A Bin, Susan MABH (2023) Green polymer nanocomposites in automotive and packaging industries. Curr Pharm Biotechnol 24:145–163

Harussani MM, Sapuan SM, Nadeem G, Rafin T, Kirubaanand W (2022) Recent applications of carbon-based composites in defence industry: a review. Def Technol 18:1281–1300

He Y, Walsh D, Shi C (2015) Fluoropolymer composite coating for condensing heat exchangers: characterization of the mechanical, tribological and thermal properties. Appl Therm Eng 91:387–398

Hirata Y, Kondo H, Ozawa Y (2014) Natural rubber (NR) for the tyre industry. In: Chemistry, manufacture and applications of natural rubber. Elsevier, pp 325–352

Hsissou R, Seghiri R, Benzekri Z, Hilali M, Rafik M, Elharfi A (2021) Polymer composite materials: a comprehensive review. Compos Struct 262:113640
Idumah CI, Nwabanne JT, Tanjung FA (2021) Novel trends in poly(lactic) acid hybrid bionanocomposites. Cleaner Mater 2:100022
Ilyas RA, Sapuan SM, Norizan MN, Atikah MSN, Huzaifah MRM, Radzi AM et al (2019) Potential of natural fibre composites for transport industry: a review. Prosiding Seminar Enau Kebangsaan, vol 2019. Persatuan Pembangunan dan Industri Enau Malaysia (PPIEM) Bahau, Negeri, pp 2–11
Ilyas RA, Sapuan SM, Harussani MM, Hakimi M, Haziq MZM, Atikah MSN et al (2021) Polylactic acid (PLA) biocomposite: processing, additive manufacturing and advanced applications. Polymers 13:1326
Ilyas RA, Zuhri MYM, Aisyah HA, Asyraf MRM, Hassan SA, Zainudin ES et al (2022) Natural fiber-reinforced polylactic acid, polylactic acid blends and their composites for advanced applications. Polymers 14:202
Isma'ila Mukhtar IM, Zulkiflle Leman ZL, Mohamad Ridzwan Ishak MRI, Edi Syams Zainudin ESZ (2016) Sugar palm fibre and its composites: a review of recent developments
Jamshidian M, Tehrany EA, Imran M, Jacquot M, Desobry S (2010) Poly-lactic acid: production, applications, nanocomposites, and release studies. Compr Rev Food Sci Food Saf 9:552–571
Jasso-Salcedo AB, Caicedo C, Fonseca-García A, Díaz-Cruz CA, Jiménez-Regalado EJ, Aguirre-Loredo RY (2023) Polylactic acid composite materials for packaging and the consumption of food products. Polylactic Acid Compos: Sustain Biocompos 18:31
John RP, Anisha GS, Nampoothiri KM, Pandey A (2009) Direct lactic acid fermentation: focus on simultaneous saccharification and lactic acid production. Biotechnol Adv 27:145–152
Johny V, Kuriakose Mani A, Palanisamy S, Rajan VK, Palaniappan M, Santulli C (2023) Extraction and physico-chemical characterization of pineapple crown leaf fibers (PCLF). Fibers 11:5
Jones RH, Giancarli L, Hasegawa A, Katoh Y, Kohyama A, Riccardi B et al (2002) Promise and challenges of SiCf/SiC composites for fusion energy applications. J Nucl Mater 307:1057–1072
Kamath SS, Sampathkumar D, Bennehalli B (2017) A review on natural areca fibre reinforced polymer composite materials. Ciência Tecnologia Dos Materiais 29:106–128
Kausar A, Ahmad I, Eisa MH, Maaza M (2023) Graphene nanocomposites in space sector—fundamentals and advancements. C 9:29
Khalid M, Ratnam CT, Walvekar R, Ketabchi MR, Hoque ME (2017) Reinforced natural rubber nanocomposites: next generation advanced material. Green Biocompos: Des Appl 309–345
Khan W, Muntimadugu E, Jaffe M, Domb AJ (2014) Implantable medical devices. Focal Control Drug Deliv 33–59
Kharbanda AK, Sinha SK, Kamalraj D (2023) Investigation of the mechanical properties of non-conventional fiber materials suitable for sportswear applications in order to achieve an innovative, sustainable, and eco-friendly design. In: AIP conference proceedings, vol 2800. AIP Publishing
Kian LK, Saba N, Jawaid M, Sultan MTH (2019) A review on processing techniques of bast fibers nanocellulose and its polylactic acid (PLA) nanocomposites. Int J Biol Macromol 121:1314–1328
Kim HS, Sun X, Lee J-H, Kim H-W, Fu X, Leong KW (2019) Advanced drug delivery systems and artificial skin grafts for skin wound healing. Adv Drug Deliv Rev 146:209–239
Kumar A, Gautam A, Dutt D (2016) Biotechnological transformation of lignocellulosic biomass in to industrial products: an overview. Adv Biosci Biotechnol 7:149–168
Kuram E (2022) Advances in development of green composites based on natural fibers: a review. Emerg Mater 5:811–831
Lips SJJ, van Dam JEG (2013) Kenaf fibre crop for bioeconomic industrial development. In: Kenaf: a multi-purpose crop for several industrial applications: new insights from the Biokenaf Project. Springer, pp 105–143
Liu H, Jian R, Chen H, Tian X, Sun C, Zhu J et al (2019) Application of biodegradable and biocompatible nanocomposites in electronics: current status and future directions. Nanomaterials 9:950

Liu S, Qin S, He M, Zhou D, Qin Q, Wang H (2020) Current applications of poly(lactic acid) composites in tissue engineering and drug delivery. Compos B Eng 199:108238
Liu Z, Song J, Kubal J, Susarla N, Knehr KW, Islam E et al (2021) Comparing total cost of ownership of battery electric vehicles and internal combustion engine vehicles. Energy Policy 158:112564
Lokesh KS, Ramachandra CG, Shrinivasa Mayya D (2023) Applications of composites—a case study. In: Structural composite materials: fabrication, properties, applications and challenges. Springer, pp 51–63
Magaña I, López R, Enríquez-Medrano FJ, Kumar S, Aguilar-Sanchez A, Handa R et al (2022) Bioelastomers: current state of development. J Mater Chem A 10:5019–5043
Maiti S, Islam MR, Uddin MA, Afroj S, Eichhorn SJ, Karim N (2022) Sustainable fiber-reinforced composites: a review. Adv Sustain Syst 6:2200258
Mallick PK (2021) Thermoplastics and thermoplastic–matrix composites for lightweight automotive structures. In: Materials, design and manufacturing for lightweight vehicles. Elsevier, pp 187–228
Martina K, Tagliapietra S, Barge A, Cravotto G (2017) Combined microwaves/ultrasound, a hybrid technology. In: Sonochemistry: from basic principles to innovative applications, pp 175–201
Misri S, Leman Z, Sapuan SM, Ishak MR (2010) Mechanical properties and fabrication of small boat using woven glass/sugar palm fibres reinforced unsaturated polyester hybrid composite. In: IOP conference series. Materials science and engineering, vol 11, IOP Publishing
Mohanty AK, Vivekanandhan S, Tripathi N, Roy P, Snowdon MR, Drzal LT et al (2023) Sustainable composites for lightweight and flame retardant parts for electric vehicles to boost climate benefits: a perspective. Compos Part C: Open Access 12:100380
Mylsamy B, Shanmugam SKM, Aruchamy K, Palanisamy S, Nagarajan R, Ayrilmis N (2024) A review on natural fiber composites: polymer matrices, fiber surface treatments, fabrication methods, properties, and applications. Polym Eng Sci
Nurazzi NM, Asyraf MRM, Khalina A, Abdullah N, Aisyah HA, Rafiqah SA et al (2021) A review on natural fiber reinforced polymer composite for bullet proof and ballistic applications. Polymers 13:646
Oladele IO, Omotosho TF, Adediran AA (2020) Polymer-based composites: an indispensable material for present and future applications. Int J Polym Sci 2020:1–12
Omerović N, Djisalov M, Živojević K, Mladenović M, Vunduk J, Milenković I et al (2021) Antimicrobial nanoparticles and biodegradable polymer composites for active food packaging applications. Compr Rev Food Sci Food Saf 20:2428–2454
Paek SW, Balasubramanian S, Stupples D (2022) Composites additive manufacturing for space applications: a review. Materials 15:4709
Palanisamy S, Murugesan TM, Palaniappan M, Santulli C, Ayrilmis N, Alavudeen A (2024) Selection and processing of natural fibers and nanocellulose for biocomposite applications: a brief review. BioResources 19
Pandey R, Singhal N, Kumar P, Choudhary K, Kumar S (2022) Graphene and its immense contribution in defense and security: a review. Anal Chem Lett 12:1–12
Pawar PA, Purwar AH (2013) Biodegradable polymers in food packaging. Am J Eng Res 2:151–164
Pozo Morales A, Güemes A, Fernandez-Lopez A, Carcelen Valero V, De La Rosa LS (2017) Bamboo–polylactic acid (PLA) composite material for structural applications. Materials 10:1286
Pradeep SA, Deshpande AM, Limaye M, Iyer RK, Kazan H, Li G et al (2024) A perspective on the evolution of plastics and composites in the automotive industry. Appl Plast Eng Handb 705–748
Qader IN, Mediha KÖK, Dagdelen F, Aydoğdu Y (2019) A review of smart materials: researches and applications. El-Cezeri 6:755–788
Ramesh M (2016) Kenaf (Hibiscus cannabinus L.) fibre based bio-materials: a review on processing and properties. Prog Mater Sci 78:1–92
Ranakoti L, Gangil B, Mishra SK, Singh T, Sharma S, Ilyas RA et al (2022) Critical review on polylactic acid: properties, structure, processing, biocomposites, and nanocomposites. Materials 15:4312

Ray SS (2014) Recent trends and future outlooks in the field of clay-containing polymer nanocomposites. Macromol Chem Phys 215:1162–1179
Reddy G, Altaf MD, Naveena BJ, Venkateshwar M, Kumar EV (2008) Amylolytic bacterial lactic acid fermentation—a review. Biotechnol Adv 26:22–34
Rossi M, Passeri D, Sinibaldi A, Angjellari M, Tamburri E, Sorbo A et al (2017) Nanotechnology for food packaging and food quality assessment. Adv Food Nutr Res 82:149–204
Saba N, Jawaid M, Hakeem KR, Paridah MT, Khalina A, Alothman OY (2015) Potential of bioenergy production from industrial kenaf (Hibiscus cannabinus L.) based on Malaysian perspective. Renew Sustain Energy Rev 42:446–459
Sabat KC (2015) Boeing's global supply chain for 787 dreamliner: a sustainable competitive advantage or lack of oversight? FIIB Bus Rev 4:32–43
Sadiku R, Ibrahim D, Agboola O, Owonubi SJ, Fasiku VO, Kupolati WK et al (2017) Automotive components composed of polyolefins. Polyolefin Fibres (Elsevier) 449–496
Samsudin H, Iñiguez-Franco F (2022) Packaging and consumer goods. In: Poly(lactic acid) synthesis, structures, properties, processing, applications, and end of life, pp 605–618
Sethulekshmi AS, Saritha A, Joseph K (2022) A comprehensive review on the recent advancements in natural rubber nanocomposites. Int J Biol Macromol 194:819–842
Sharma S, Sudhakara P, Singh J, Ilyas RA, Asyraf MRM, Razman MR (2021) Critical review of biodegradable and bioactive polymer composites for bone tissue engineering and drug delivery applications. Polymers 13:2623
Shirke N, Ghase V, Jamdar V (2024) Recent advances in stealth coating. Polym Bull 1–30
Sienkiewicz M, Kucinska-Lipka J, Janik H, Balas A (2012) Progress in used tyres management in the European Union: a review. Waste Manag 32:1742–1751
Silvestre C, Duraccio D, Cimmino S (2011) Food packaging based on polymer nanomaterials. Prog Polym Sci 36:1766–1782
Simon J, Flahaut E, Golzio M (2019) Overview of carbon nanotubes for biomedical applications. Materials 12:624
Singh NB, Agarwal S (2016) Nanocomposites: an overview. Emerg Mater Res 5:5–43
Singha NR, Mahapatra M, Karmakar M, Chattopadhyay PK (2019) Processing, characterization and application of natural rubber based environmentally friendly polymer composites. Sustain Polym Compos Nanocompos 855–897
Singhal R, Sarangi MK, Rath G (2024) Injectable hydrogels: a paradigm tailored with design, characterization and multifaceted approaches. Macromol Biosci 2400049
Sreejith M, Rajeev RS (2021) Fiber reinforced composites for aerospace and sports applications. In: Fiber reinforced composites. Elsevier, pp 821–859
Stevenson A (1992) Engineering properties and applications of rubber. Dev Crop Sci 23:487
Stodolak E, Gumula T, Leszczynski R, Wieczorek J, Blazewicz S (2010) A composite material used as a membrane for ophthalmology applications. Compos Sci Technol 70:1915–1919
Teh TW, Tan CG, Jumaat MZ (2022) Feasibility evaluation of novel high-damping rubbers as energy-dissipation material under axial dynamic load for damper devices. Buildings 12:1917
Teo AJT, Mishra A, Park I, Kim Y-J, Park W-T, Yoon Y-J (2016) Polymeric biomaterials for medical implants and devices. ACS Biomater Sci Eng 2:454–472
Tertyshnaya YV, Karpova SG, Podzorova MV, Khvatov AV, Moskovskiy MN (2022) Thermal properties and dynamic characteristics of electrospun polylactide/natural rubber fibers during disintegration in soil. Polymers 14:1058
Tertyshnaya YV, Podzorova MV, Varyan IA, Tcherdyntsev VV, Zadorozhnyy MY, Medvedeva EV (2023) Promising agromaterials based on biodegradable polymers: polylactide and poly-3-hydroxybutyrate. Polymers 15:1029
Thiyagu C, Narendrakumar U (2024) Mechanical and ballistic performance of ramie fiber-reinforced polymer composites: a review. Proc Inst Mech Eng Part C: J Mechan Eng Sci 09544062241230224

Thongchul N (2013) Production of lactic acid and polylactic acid for industrial applications. In: Bioprocessing technologies in biorefinery for sustainable production of fuels, chemicals, and polymers, pp 293–316

Tripathi N, Misra M, Mohanty AK (2021) Durable polylactic acid (PLA)-based sustainable engineered blends and biocomposites: recent developments, challenges, and opportunities. ACS Eng Au 1:7–38

Urtekin G, Ullah MS, Yildirim R, Ozkoc G, Kodal M (2023) A comprehensive review of the recent developments in thermoplastics and rubber blends-based composites and nanocomposites. Polym Compos 44:8303–8329

Utrera-Barrios S, Verdejo R, López-Manchado MÁ, Santana MH (2023) Self-healing elastomers: a sustainable solution for automotive applications. Eur Polym J 112023

Varma M, Chandran S, Vijay Kumar V, Suyambulingam I, Siengchin S (2024) A comprehensive review on the machining and joining characteristics of natural fiber-reinforced polymeric composites. Polym Compos

Vignesh Kumar B, Muthumari B, Kavitha M, John Praveen Kumar JK, Thavamurugan S, Arun A et al (2022) Studies on optimization of sustainable lactic acid production by Bacillus amyloliquefaciens from sugarcane molasses through microbial fermentation. Sustainability 14:7400

Wang F, An C, Jia Q (2021) Effect of multi-phase STF on the high-speed impact performance of shear-thickening fluid (STF)-impregnated Kevlar composite fabrics. J Phys: Conf Ser (IOP Publishing) 1759:12015

Wang Y, Chen Z, Lu Y, Yang L, Xu T, Wu H et al (2023) A review of application, modification, and prospect of melamine foam. Nanotechnol Rev 12:20230137

Wazeer A, Das A, Abeykoon C, Sinha A, Karmakar A (2023) Composites for electric vehicles and automotive sector: a review. Green Energy Intell Transp 2:100043

Williams JC, Starke EA Jr (2003) Progress in structural materials for aerospace systems. Acta Mater 51:5775–5799

Xu K (2004) Nonaqueous liquid electrolytes for lithium-based rechargeable batteries. Chem Rev 104:4303–4418

Yang Z, Chen Y, Zhou Y, Wang Y, Dai P, Zhu X et al (2017) Microscopic interference full-color printing using grayscale-patterned Fabry-Perot resonance cavities. Adv Opt Mater 5:1700029

Yang MFM, Hamid H, Abdullah AM (2018) Potential use of cellulose fibre composites in marine environment—a review. Eng Appl New Mater Technol 25–55

Yuan S, Shen F, Chua CK, Zhou K (2019) Polymeric composites for powder-based additive manufacturing: materials and applications. Prog Polym Sci 91:141–168

Zhang X, Chen L, Lim KH, Gonuguntla S, Lim KW, Pranantyo D et al (2019) The pathway to intelligence: using stimuli-responsive materials as building blocks for constructing smart and functional systems. Adv Mater 31:1804540

Zhang Q, Ying Y, Ping J (2022) Recent advances in plant nanoscience. Adv Sci 9:2103414

Chapter 8
Bacterial Cellulose/BC/Rubber Nanocomposites

Mukesh Kumar, Madhu Yadav, Rahul, Ashutosh Sharma, Sivasubramanian Palanisamy, and Arup Kar

Abstract BC is an incredibly robust material, with an estimated modulus of over 100 GPa and a strength exceeding 1.5 GPa. It is truly one of nature's strongest creations. Research has shown the effectiveness of using BC as a nano-additive in polymeric materials leading to the formation of lightweight and superior strength nanocomposites at high fiber fractions of BC in the matrices. This chapter explores the various uses of BC in polymers and rubber nanocomposites. These materials consist of nanocomposites that are transparent to light and composites that have been reinforced with BC and natural fibres, creating a hierarchical structure. It covers the synthesis, properties, and applications of these advanced materials. Emphasis is placed on the synergy between BC and rubber matrices, the techniques employed for their fabrication, and their enhanced mechanical, thermal, and biocompatible properties. The potential applications of BC/rubber composites in various industries, including biomedical, automotive, and electronics, are discussed. The authors have explored the application of BC in various cellulosic and biomimetic composites. This chapter also highlights then the usage of BC in the production of BC/rubber nanocomposites.

M. Kumar · M. Yadav · Rahul
Department of Physics, Faculty of Applied and Basic Sciences, Shree Guru Gobind Singh Tricentenary University, Gurugram 122505, India

A. Sharma
Department of Materials Science and Engineering, Ajou University, Suwon 16499, Korea

Present Address:
Amity Institute of Applied Sciences, Amity University Jharkhand, Ranchi 834002, India

A. Sharma
e-mail: ashu.materials@gmail.com

S. Palanisamy (✉)
Department of Mechanical Engineering, P T R College of Engineering and Technology, Thanapandiyan Nagar, Austinpatti, Madurai-Tirumangalam Road, Madurai 625008, Tamil Nadu, India
e-mail: sivaresearch948@gmail.com

A. Kar
Department of Physics, Dibrugarh University, Dibrugarh 786004, Assam, India

© The Author(s), under exclusive license to Springer Nature Switzerland AG 2025
P. M. Visakh (ed.), *Rubber Based Bionanocomposites*, Advanced Structured Materials 227, https://doi.org/10.1007/978-3-031-78557-3_8

8.1 Introduction

Cellulose finds applications in various industries such as paper, pharmaceuticals, and cosmetics. It is also being studied as a potential additive for polymeric and rubbery materials (Sczostak 2009; Klemm et al. 2005; Kamel et al. 2008; Pandey et al. 2010; Satyanarayana et al. 2009; Blaker et al. 2011). There are also a variety of products that come from cellulose, including technical textile fibers like viscose and Lyocell, as well as thermoplastic polymers like cellulose acetate (Harms 2003; Abdelhamid and Mathew 2022). The various applications of BC are summarized in Fig. 8.1.

BC is a biopolymer produced by certain bacteria, notable for its high purity, crystallinity, and mechanical strength. When combined with rubber, a versatile elastomeric material, BC/rubber composites exhibit improved properties suitable for a range of applications. This chapter explores the interdisciplinary nature of BC/rubber composites, bridging materials science, biotechnology, and engineering. Currently, there is a significant amount of research and attention dedicated to isolating and producing nanoscale cellulose reinforced rubber and polymeric materials (Klemm et al. 2011; Siro and Plackett 2010). The fascination with nanocellulose arises from the unique combination of physical and chemical characteristics it possesses. Nanoscale cellulose not only exhibits properties like hydrophilicity and easy chemical modification, but also boasts high specific volume and aspect ratio.

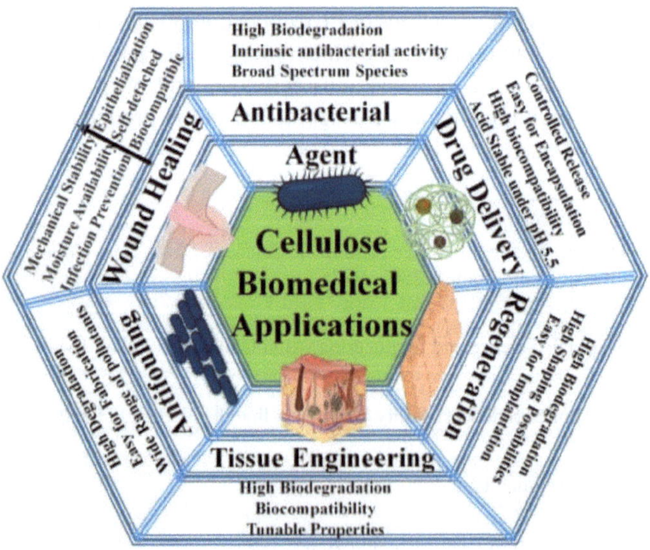

Fig. 8.1 Applications of cellulose in biomedicine (Abdelhamid and Mathew 2022)

8.2 Synthesis

In literature, two approaches are available to obtain nanocellulosic materials: top-down and bottom-up. Through the top-down approach, (ligno)cellulose, like wood fibers are broken down to nanofibers. A study conducted by Wuhrmann et al. (1946) introduced a novel technique involving ultrasonication to extract nanoscale cellulose from ramie, hemp, and cotton fibers. Wood pulp is transformed into nanocellulose by passing it through a homogenizer at high pressure, which effectively reduces the diameter of the fibers to the nano regime (Herrick et al. 1983; Brown 1886). Grinding is a more recent approach to produce nanocellulose from plant cellulose. In this method, pulpwood is thinned down by the intense grinding force created when it passes through the narrow gap between a stationary and moving grindstone.

In the bottom-up method, fermentation of low molar mass sugar is employed with cellulose-synthesising bacteria obtained from the Acetobacterium species. This method results in the synthesis of nanoscale cellulose, also called microbial or bacterial cellulose (BC) (Jonas and Farah 1998; Ross et al. 1991; Iguchi et al. 2000). BC is composed of only the cellulosic material which is free from hemicellulose, pectic polysachhrides or lignin-polymers. The bacteria release cellulose into a liquid culture medium forming nanofibers with size varying from 25 to 100 nm. These nanofibers are responsible for forming a thin cellulosic skin the culture (Jonas and Farah 1998; Ross et al. 1991; Iguchi et al. 2000; Gindl and Keckes 2004). BC was originally utilized as a nano additive in polymeric matrix (Gindl and Keckes 2004). The authors experimented with different loading fractions of BC to embed cellulose acetate butyrate (CAB). The tensile properties of the resultant polymer composites showed up to five times better strength than the pristine material. Over the years, there has been a significant increase in studies exploring BC as a nano-additive in polymeric and rubbery matrices. One of the main reasons behind this is the exciting possibility of harnessing the exceptional rigidity of cellulose crystals. The cellulose crystals are reported to with 100–160 GPa stiffness which makes it an extremely desirable reinforcement for polymer matrices (Matsuo et al. 1990; Hsieh et al. 2008; Eichhorn et al. 2010; Yano et al. 2005; Nogi and Yano 2008; Fernandes et al. 2009; Martins et al. 2009; Lee et al. 2009; Quero et al. 2010; Gu et al. 2010; Li et al. 2010; Trovatti et al. 2010; Retegi et al. 2012). However, the addition of nanoscale cellulose is found to enhance the mechanical performance of these polymeric materials. Table 8.1 presents the tensile characteristics of polymer nanocomposites reinforced with BC.

Table 8.1 Tensile strength and fracture strain of BC/CAB nanocomposites (Gindl and Keckes 2004)

Samples	Fiber fraction (%)	Tensile strength (MPa)	Fracture strain (%)
CAB	0	25.9	3.5
BC/CAB	10	25.9	3.5
BC/CAB	32	128.9	3.6

A number of research activities have been done in this area of BC/polymer nanocomposites as reported by previous studies (Eichhorn and Davies 2006; Khalil et al. 2012; Blaker et al. 2011). BC, also known as bacterial cellulose is generated by certain bacteria (Volova et al. 2022). With its outstanding mechanical capabilities, elevated crystallinity, and a nanofiber network structure, this material finds utility in widespread applications, such as biomedical, food, and material sciences (Hou et al. 2023). The predominant bacterial species utilized for cellulose synthesis are derived from the genus Gluconacetobacter (formerly known as Acetobacter), specifically *Gluconacetobacter xylinus.* Additional species are *Gluconacetobacter hansenii, Gluconacetobacter europaeus,* and *Sarcina ventriculi* (Walling et al. 2023). BC production may be accomplished with many growing methods.

8.2.1 Static Culture

It refers to a culture that remains unchanged or unaltered across time. The bacteria are cultivated in a liquid media under static conditions (Yang et al. 2024). The bacteria are introduced into a static container containing a thin coating of nutritional material. Over the course of many days, the bacteria create a gelatinous BC pellicle on the liquid's surface (Mauro et al. 2024). The setup is straightforward and inexpensive, but the diffusion of oxygen is restricted, which might hinder the yield of BC.

8.2.2 Disturbed Culture

The bacteria are cultivated in a culture that is either agitated or shaken. The culture is sustained by continuous movement utilizing orbital shakers or fermenters, which improve the transport of oxygen and mixing of nutrients (Uchimura and Tsumori 2024). Increased biomass output and accelerated manufacturing process as a result of enhanced oxygenation and nutrient accessibility. Nevertheless, BC is generated as irregular particles rather than a cohesive film.

8.2.3 Bioreactor Cultivation

Massive production utilizing bioreactors equipped with regulated environmental conditions. The options consist of stirred-tank reactors, airlift reactors, and rotating disc reactors. Accurate manipulation of variables such as pH, temperature, and oxygen provision, resulting in reliable and superior synthesis of BC (Öz and Kalender 2023). Drawbacks: Elevated operations and maintenance expenses.

Table 8.2 Hestrin-Schramm medium

Chemical	Amount (g/L)
Glucose	20.00
Yeast extract	5.00
Peptone	5.00
Disodium phosphate	2.70
Citric acid	1.15

8.2.4 Culture Media

The formulation of the nutrition medium is essential for achieving optimal BC production. The common constituents consist of diverse carbon sources such as glucose, fructose, sucrose, mannitol, and glycerol (Tang et al. 2024). The sources used in this study are yeast extract, peptone, maize steep liquor, and ammonium sulfate. Additional nutrients include phosphates, magnesium sulfate, and several trace elements. Table 8.2 displays an example medium formulation called Hestrin-Schramm medium.

8.2.5 Rubber Matrix

Rubber matrices are materials made from rubber that are used for many purposes. An examination of natural rubber (NR), synthetic rubber (such as SBR and EPDM), and their respective characteristics. Rubber is processed by vulcanization, compounding, and mixing techniques.

8.2.6 Methods for Creating Composite Materials

The process involves dispersing black carbon (BC) in rubber solutions, followed by casting and evaporating the solvent (Hamedi et al. 2024). In situ polymerization refers to the process of polymerizing rubber monomers in the presence of BC (Binti Ismail et al. 2024). The process of incorporating BC into rubber matrices is achieved through the use of melt mixing processes (Thomas et al. 2024).

8.3 Characteristics of BC/Rubber Composites

Bacterial cellulose (BC) is used to strengthen rubber composites, resulting in materials that exhibit improved performance due to the combination of BC's distinctive characteristics and rubber's qualities. BC gives a high level of tensile strength, exceptional heat stability, and biocompatibility, whilst rubber offers flexibility and elasticity. This combination leads to composite materials that have a wide range of uses, including biomedical devices and industrial components.

8.3.1 Tensile Strength

By using BC, the tensile strength of rubber composites is greatly enhanced. The boost is attributed to the extremely crystalline structure and strong hydrogen bonding of BC. Tensile testing demonstrates that BC/rubber composites display superior tensile strength in comparison to pure rubber. The extent of enhancement relies on the BC concentration and the quality of its dispersion (Singha et al. 2019).

8.3.2 Modulus of Elasticity

BC enhances the rigidity of rubber composites. The composite material has a larger Young's modulus compared to the unreinforced rubber, suggesting increased rigidity. The stress–strain curves of BC/rubber composites often exhibit a greater initial slope, indicating an enhanced stiffness (Jain and Pradhan 2016).

Elongation at Break Ductility

Although BC improves strength and stiffness, it can decrease the elongation at break because of the inflexible characteristics of cellulose fibers. Optimizing the content of BC is essential for achieving a suitable equilibrium between strength and flexibility (Li et al. 2024).

Tear Resistance

The capacity of a material to bear tearing or ripping forces. BC/rubber composites frequently demonstrate enhanced tear resistance, a critical characteristic for applications necessitating long-lasting materials (Ping et al. 2024).

8.3.3 Thermal Stability

Thermogravimetric analysis (TGA) reveals that BC/rubber composites have elevated breakdown temperatures in comparison to pure rubber. The improvement is attributed to the excellent heat stability of BC. BC, or black carbon, has the effect of prolonging the time it takes for thermal deterioration to occur. This leads to an increased lifespan of the material when exposed to thermal stress, as stated in reference (Smejda-Krzewicka et al. 2024).

8.3.4 Thermal Conductivity.

It indicates the capacity of a material to conduct heat energy. Thermal conductivity of BC can enhance the dissipation of heat in rubber composites, making it advantageous for applications that include thermal cycling or exposure to high temperatures (Zhou et al. 2015).

8.3.5 Glass Transition Temperature (T_g)

The T_g of rubber composites with BC addition may be determined using Dynamic Mechanical Analysis (DMA), which detects changes in T_g. The T_g may be modified by either increasing or decreasing the contact between BC and the rubber matrix (Zhang et al. 2024a).

8.3.6 Density

The concept of density refers to the measure of mass per unit volume.

The density of BC/rubber composites may see a modest rise as a result of the higher density of BC in comparison to rubber. Nevertheless, the rise is often moderate (Kitsawat et al. 2024). BC establishes an interconnected network inside the rubber matrix, which has the potential to affect the porosity of the composite. Controlled porosity can provide advantages in specific applications, such as biomedical implants (Kumineck Junior et al. 2023).

8.3.7 Chemical Characteristics

BC/rubber composites generally exhibit improved resistance to solvents and chemicals as a result of the stable structure of BC. BC's resistance to degradation enhances the overall longevity of the composite in challenging chemical conditions (Kumineck Junior et al. 2023). BC, or carbon black, has been found to have the ability to decrease the rate of oxidation in rubber composites, hence prolonging their lifespan (Raut et al. 2023; Archibong et al. 2023).

8.3.8 Evaluation of Compatibility with Living Systems

In vitro studies assess the cytotoxicity of BC/rubber composites by employing conventional cell viability tests. The results often indicate that these composites are devoid of toxicity and well-suited for biological applications.

BC improves the hemocompatibility of rubber composites, rendering them appropriate for devices that come into touch with blood (Hou et al. 2023).

8.3.9 Biodegradability

Biodegradability refers to the ability of a substance to be broken down by natural processes into simpler, harmless components. BC is biodegradable, which means it can contribute to the ecologically friendly properties of the composite. This feature is especially advantageous for applications in items that are used temporarily or for a limited period of time (Hou et al. 2023).

8.3.10 Morphological Characteristics

The microstructure study demonstrates the distribution and interaction of BC particles inside the rubber matrix. Optimal dispersion results in enhanced mechanical characteristics. The contact between BC fibers and the rubber matrix plays a key role in determining the overall performance of the composites (Lu et al. 2023).

8.3.11 Electrical Characteristics

BC, being a non-conductive substance, means that the electrical characteristics of BC/rubber composites are mostly determined by the rubber matrix and the presence of any conductive fillers (Shiva 2024). By including conductive additives, the composite may be customized to suit certain electrical applications (Wloch et al. 2023).

8.4 BC Nanocomposites

Polymers that are both optically transparent and flexible have garnered considerable attention, thanks to the booming electronic industries. On the other hand, flexible polymers have significant coefficient of thermal expansion (CTE), typically around $200 \times 10^{-6}/K^{-1}$). Due to the difference in CTE, any functional material coated onto the substrates could potentially be destroyed. Therefore, it is crucial to create innovative materials that are both flexible and lightweight, while also having a CTE similar to glass (around $8 \times 10^{-6}/K^{-1}$) while maintaining optical transparency. Utilizing BC to strengthen polymers and create nanocomposites is a promising direction in bio nanocomposites. A low CTE of BC approximately $0.1 \times 10^{-6}/K^{-1}$ and a high Youngs modulus of 114 GPa (Gindl and Keckes 2004). In a groundbreaking study, Yano et al. (2005) showcased the fabrication of optically transparent BC/acrylic resin nanocomposites. The authors varied the BC fiber fractions up to 70 wt% in resin matrix. The resulting materials are optically transparent due to their small dimensions compared to the wavelength of radiation and the matching refractive index of the materials. This prevents light scattering in these nanocomposites (Novak 1993). What's even more impressive is that the material has an incredibly low CTE of just $(3–6) \times 10^{-6}/K^{-1}$ with Youngs modulus of 21 GPa and a tensile strength of 325 MPa.

8.5 Graded/Hierarchical BC Nanocomposites

Traditional natural fibers are the choice of materials for more than a decade for various composites, but they often face challenges of shape uniformity and variable mechanical strength (Nogi et al. 2006; Mohanty et al. 2000). In addition, the use of natural fibers as the reinforcing phase is undesirable due to the inconsistent polymer composite properties (Bismarck et al. 2005; Cichocki and Thomason 2002; Baley et al. 2006). One significant factor is the inherent anisotropic properties of natural fibers, high linear CTE, which leads to a delamination of the matrix interface (Thomason 2009).

Depositing BC over natural fibers has been shown to have a promising effect when dealing with thermal processing of composites. BC coated natural fiber has the potential to bridge the CTE mismatch of the fibers and the matrix. According to the previous studies, through the cultivation of cellulose-releasing bacteria with natural fibers in a suitable culture, BC can be selectively deposited onto the fiber surfaces (Cichocki and Thomason 2002; Baley et al. 2006). Mass gain experiments revealed that around 5–6 wt% of BC was coated on the fiber surfaces. Additionally, the mechanical performance of the natural fibers after BC coating are impacted by

Table 8.3 The mechanical performance of natural fibers enhanced by BC (Thomason 2009; Juntaro 2009)

Natural fiber	Tensile strength (MPa)	Fracture strain (%)
BC/sisal	342	2.9
Neat hemp	324	4.5
BC/hemp	171	2.9

the specific type of natural fiber employed. As demonstrated in Table 8.3, the coating process had no impact on the mechanical characteristics of sisal fibers, it had a major effect on the mechanical strength of hemp fibers. Such BC-coated sisal and hemp fibers are also utilized in the synthesis of monodirectional natural fiber/CAB and fiber/polylactides (PLLA) (Baley et al. 2006).

The study also examined the tensile characteristics of composites made from randomly oriented short sisal fibers reinforced with BC-coated PLLA (Jain and Pradhan 2016). Two distinct categories of graded composites were fabricated: (a) BC/sisal/PLLA and (b) BC/sisal/PLLA/BC composites. The initial material consisted of only BC deposited sisal fibers, while the subsequent composites had BC coated fibers over the PLLA material. The tensile moduli of these composites increased when sisal fibers, coated with BC. This is believed to be a result of the matrix being rigid owing to the presence of BC (Lee et al. 2009). By dispersing BC throughout the matrix as well as the fibers, both the matrix and the fiber bonding strength can be enhanced. The tensile performance of the graded composites exhibited a marginally distinct pattern in comparison to the tensile modulus. A reduction in the ability of PLLA to withstand stretching forces was noted when sisal fibers, coated with BC, were used as reinforcement, without any BC particles distributed throughout the material. The tensile characteristics of the BC/PLLA system increased by 11% over pure PLLA and 21% over BC/sisal/PLLA free from BC in the matrix. This process may be related to the improved bonding at the interface of the fiber and matrices (Fig. 8.2).

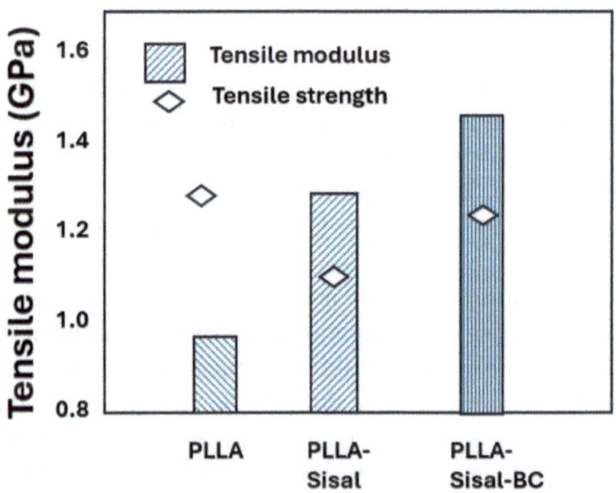

Fig. 8.2 Tensile properties of hierarchical BC composites (Juntaro et al. 2007)

8.5.1 BC Fiber Preforms for Composites

Hornification is a physical phenomenon that occurs during the production of paper. It involves the irreversible creation of hydrogen bonds as wood pulp, which is mostly composed of cellulose, dries. When this cellulose, which has been transformed into a horn-like structure, is mixed with water again, it will not return to its former state of being well-dispersed in water. Likewise, BC, composed only of cellulose, will likewise experience hornification, resulting in the synthesis of papers with exceptional stiffness and strength (Juntaro et al. 2007). The phenomenon of hornification may be integrated with the old slurry immersion technique to produce non-woven fiber preforms (Pommet et al. 2008). Apart from the immersion process, the sisal/BC mixture was subjected to vacuum filtration, pressing, and drying to create firm fiber preforms. These preforms utilize BC as a binding agent for the manufacturing of composite materials. A tensile strength of around 13 kN m^{-1} can be reached by using 10 wt%BC/sisal fiber binding. Nevertheless, the tensile performance of pristine sisal preforms free from BC was not determined due to weaker bonding of sisal fibers. The enhanced mechanical properties of BC/sisal preform may be ascribed to the BC acting as the adhesive which facilitates the transmission of stress between fibers (Juntaro 2009). The mechanical properties of the created BC/sisal preform was achieved by the high mechanical strength of the fiber network between sisal fibers.

In another study, acrylated epoxidised soya oil (AESO) was used to infuse the natural fiber preforms, which were then polymerized to create BC/sisal fiber composites (Lee et al. 2012a; Diniz et al. 2004). The fiber fraction of sisal/polyAESO and BC/sisal/polyAESO was fixed at 4%. The use of sisal fibers in polyAESO resulted in a substantial enhancement of the tensile mod from 400 MPa in pristine polyAESO to 3200 MPa in polyAESO composites reinforced with 4% sisal fibers. The use of BC as the binding agent for the natural fiber preform led to a considerable enhancement in the tensile mod from 3200 to 5600 MPa. This phenomenon is believed to occur because of the increased rigidity of polymer matrix due to a densified network of BC. The notable enhancement observed in hierarchical composites created with BC/sisal preform may be due to two factors: the improved fiber–matrix interfacial contact, and the greater transmission of stress between fibers. The BC acting as a binding agent for the fibers led to the creation of a continuous and dense BC web, which enveloped the sisal web while binding them together. It is inferred that this improves the transmission of stress between fibers in comparison to preforms made exclusively of sisal fibers, where the fibers are generally separate. Furthermore, research has demonstrated that the utilization of BC as a binding agent significantly improves the mechanical characteristics of BC/sisal fiber preform. This results in an enhanced tensile modulus of BC/sisal/polyAESO composite.

8.5.2 Self-reinforced Cellulosic Nanocomposites

Recent developments in polymer science have specifically emphasized the engineering of designs using only one kind of material, resulting in the creation of polymer composites which are self-reinforced. The fundamental concept behind self-reinforced polymers is the compaction process (Lee et al. 2012a, b, c; Diniz et al. 2004). This involves melting a portion of the fiber surface at a low contact pressure and then consolidating the material at a greater stress for a short instant of time to solidify (Hine et al. 1993). When the material cools down, the polymer undergoes recrystallization and forms a strong bond with the fibers, like how resin works in a fiber-reinforced composite system. Composites made entirely of polyethylene with a composition of 56–60% and entirely of polypropylene with a composition of 60% have been successfully created and have achieved significant commercial success. Self-reinforced cellulose composites may be made by applying a similar principle. However, it is important to mention that cellulose does not have melting temperatures (T_m) like thermoplastics do, as it breaks down before reaching its glass transition temperature (T_g). There are two possible methods: (i) saturating cellulose fibers in cellulose matrix, then regenerating it, and (ii) selectively dissolving the surface of the cellulosic fibers to join them together (Eichhorn et al. 2010). For approach (i), the method proposed by Nishino and his co-workers (Kabeel et al. 1994, 1995) shows the etching of Kraft pulp in 8 wt% LiCl+ dimethylacetamide (DMA). Next, the solution was added to ramie fibers followed by solvent evaporation and air-drying at ambient temperature. This results in the formation of a biocomposite where both the matrix as well as the fiber consist of cellulose. Additionally, Nishino et al. also showed approach (ii) by immersing filter papers successively in deionised water, acetone, and DMA. The sheets were then submerged in 8 wt% LiCl in DMA to selectively etch out the fibers followed by evaporation and drying. This notion bears a resemblance to the self-reinforced polymer that was mentioned before (Olley et al. 1993; Ward and Hine 1997; Nishino et al. 2004; Nishino and Arimoto 2007; Soykeabkaew et al. 2009). The cellulose fiber surface undergoes selective breakdown, resulting in the formation of a matrix by cellulose regeneration. Meanwhile, the fiber core is preserved and can be used as reinforced phase for the new cellulosic matrix. Such cellulose only nanocomposite may be made utilizing BC and new cellulose fibers, following a similar method.

The results indicate that a dissolving time of 10 min is the most favorable, resulting in the greatest tensile strength of 411 MPa and a satisfactory Youngs mod of 18,000 MPa. The optimal point refers to the ideal level of BC dissolution that allows the fibers to join without causing any disruption to the fiber network (Olley et al. 1993). During a short etching time, the BC surface was not fully etched to create the cellulosic matrix. However, over a prolonged dissolving interval, the crystals of BC completely dissolve, leading to a reduction in mechanical characteristics of these composites.

In comparison to all cellulose nanoscale composites to all-cellulose composites manufactured using membrane filters, the nanocomposites have greater tensile

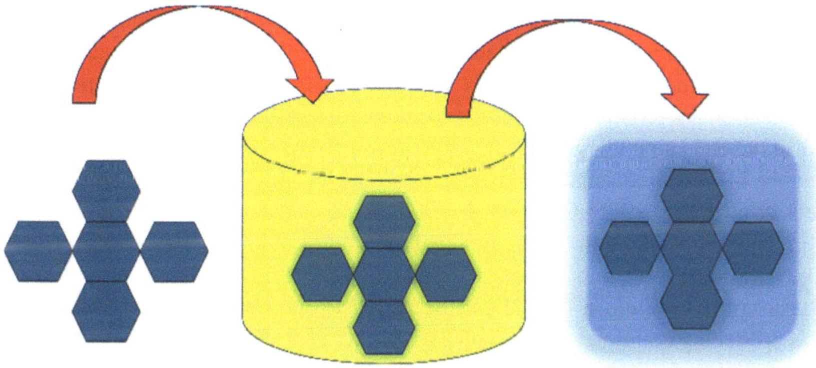

Fig. 8.3 Selective etching of cellulose fiber to prepare self-reinforced cellulose nanocomposites

modulii at a shorter dissolving time than that of the composite matrix. The presence of a rigid BC web in the nanocomposites is responsible for this effect. Conversely, there is a notable disparity in the tensile strength trend is also observed (Fig. 8.2). The nanocomposite tensile strength falls as the etching time grows. However, the composite tensile strength improves to 50–250 MPa as the dissolution time increases. Filter membranes consist of a loosely connected network of fibers with significant empty spaces. It is concluded that the dissolving process resulted in the full occupancy of these empty spaces and enhanced the tensile performance of these nanocomposites (Nishino et al. 2004) (Fig. 8.2 and 8.3).

8.5.3 *Biomimetic BC-Reinforced Nanocomposites*

Nature can serve as a source of inspiration for the preparation of BC/cellulose nanocomposites with exceptional performance. Pulpwood is a biocomposite composed of cellulose mass, hemicellulose, lignin web, and pectic polysachhrides. Cellulose functions as the strengthening component, whereas lignin forms the outer structure. Hemicellulose protects the cellulose within the cell wall and provides an adhesive link between cellulosic space and lignin polymer (Abbott and Bismarck 2010). This arrangement imparts pulpwood with its mechanical stiffness. Nevertheless, even when we comprehend the composition of a woody structure, the task of converting it into engineering materials continues to be a formidable obstacle. The mechanical performance of biomimetic cellulose coatings is sometimes compromised due to the propensity of cellulose to agglomerate. Additionally, the expense associated with extracting nanocellulose from wood materials by high energy fibrillation or homogenization procedures. This, in turn, hampers the advancement of biomimetic nanocellulose coatings. Recent findings indicate that nanocellulose derived from woody sources exhibits comparable reinforcement capabilities in composite structures to BC (Pommet et al. 2008). It is unexpected because the nanocellulose kinds have significantly varying levels of crystallinity and predicted mechanical characteristics. In contrast to nanoscale cellulose derived from wood, the fibrils of BC can undergo modification during biosynthesis, allowing to produce genuinely bio-inspired engineered materials (Lee et al. 2012c; Hine et al. 1993; Kabeel et al. 1994, 1995; Olley et al. 1993; Ward and Hine 1997; Nishino et al. 2004; Nishino and Arimoto 2007; Soykeabkaew et al. 2009; Abbott and Bismarck 2010; Huber et al. 2012a, b, c).

Hydroxyethyl cellulose (HEC) can be added to the culture when Acetobacter aceti is producing cellulose via biosynthesis (Huber et al. 2012c; Kretschmann 2003; Seifert et al. 2004; Tokoh et al. 2002; Yamamoto et al. 1996; Haigler et al. 1982; Zhou et al. 2009). After incubating for a period of 7 days, the HEC (hydroxyethyl cellulose) including BC (bacterial cellulose) was separated, purified using a weak alkaline solution, and compressed to produce a composite material reinforced with BC and HEC. This nanocomposite is biomimetic and consists of 80% by weight of BC. The structural analysis revealed the widening of the 110, $1\bar{1}0$, and 200 peaks, which suggests a reduction in the size of the crystallites. These findings indicate that the introduction of HEC in the culture media leads to a partial aggregation of cellulosic fluid. HEC is hypothesized to decrease the level of co-crystallization of individual cellulose fibers. It is seen that the BC/HEC composites had higher tensile modulus and strength with 80% BC fraction. The exceptional tensile capabilities of BC/HEC are related to the enhanced dispersion of BC fibrils inside the matrix, which is achieved by encapsulating individual nanoscale BC fibrils via HEC.

Polyvinyl alcohol (PVA), which is another type of water-soluble polymer, has also been investigated. PVA has garnered significant attention because of its ability to dissolve in water, favorable mechanical characteristics, and compatibility with living organisms. Gea et al. (2010), conducted a study where they cultivated BC in the presence of PVA and compared it to BC that had been infused with PVA. Following the purification process, the BC loaded PVA had 96.3 wt%, whereas the BC infused PVA had 98.6 wt% loading of BC. The tensile mod of BC loaded PVA was lower as compared to that of BC infused PVA due to the reduced loading volume in the matrix made in situ. The tensile performance of BC loaded PVA was close to 210 MPa, which is closer to that of the pristine BC, compared to the 110 MPa of BC infused PVA. This can be attributable to the factors outlined before. Significantly, the researchers discovered that BC infused PVA formed at its original location exhibits additional properties like PVA such as plastic deformation, followed by a work hardening modulus of 14,700 MPa. It is inferred that the co-culturing of BC and PVA resulted in enhanced deposition of individual fibres to some extent. The breakdown of the H-bonding between BC caused softening of the BC web.

8.5.4 Macroporous BC/Polymer Nanoscale Composites

BC has several applications, including its usage as a thickening and stabilizing agent in the manufacturing of polymeric foam. Blomfeldt and his co-workers (Blomfeldt et al. 2010, 2011) included 1 wt% of BC added to foam made from wheat gluten. The foam was created by dispersing gluten and water followed by cooling the mixture at −25 °C for a duration of 6 h, and finally undergoing freeze-drying. The addition of BC serves to scale up the solution viscosity and minimize fluctuations in foam density. Furthermore, the BC serves as a reinforced phase, augmenting the mechanical strength of the nanofoam composites. BC considerably enhanced the bulk mod up to 50% (from 2.1 MPa for the foam without BC to 3.1 MPa for the foam with 1 wt% BC) (Blomfeldt et al. 2010, 2011; Lee et al. 2011). BC was employed concomitantly as a thickener and reinforced phase for macroporous polymers derived from soybean oil (Lee et al. 2011). Air was introduced into the system by physically foaming the viscous AESO, resulting in the production of oil foams. The foam template was then polymerized with the use of microwaves. When BC was added to AESO at a minimum fraction of 0.5 wt%, the stability index of the oil foam showed a considerable increase. No visible separation of phases was seen for a period of 2 months after frothing AESO.

The BC free AESO exhibited a high degree of phase separation around one month after it was prepared. It is hypothesized that BC (bubble clusters) in the liquid foam gather at the edges of the foam structure, impeding the movement of liquid between the layers of the foam. This finally obstructs the merging of the froth in the fluid foam. The reduction in the modulus can be related to the inadequate macroporosity of the polymer that has a sufficient BC fraction. This poor pore structure is a direct consequence of the challenges in creating froth in AESO (Lee et al. 2011).

8.5.5 Carbon Nanofillers in Rubber

Carbon black (CB) is the predominant filler material used in natural rubber composites. Nevertheless, a substantial quantity of carbon black is essential to achieve the intended mechano-chemical and conducting characteristics of the composite materials. Currently, creative research is mostly centered around alternative carbon structures such as carbon nanotubes (CNTs), graphite, graphene and graphene oxide (GO). These carbon entities have tubular, lamellar, or planar morphology, have gained significant interest as fillers for rubbery compounds due to their inherent mechano-electrical characteristics (Galimberti et al. 2014; Sadasivunia et al. 2014; Ponnamma et al. 2014a). The high aspect ratio of these conducting fillers is particularly noteworthy. When the fillers are optimally dispersed in the rubber matrix, they can achieve the percolation threshold at extremely low filler fraction. However, the alignment and adhesion between the rubber and polymer molecules are also important factors in reinforcing rubber matrices.

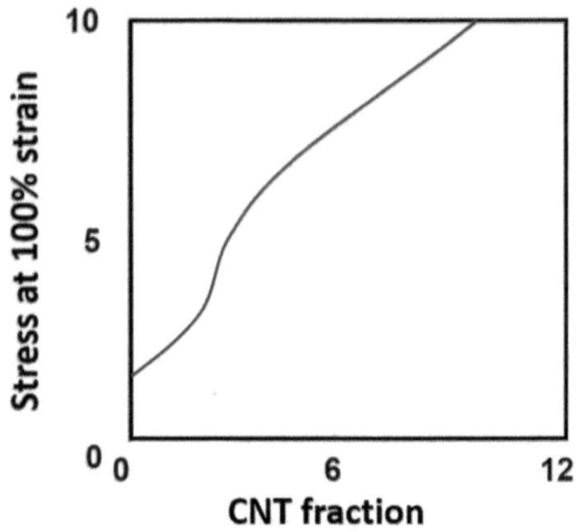

Fig. 8.4 Stress at 100% strain for rubber nanocomposites embedded with CNTs (Bokobza 2012)

CNTs have attracted significant attention due to their distinctive features resulting from their one-dimensional nature, among other carbon-based nanofillers. Significant enhancements in rigidity have been documented with the use of multiwall CNTs in rubber (Fakhru'l-Razi et al. 2006; Bhattacharyya et al. 2008; Bokobza 2012). Regarding the pure polymer, the addition of 1 part per hundred (phr) of multi-walled CNTs to natural rubber via the sonication method results in a 65% enhancement in stress when subjected to a strain of 100%. At a concentration of 10 phr, there is a significant rise of over 700% (Fig. 8.4). Additionally, the point at which electrical percolation occurs is achieved at a concentration of less than 1 phr (Bokobza 2012). However, the CNT's capacity to reach its maximum potential is significantly restricted owing to the existence of bundles caused by the van der Waals interaction between the tubes. The nanotube loading is responsible for reducing the strain at rupture using these CNT bundles.

Varying degrees of improved diverse electrical conductivities are achieved through the use of different processing methods in the production of nanocomposites (Bokobza and Kolodziej 2006; Kolodziej et al. 2007). According to Maa et al. (2010), the addition of CNTs in a rubber matrix is more challenging compared to other reinforcements due to their large surface area and aspect ratios. The research discusses several approaches such as ultrasonic treatments, mixing, milling, and extrusion, which aim to distribute the CNTs into a rubber matrix. An alternative method to achieve the separation and even distribution of CNTs in matrix materials involves the utilization of surfactant molecules that adhere to the surface of the tubes through non-covalent interactions (Vaisman et al. 2006; Tkalya et al. 2012). Unlike the functionalizing of CNTs by chemical bonding, this approach doesn't degrade the electrical characteristics of the CNTs. Ponnamma et al. (2014b) conducted a study to examine how the rheology of natural rubber latex nanocomposites is affected by the non-covalent functionalization of CNTs using different surfactants. The study demonstrated that the anionic sodium dodecyl sulphate achieved the most effective dispersion which is attributed to the establishment of a 3D network. It is worthy to note that in traditional composites, a high level of filler dispersion is noted. Conversely, anisotropic fillers that can form a connected network at very low filler fraction will show the opposite behavior if they are uniformly dispersed in the rubber. According to Galimberti et al. (2018), the significant hysteresis observed in CNTs might be harmful for dynamic loading applications. The dispersion of CNTs in elastomers has been enhanced by introducing an additional form of filler, such as CB, clays, or reduced graphene oxide. This is due to the synergistic effects that occur between the two distinct filler morphology. The use of dual filling results in enhanced mechano-electrical characteristics, surpassing the percolation threshold achieved by composites filled with a single filler (Bokobza et al. 2008; Zhan et al. 2011; Rooj et al. 2015; Ponnamma et al. 2013).

Weng et al. (2010) demonstrated that the presence of multi-walled CNTs in natural rubber results in increased crystallinity and a lower temperature at which crystallization begins, as shown using synchrotron wide-angle X-ray diffraction. Additional information may be found in the cited reference (Jasse and Koenig 1979).

Graphene consists of a solitary layer of atomic carbon that are linked together in a sp2 configuration and arranged in a two-dimensional honeycomb pattern. Recently, it has gained significant attention as a nanofiller due of its extraordinary features, including its superb modulus and astonishingly high electrical and thermal conductivities. The unique characteristics inherent in the single-layer structure of these materials, surpassing those of existing carbon nanomaterials, have paved the way for the development of ultra-lightweight, low-cost advanced composite (Kuilla et al. 2010; Singh et al. 2011). However, similar to other nanomaterials, the dispersion state of graphene plays a vital role in determining the physical properties of the elastomers. The primary obstacle in harnessing the distinctive characteristics of individual graphene sheets is preventing the reassembly of the monolayers through van der Waals contacts during the production of graphene-filled composites.

Various methods have been employed to acquire graphene nanosheets, but the most effective approach for producing a significant quantity of monolayers or graphene nanoflakes formed by multiple stacked graphene sheets appears to be the graphite oxide route. This method, as extensively explained by Kuilla et al. (2010), involves oxidation of graphite, followed by exfoliation and reduction to obtain graphene. However, this procedure, designed to produce significant quantities of monolayers, creates structural flaws that interfere with the electrical structure of graphene, as noted by Singh et al. (2011). Raman spectroscopy, a very effective tool for analyzing carbon-based materials, definitively detects the existence of defects (Bokobza et al. 2015, 2017; Stankovich et al. 2007; Eda et al. 2008).

8.6 Carbon Nanofillers/Rubber Nanocomposites

The elastic modulus of rubber materials quantifies the amount of recoverable strain in a sheared sample. On the other hand, the loss factor is associated with the energy dissipated as heat during the process of energy damping. Consequently, an investigation has been conducted on the dynamic mechanical characteristics of several rubber materials that incorporate carbon nanofillers. The storage modulus, loss modulus, and T_g exhibited an increase in the case of all multi-walled CNTs reinforced natural rubber (Tarawneh et al. 2013). The use of CNT at a minimal concentration significantly improves the storage modulus of natural rubber blend nanocomposites produced by latex compounding, as demonstrated in a separate study (Boonmahitthisud and Chuayjuljit 2012).

Boonmahitthisud and Chuayjuljit (2012) examined the changes in storage modulus and loss tan δ with temperature for plain natural rubber blends, as well as nanocomposites including different amounts of CNT. The results showed that the addition of CNT to rubber blends led to an increase in the E' (elastic modulus) in a dose-dependent manner. This increase was attributed to the high stiffness of the CNT, which restricts the movement of the rubber chains. In addition, the T_g of each rubber in these nanocomposites rose as the loading of CNT increased. Thomas et al. demonstrated that the T_g of NR/phenol functionalized CNT nanocomposites was increased, moving to a higher temperature (Thomas et al. 2012). The loss angle exhibited a reduction in the presence of CNT, with a more noticeable impact observed when using phenol-functionalized CNT. Sui et al. (2008) found that the elastic modulus of the CNT/rubber nanocomposites is much higher than that of pure rubber and CB/rubber composites. The T_g of natural rubber/ethylene propylene nanocomposites gradually increases when the amount of multiwalled CNT is increased (Jahed et al. 2016). Xu et al. (2016) conducted dynamic mechanical tests on composites of natural rubber with CB, as well as with CB and multi-walled CNTs at various CNT loadings. The thermoplastic rubber blend filled with acidified CNTs exhibited the maximum storage modulus at −90 °C, whereas the rubber containing untreated CNTs had the lowest storage modulus (Boonmahitthisud and Chuayjuljit 2012). This is likely because the acid treated CNTs are evenly distributed in the TPNR matrix. Bhattacharya et al. found that rubber nanocomposites exhibited better dynamic mechanical characteristics compared to clay-based nanocomposites (Bhattacharya et al. 2010). The dynamic characteristics of rubber/graphene composites were enhanced by direct mechanical mixing (Al-Hartomy et al. 2015). The elastic modulus of natural rubber at ambient temperature rises, whereas the maximum loss tangent and the related T_g of composites drop as the quantity of multi-walled CNT increases (Zhan et al. 2011). The inclusion of graphene oxide (GO) in the rubber/GO nanocomposites led to a significant enhancement in the storage modulus. This suggests that GO effectively reinforced the NR material (Cecen et al. 2017). The modulus at 300% for rubber composites with 0.9 phr graphene was raised by 154.9% compared to pure rubber, according to reference (Wu et al. 2015). The enhancements in the dynamic mechanical characteristics were attained with a low level of replacement of GO or decreased graphene nanosheets for CB in natural rubber.

The rolling resistance characteristics of styrene-butadiene rubber (SBR) are highly advantageous for their use, particularly in the tire industry. Therefore, the investigation of the mechanical properties of SBR reinforced with carbon-based nanoparticles has garnered interest (Srivastava and Bhuyan 2018). Peddini et al. (2014) conducted an experiment to assess the storage and loss modulii, and tan δ as a function of temperature for several dilutions using SBR. It was observed that the storage and loss modulii both rise when CNT is added to the composite. The T_g of the cured plaques, as measured by dynamic mechanical thermal analysis (DMA), increases by 4.5 °C when the SBR-12.3 wt% CNT, is diluted. Pedroni et al. (2012) conducted a study where they examined the DMA results of SBR/CNTs composites produced using the solution casting and melting. The storage modulus of SBR rises proportionally with the quantity of filler used, regardless of the process employed. This increase is mostly attributed to the reinforcing action of the filler. However, the impact of this effect is more pronounced in composites created using the casting method. The T_g of the polybutadiene in the composites, which were manufactured using the casting process, changes from −88 to −80 °C when the content of multi-walled CNTs increases. However, the composites formed via extrusion did not show any notable alteration in the T_g of the PB blocks. The incorporation of CNTs into a blend of SBR and butadiene rubber in a 50:50 ratio has an impact on the glass transition behavior, as reported in reference (Das et al. 2012). Adohi et al. discovered that the storage modulus of the SBR/CB and CNTs composites is almost identical to that of pure SBR when the frequency is less than 0.1 (Adohi et al. 2010). The modified graphite/ SBR nanocomposites exhibited enhanced DMA characteristics compared to the control samples (Malas et al. 2014). Zhang et al. conducted a study on the dynamic mechanical characteristics of SBR/CB and SBR/CB-RG hybrid composites. They found that the T_g increased marginally when the amount of CB added to SBR increased from 10 to 13 phr (Zhang et al. 2015). The T_g moved to a higher temperature with equal total filler loading in SBR SBR/CB-RG (100/10:3) as compared to that had a greater T_g compared to SBR/CB (100/13). Tang and colleagues observed that the combination of tubular clay (HNT) and tannic acid/ graphene hybrid at a ratio of 22 phr exhibited a greater modulus compared to using 20 phr of HNT alone or 2 phr of TAG alone. This finding was reported in reference (Tang et al. 2013). The addition of liquid polyisoprene and MgO had a significant impact on the DMA characteristics of the SBR/CNT vulcanizate. Specifically, it led to an increase in the elastic modulus, T_g, and tan δ at 0 °C (Wang and Zhang 2018).

Several researchers have investigated the mechanical characteristics of various nanocomposites made from natural rubber and carbon fillers. Tarawneh et al. (2011) examined the strength characteristics of rubber composites that were strengthened by multiwalled CNTs. The study revealed that incorporating 3 wt% of CNTs into rubber led to a significant enhancement of approximately 39% and 30% in strength and elastic modulus, respectively. Furthermore, it was shown that the elongation at break reduced as the fraction of CNTs increased. The highest impact strength was seen at a concentration of 5 wt% of CNTs, resulting in a 74% increase compared to a pure thermoplastic natural rubber. Ponnamma and colleagues (2015) presented stress–strain data for both pure natural rubber and CNT composites. Rubber demonstrates a significant rise in stress at greater deformations because of work-hardening. Conversely, the strain at which rupture occurs is decreased for its CNT composites. The mechanical data of rubber/CNT composites indicates that the elastic modulus of rubber increases (~0.50 MPa) as the CNT fraction rise. The greatest value of ~1.10 MPa was observed when the maximal loading of CNTs reached 5 wt%. This phenomenon can be related to the unevenness of the filler structures, its ability to promote nucleation, and the dispersion of the filler as well as the interactions between the filler and the polymer. Additionally, they noted that the greatest enhancement in tensile strength occurred when the loading of CNTs in natural rubber reached around 1 wt%. However, the addition of CNT in rubber did not result in any increase in elongation at break. Tarawneh et al. (2013) investigated the impact of sonication on the mechanical characteristics of rubber composites that were strengthened by the addition of 1 wt% of CNT. After an appropriate sonication period of 1 h, the elastic modulus, tensile strength, elongation at break, and impact strength of the material rose by approximately 11, 21, 43, and 50% correspondingly, compared to its original state. This improvement may be attributed to the successful dispersion obtained.

Nakaramontri et al. (2015) applied ex-situ and in-situ surface functionalized CNTs as a filler in natural rubber and investigated their mechanical characteristics. The inclusion of CNTs led to a doubling of the modulus and a reduction in elongation, mostly owing to the reinforcing impact of multi-walled CNTs. The tensile strength of the rubber reduced when either raw CNTs or ex-situ functionalized CNTs were added. On the other hand, the tensile strength of in-situ functionalization was shown to be greater than that of matrix. The elongation values of natural rubber composites filled with either raw or ex situ-functionalized multi-walled CNTs were determined to be lower than those of matrices. Ali and Ahmad conducted a study on the mechanical characteristics of polylactic acid (PLA)/rubber that was filled with CNTs. The PLA/rubber/CNT composite was created using the melt blending process. The findings indicate that the PLA/rubber nanocomposite, including 3.5 wt% of CNTs, demonstrated a 22% increase in strength and a 20% increase in modulus of elasticity. These enhancements might be focused on achieving a uniform dispersion of CNTs throughout the PLA/rubber matrix. The experiments also revealed a reduction in the elongation of PLA/rubber as the number of CNTs increased. The impact strength of PLA/rubber/CNTs (3.5 wt%) rose by approximately 42% because of the improved dispersion of CNTs in the matrix, which led to a considerable toughening effect. Anand et al. (2009) reported a notable improvement in the mechanical characteristics of rubber/CNTs nanocomposites synthesized using a latex stage mixing approach. The addition of 2.0 phr of rubber/CNTs resulted in a significant increase in both tensile strength and tensile modulus compared to pure rubber with CNTs. Specifically, the tensile strength was 56% higher and the tensile modulus was 63% higher.

8.6.1 Uses of BC/Rubber Composites

BC/rubber composites have the mechanical strength and ability to interact with living organisms of BC, while also exhibiting the stretchability and adaptability of rubber. The combination of these elements creates a composite material that has a diverse variety of possible uses in several industries, such as biomedical, automotive, electronics, and consumer products.

Applications in the Field of Biomedicine

BC's exceptional water retention capacity and biocompatibility make it highly suitable for wound dressings, since it creates a moist environment that promotes healing. By using rubber, the dressings become more flexible, allowing them to adhere better to the shape of the body. BC/rubber composites act as supportive structures for tissue engineering, facilitating the attachment and growth of cells. BC exhibits robust mechanical strength, which facilitates tissue regeneration, while rubber's flexibility closely imitates the inherent characteristics of soft tissues. Vascular grafts: BC/rubber composites possess the necessary flexibility and biocompatibility to be well-suited for use in vascular grafts. By manipulating the material, it is possible to adjust its mechanical characteristics to closely resemble those of normal blood arteries, hence minimizing the likelihood of difficulties (Mukhopadyay and De 1979).

BC/rubber composites have the potential to be utilized for the purpose of controlled medication delivery. BC's porous structure enables drug loading, while rubber's elasticity facilitates controlled deformation and release kinetics (Das et al. 2024). The biocompatibility and mechanical features of these composites offer benefits for implanted drug delivery devices that need both flexibility and durability (Azka et al. 2024). BC/rubber composites provide exceptional strength and flexibility, making them suitable for application in bone implants. The material possesses the ability to facilitate bone formation while also effectively absorbing impact, therefore imitating the characteristics of genuine bone (Kumar et al. 2024).

Applications in the Automotive and Aerospace Industries

BC, when added to rubber, improves the mechanical characteristics of tires, such as tensile strength and resistance to wear. As a result, the tires have increased durability and enhanced performance. Vibration Damping: BC/rubber composites exhibit enhanced stiffness and damping qualities, making them highly useful for applications involving vibration dampening, such as engine mounts and suspension bushings (Guajardo and Andler 2024).

Seals and Gaskets

These components used to prevent the leakage of fluids or gases in a system.

BC/rubber composites possess improved mechanical and thermal stability, rendering them appropriate for use in automotive and aerospace applications as seals and gaskets. These materials have exceptional durability, enduring challenging circumstances such as severe temperatures and chemical exposure. The enhanced tear resistance and suppleness contribute to the preservation of a secure seal, hence reducing leakage in diverse mechanical systems (Low and Tang 2024).

Low-Weight Components

Weight Loss: By using BC, the weight of rubber components may be reduced without sacrificing strength, hence enhancing fuel efficiency in cars and airplanes. BC/rubber composites offer the essential structural integrity for lightweight components utilized in the automotive and aerospace sectors (Zhang et al. 2024b).

Electronics and Wearable Devices

The combination of flexibility and durability makes BC/rubber composites ideal for use in flexible electrical circuits. These materials exhibit high resilience to repeated bending and stretching (Nargatti et al. 2024). BC/rubber composites are very suitable for wearable sensors that measure physiological characteristics including heartbeat, body temperature, and movement because of their biocompatibility and flexibility (Feng et al. 2024).

Conductive Composites

It refers to materials that have the ability to conduct electricity. By integrating conductive fillers like carbon nanotubes or graphene into BC/rubber composites, it is possible to enhance their electrical conductivity, hence broadening their potential use in flexible and elastic electronic devices. Touchscreens and displays: The favorable properties of BC/rubber composites, such as transparency and flexibility, make them suitable for touchscreens and flexible display technologies (Ye et al. 2024).

Consumer and Sports

BC/rubber composites provide exceptional strength and flexibility, rendering them very ideal for sports equipment, including protective gear, balls, and athletic shoes. These composites enhance performance and safety to a significant extent. Improved durability guarantees the longevity of sporting equipment, enabling it to endure demanding usage (Adak 2024). BC/rubber composites are suitable for enhancing the cushioning, flexibility, and durability of shoe soles. The material's capacity to adapt to the contour of the foot promotes comfort. BC's biodegradability renders these composites a sustainable choice for manufacturing footwear (Pandit and Avachat 2024).

Household Things

Versatile and long-lasting products: Gloves, mats, and cushions can experience greater durability and comfort by utilizing BC/rubber composites, which possess enhanced characteristics. BC/rubber composites include biocompatibility and chemical resistance, making them appropriate for food-contact goods such as kitchen utensils and storage containers (Vasanth Kumar et al. 2024).

Applications in Industry

BC/rubber composites are suitable for incorporation into conveyor belts to enhance tensile strength and flexibility, hence improving efficiency and durability in industrial applications. Improved durability results in extended lifespan and less maintenance expenses (Kalaee et al. 2024).

Coatings for Protection

BC/rubber composites have the ability to serve as protective coverings for machinery and structures, effectively preventing corrosion and chemical damage. BC/rubber composites provide thermal qualities that can enhance insulation for equipment and pipes, hence enhancing energy efficiency (Sperotto et al. 2021).

Materials Used for Sealing and Insulation

BC/rubber composites are highly useful for sealing applications in a wide range of industrial settings because of their enhanced flexibility and durability. Thermal and acoustic insulation refers to materials or techniques used to prevent the transfer of heat and sound. These composites have the ability to be utilized for thermal and acoustic insulation, offering defense against high temperatures and noise in industrial settings (Shokatayeva et al. 2019).

Applications in the Field of Environmental Science and Conservation

The BC/rubber composites' biodegradability renders them an eco-friendly substitute for disposable items and packaging materials, hence mitigating plastic waste (Tang et al. 2019). These composites may be produced using sustainable methods, so helping to minimize their environmental footprint. BC/rubber composites have the potential to be utilized in the production of filtration membranes for water purification. This is due to the advantageous properties of BC's nanofiber network, which enables efficient filtration, and rubber's flexibility, which ensures the durability of the membranes (Kong et al. 2023).

8.7 Obstacles and Prospects for the Future

8.7.1 Production and Scale-Up Bacterial Cellulose Production

Expanding the production of BC is a substantial difficulty. Industrial-scale manufacturing necessitates achieving constant and high yields, which can be challenging due to the sluggish growth rate of cellulose-producing bacteria and the requirement for meticulous regulation of fermentation conditions (Blanco Parte et al. 2020).

Price: The manufacturing expenses of BC might be elevated as a result of the costly culture media and the requirement for aseptic conditions. Discovering cost-efficient and environmentally friendly methods to manufacture BC is essential for its economic feasibility (Jabbari and Babaeipour 2023).

8.7.2 Dispersion and Homogeneity

Uniform dispersion is crucial for achieving excellent composite characteristics by ensuring that BC is evenly distributed inside the rubber matrix. Inadequate dispersion can cause particles to clump together, which can create areas of weakness in the composite material (Almihyawi et al. 2024).

Interfacial adhesion is crucial for optimal stress transmission between the BC and rubber matrix. Inadequate interfacial adhesion may undermine the mechanical characteristics of the composite (Srivastava and Mishra 2018).

8.7.3 Optimization of Mechanical Properties

BC enhances the tensile strength and modulus of rubber, but it can decrease elongation at break and flexibility. Optimizing the composition of BC and the manufacturing of composites is necessary to achieve the ideal balance between strength and flexibility.

8.7.4 Techniques for Processing

Complex Processing: The processing methods for BC/rubber composites can be intricate and may necessitate specialist equipment. Precise control is necessary for achieving desirable qualities in both melt blending and in situ polymerization.

Scalability of Processing Methods: Techniques that exhibit high efficacy in a laboratory setting may encounter challenges when attempting to be scaled up to industrial proportions. It is crucial to develop processing technologies that can be easily expanded and are economical for use in commercial applications.

8.7.5 *Biodegradability and Recycling*

Regulating Biodegradability: Although biodegradability is beneficial, it must be carefully managed to prevent premature degradation of the composite materials throughout their intended usage. Creating composites with known and regulated rates of deterioration is a difficult task. Efforts should be made to develop efficient recycling methods for BC/rubber composites to enhance their sustainability and environmental friendliness.

8.8 Conclusions

Bacterial cellulose/rubber composites represent a promising class of materials with enhanced mechanical, thermal, and biocompatible properties. Their potential applications span various industries, driven by ongoing advancements in synthesis, fabrication, and functionalization techniques. Future research will likely focus on overcoming current challenges and exploring new frontiers in this dynamic field.

The application of BC in polymer composite applications has been stimulated by recent developments in nanomaterials. Evidence demonstrates that BC has the capability to enhance polymers, resulting in the production of nanocomposites that possess enhanced characteristics. This chapter highlights the many uses of BC in polymeric matter. BC may be utilized to manufacture nanocomposites that have both high modulus and strength, while yet being optically transparent. Furthermore, BC has demonstrated the ability to enhance the mechanical characteristics of traditional natural polymer composites through mechanisms like mechanical key lock mechanisms. Additionally, it can serve as a binding agent to consolidate free fibres into a dense fibre network, enhancing the strength of downstream handling in the fabrication of composites.

The nanocomposites demonstrate exceptional mechanical performance by fully using the modulus of the BC. BC/biomimetic composites may be produced, drawing inspiration from nature. These biocomposites are created by adding water-dispersible polymers to the culture while producing BC via biosynthesis. This causes the formation of fibrillated BC inside the nanoweb matrix, resembling a natural biomaterial such as pulpwood. Furthermore, BC can serve as a stabilizer and thickening agent in the production of macroporous polymer, expanding its application beyond its traditional employment in polymer biocomposites. The promises of BC are boundless and constrained solely by the scientist's creativity. BC is considered a very promising material for producing lightweight yet structurally stable renewable composites, which are expected to be used in the fabrication of the future generation of materials.

In addition to their inherent high rigidity and durability, carbon fillers offer a significant benefit over others by imparting electrical characteristics to polymers, which are often regarded as electrical insulators. The attainment of the electrical percolation threshold, which signifies the establishment of a network of fillers, occurs at minuscule concentrations of nanofillers in comparison to typical CB particles.

The addition of CNTs to the rubber matrix results in a significant increase in stiffness. However, it also leads to a substantial decrease in the strain at which rupture occurs, which is attributed to the existence of agglomerates. Currently, a significant obstacle that persists is the need to address the issue of CNTs agglomeration caused by van der Waals forces acting between the individual CNT strands.

An effective method to enhance the dispersion of CNTs is to introduce a secondary filler into the matrix, since it might potentially lead to synergistic effects resulting from the combination of two distinct filler morphologies. The dual filling strategy seeks to merge the advantages of each reinforcing agent to further augment the features of the matrix. A fascinating use of hybrid filler involves adding a tiny quantity of carbon nanotubes to silica-filled systems in order to provide electrical conductivity to the insulating silica-embedded rubber. Similar to other fillers, the degree of enhancement in the graphene composite's properties relies on the level of dispersion. However, the effectiveness of this exceptional filler is still constrained by uneven distributions. Efficiently addressing the difficulties in processing and dispersion is crucial for fully harnessing the promise of graphene.

8.9 Future Directions

8.9.1 Enhanced Production Methods

Genetic Engineering: Genetic modification of cellulose-producing bacteria could enhance the yield and quality of BC. Strains with faster growth rates and higher cellulose output could make BC production more efficient and cost-effective (Srivastava and Mishra 2018).

Alternative Feedstocks: Using alternative, low-cost feedstocks for BC production, such as agricultural waste or byproducts, could reduce production costs and improve sustainability.

8.9.2 Advanced Dispersion Techniques

Nano-Dispersion Methods: Advanced dispersion techniques, such as ultrasonic treatment or high-shear mixing, can improve the uniformity of BC dispersion in rubber matrices.

Surface Functionalization: Functionalizing BC surfaces with compatible chemical groups can enhance interfacial adhesion with rubber matrices. This can be obtained via chemical treatments or the use of coupling agents.

8.9.3 Hybrid Composites

Multi-Component Systems: Incorporating additional reinforcing agents, such as nanoparticles (e.g., carbon nanotubes, graphene), can create hybrid composites with superior properties. These multi-component systems can leverage the benefits of each component to achieve enhanced performance.

Functional Additives: Adding functional additives, such as antioxidants, UV stabilizers, or antimicrobial agents, can extend the application range of BC/rubber composites.

8.9.4 Tailoring Mechanical Properties

Optimized Formulations: Developing optimized formulations and processing conditions to achieve the desired balance of mechanical properties, such as tensile strength, elasticity, and toughness.

Hierarchical Structures: Designing composites with hierarchical structures at different scales (nano, micro, and macro) can enhance mechanical performance by improving load distribution and crack resistance.

8.9.5 Sustainable and Eco-friendly Approaches

Green Chemistry: Utilizing green chemistry principles in the synthesis and processing of BC/rubber composites can reduce environmental impact and improve sustainability.

Recycling and Reuse: Developing efficient recycling processes for BC/rubber composites to recover and reuse materials, reducing waste and promoting circular economy principles.

8.9.6 Biomedical Innovations

Smart Materials: Developing BC/rubber composites with smart properties, such as responsiveness to environmental stimuli (e.g., pH, temperature, humidity), for advanced biomedical applications.

Personalized Medicine: Using 3D printing and other advanced manufacturing techniques to create personalized medical devices and implants tailored to individual patient needs.

8.9.7 Industrial and Commercial Applications

Automotive and Aerospace: Expanding the use of BC/rubber composites in automotive and aerospace industries by optimizing properties for specific applications, such as lightweight components and high-performance parts.

Consumer Goods: Developing new consumer products, such as eco-friendly footwear, sports equipment, and household items, leveraging the unique properties of BC/rubber composites.

8.9.8 Research and Development

Interdisciplinary Research: Encouraging interdisciplinary research collaborations to explore new applications and improve the understanding of BC/rubber composite behavior.

Standardization and Testing: Establishing standardized testing methods and protocols to evaluate the performance and durability of BC/rubber composites, ensuring consistency and reliability in commercial products.

References

Abbott A, Bismarck A (2010) Self-reinforced cellulose nanocomposites. Cellulose 17(4):779–791

Abdelhamid HN, Mathew AP (2022) Cellulose-based nanomaterials advance biomedicine: a review. Int J Mol Sci 23(10):5405

Adak S (2024) Development of nanocellulose-based nanocomposites and its properties. Nanocellulose 113–140

Adohi BJ-P, Mdarhri A, Prunier C, Haidar B, Brosseau C (2010) A comparison between physical properties of carbon black-polymer and carbon nanotubes-polymer composites. J Appl Phys 108:074108

Al-Hartomy O, Al-Ghamdi A, Al Said SF, Dishovsky N, Mihaylov M, Ivanov M, Zaimova D (2015) Comparison of the dielectric thermal properties and dynamic mechanical thermal properties of natural rubber-based composites comprising multiwall carbon nanotubes and graphene nanoplatelets. Full Nanotub Carbon Nanostruct 23:1001–1007

Almihyawi RH, Musazade E, Alhussany N, Zhang S, Chen H (2024) Production and characterization of bacterial cellulose by Rhizobium sp. isolated from bean root. Sci Rep. https://doi.org/10.1038/s41598-024-61619-w

Anand KA, Jose TS, Alex R, Joseph R (2009) Natural rubber-carbon nanotube composites through Latex compounding. Int J Polym Mater 59:33–44

Archibong FN, Orakwe LC, Ogah OA, Mbam SO, Ajah SA, Okechukwu ME, Igberi CO, Okafor KJ, Chima MO, Ikelle II (2023) Emerging progress in montmorillonite rubber/polymer nanocomposites: a review. J Mater Sci 58:2396–2429

Azka MA, Adam A, Ridzuan SM, Sapuan SM, Habib A (2024) A review on the enhancement of circular economy aspects focusing on nanocellulose composites. Int J Biol Macromol 269:132052

Baley C, Perrot Y, Busnel F, Guezenoc H, Davies P (2006) Transverse tensile behaviour of unidirectional plies reinforced with flax fibres. Mater Lett 60(24):2984–2987

Bhattacharya M, Maiti M, Bhowmick AK (2010) Synergy in carbon black filled natural rubber nanocomposites. Part I: Mechanical dynamical properties, and morphology. J Mater Sci 45:6126–6138

Bhattacharyya S, Sinturel C, Bahloul O, Saboungi M-L, Thomas S, Salvetat J-P (2008) Improving reinforcement of natural rubber by networking of activated carbon nanotubes. Carbon 46:1037–1045

Binti Ismail NS, Hakhiri SMABSM, Binti Mustapha SNH, Bin a Razak AHB, Bin Ab. Rahim MohdH, Bin Zakaria S (2024) Confinement of the permittivity enhancing fillers in bacterial cellulose for dielectric elastomer applications. Solid State Phenom 355:141–152

Bismarck A, Mishra S, Lampke T (2005) In: Mohanty AK, Misra M, Drzal LT (eds) Natural fibers, biopolymers and biocomposites, 1st edn, Chap 2. CRC Press, Boca Raton

Blaker JJ, Lee KY, Bismarck A (2011a) Hierarchical composites made entirely from renewable resources. J Biobased Mater Bioenergy 5(1):1–16

Blanco Parte FG, Santoso SP, Chou C-C, Verma V, Wang H-T, Ismadji S, Cheng K-C (2020) Current progress on the production, modification, and applications of bacterial cellulose. Crit Rev Biotechnol 40:397–414

Blomfeldt TOJ, Olsson RT, Menon M, Plackett D, Johansson E, Hedenqvist MS (2010) Novel foams based on freeze-dried renewable vital wheat gluten. Macromol Mater Eng 295(9):796–801

Blomfeldt TOJ, Kuktaite R, Johansson E, Hedenqvist MS (2011) Mechanical properties and network structure of wheat gluten foams. Biomacromolecules 12(5):1707–1715

Bokobza L (2012) Multiwall carbon nanotube-filled natural rubber: electrical and mechanical properties. eXPRESS Polym Lett 6:213–223

Bokobza L, Kolodziej M (2006) On the use of carbon nanotubes as reinforcing fillers for elastomeric materials. Polym Int 55:1090–1098

Bokobza L, Rahmani M, Belin C, Bruneel J-L, El Bounia N-E (2008) Blends of carbon blacks and multiwall carbon nanotubes as reinforcing fillers for hydrocarbon rubbers. J Polym Sci Part B Polym Phys 46:1939–1951

Bokobza L, Bruneel J-L, Couzi M (2015) Raman spectra of carbon-based materials (from graphite to carbon black) and of some silicone composites. C J Carbon Res 1:77–94

Bokobza L, Couzi M, Bruneel J-L (2017) Raman spectroscopy of polymer–carbon nanomaterial composites. Rubber Chem Technol 90:37–59

Boonmahitthisud A, Chuayjuljit S (2012) Effects of carbon nanotube on tensile and dynamic mechanical properties of NR/SBR and NR/XSBR nanocomposites prepared by latex compounding. Adv Mater Res 488–489:612–616

Brown AJ (1886) The chemical action of pure cultivations of bacterium aceti. J Chem Soc Trans 49:172–187

Cecen V, Thomann R, Mülhaupt R, Friedrich C (2017) Thermal conductivity, morphology and mechanical properties for thermally reduced graphite oxide-filled ethylene vinylacetate copolymers. Polymer 132:294–305

Cichocki FR, Thomason JL (2002) Thermoelastic anisotropy of a natural fiber. Compos Sci Tech 62(5):669–678

Das A, Kasaliwal GR, Jurk R, Boldt R, Fischer D, Stöckelhuber KW, Heinrich G (2012) Rubber composites based on graphene nanoplatelets, expanded graphite, carbon nanotubes and their combination: a comparative study. Compos Sci Technol 72:1961–1967

Das M, Lalsangi S, Santra S, Banerjee R (2024) Nanocellulose as a carrier for improved drug delivery: progresses and innovation. J Drug Deliv Sci Technol 97:105743
Diniz J, Gil MH, Castro J (2004) Hornification—its origin and interpretation in woodpulps. Wood Sci Technol 37(6):489–494
Eda G, Fanchini G, Chhowalla M (2008) Large-area ultrathin films of reduced graphene oxide as a transparent and flexible electronic material. Nat Nanotechnol 3:270–274
Eichhorn SJ, Davies GR (2006) Modelling the crystalline deformation of native and regenerated cellulose. Cellulose 13(3):291–307
Eichhorn SJ, Dufresne A, Aranguren M, Marcovich NE, Capadona JR, Rowan SJ, Weder C, Thielemans W, Roman M, Renneckar S, Gindl W, Veigel S, Keckes J, Yano H, Abe K, Nogi M, Nakagaito AN, Mangalam A, Simonsen J, Benight AS, Bismarck A, Berglund LA, Peijs T (2010) Review: current international research into cellulose nanofibres and nanocomposites. J Mater Sci 45(1):1–33
Fakhru'l-Razi A, Atieh MA, Girun N, Chuah TG, El-Sadig M, Biak DRA (2006) Effect of multi-wall carbon nanotubes on the mechanical properties of natural rubber. Compos Struct 75:496–500
Feng L, Cao X, Wang ZL, Zhang L (2024) A transparent and degradable bacterial cellulose-based film for triboelectric nanogenerator: efficient biomechanical energy harvesting and human health monitoring. Nano Energy 120:109068
Fernandes SCM, Oliveira L, Freire CSR, Silvestre AJD, Neto CP, Gandini A, Desbrieres J (2009) Novel transparent nanocomposite films based on chitosan and bacterial cellulose. Green Chem 11(12):2023–2029
Galimberti M, Cipolletti V, Kumar V (2014) Nanofillers in natural rubber. In: Thomas S, Maria HJ, Joy J, Chan CH, Pothen LA (eds) Chapter 2 Natural rubber materials. Composites and nanocomposites, vol 2. Royal Society of Chemistry, Cambridge, pp 34–72
Galimberti M, Infortuna G, Guerra S, Barbera V, Agnelli S, Pandini S (2018) sp2 carbon allotropes in elastomer matrix: from master curves for the mechanical reinforcement to lightweight materials. eXPRESS Polym Lett 12:265–283
Gea S, Bilotti E, Reynolds CT, Soykeabkeaw N, Peijs T (2010) Bacterial cellulose poly(vinyl alcohol) nanocomposites prepared by an in-situ process. Mater Lett 64(8):901–904
Gindl W, Keckes J (2004) Tensile properties of cellulose acetate butyrate composites reinforced with bacterial cellulose. Compos Sci Tech 64(15):2407–2413
Gu RJ, Kokta BV, Frankenfeld K, Schlufter K (2010) Bacterial cellulose reinforced thermoplastic composites: preliminary evaluation of fabrication and performance. BioResources 5(4):2195–2207
Guajardo C, Andler R (2024) Challenges and perspectives in enzymatic polymer fragmentation: the case of rubber and polyethylene terephthalate. J Clean Prod 450:141875
Haigler CH, White AR, Brown RM, Cooper KM (1982) Alteration of in vivo cellulose ribbon assembly by carboxymethyl cellulose derivatives. J Cell Biol 94(1):64–69
Hamedi S, Mahmoodi-Barmesi M, Kermanian H, Ramezani O, Razmpour Z (2024) Investigation of physicochemical and biological properties of bacterial cellulose & zein-reinforced edible nanocomposites based on flaxseed mucilage containing Origanum vulgare L. essential oil. Int J Biol Macromol 254:127733
Harms H (2003) Lenzing LYOCELL—chances of a new generation of manmade fibres. Materialwiss Werkstofftech 34(3):267–271
Herrick FW, Casebier RL, Hamilton JK, Sandberg KR (1983) Microfibrillated cellulose: morphology and accessibility. New York
Hine PJ, Ward IM, Olley RH, Bassett DC (1993) The hot compaction of high modulus melt-spun polyethylene fibers. J Mater Sci 28(2):316–324
Hou S, Xia Z, Pan J, Wang N, Gao H, Ren J, Xia X (2023) Bacterial cellulose applied in wound dressing materials: production and functional modification—a review. Macromol Biosci. https://doi.org/10.1002/mabi.202300333
Hsieh YC, Yano H, Nogi M, Eichhorn SJ (2008) An estimation of the Young's modulus of bacterial cellulose filaments. Cellulose 15(4):507–513

Huber T, Pang S, Staiger MP (2012a) All-cellulose composite laminates. Compos Pt A-Appl Sci Manuf 43(10):1738–1745
Huber T, Muessig J, Curnow O, Pang S, Bickerton S, Staiger MP (2012b) A critical review of all-cellulose composites. J Mater Sci 47(3):1171–1186
Huber T, Bickerton S, Muessig J, Pang S, Staiger MP (2012c) Solvent infusion processing of all-cellulose composite materials. Carbohydr Polym 90(1):730–733
Iguchi M, Yamanaka S, Budhiono A (2000) Bacterial cellulose—a masterpiece of nature's arts. J Mater Sci 35(2):261–270
Jabbari F, Babaeipour V (2023) Bacterial cellulose as a potential biopolymer for wound care. A review. Int J Polym Mater Polym Biomater 73:455–477
Jahed M, Naderi G, Mir HRG (2016) Microstructure, mechanical, and rheological properties of natural rubber/ethylene propylene diene monomer nanocomposites reinforced by multi-wall carbon nanotube. Polym Compos 39:E745–E753
Jain M, Pradhan MK (2016) Morphology and mechanical properties of sisal fiber and nano cellulose green rubber composite: a comparative study. Int J Plast Technol 20:378–400
Jasse B, Koenig JL (1979) Orientational measurements in polymers using vibrational spectroscopy. J Macromol Sci Part C Polym Rev 17:61–135
Jonas R, Farah LF (1998) Production and application of microbial cellulose. Polym Degrad Stabil 59(1–3):101–106
Juntaro J, Pommet M, Mantalaris A, Shaffer M, Bismarck A (2007) Nanocellulose enhanced interfaces in truly green unidirectional fibre reinforced composites. Compos Interface 14(7–9):753–762
Juntaro J (2009) PhD thesis, Imperial College London
Kabeel MA, Bassett DC, Olley RH, Hine PJ, Ward IM (1994) Compaction of highmodulus melt-spun polyethylene fibers at temperatures above and below the optimum. J Mater Sci 29(18):4694–4699
Kabeel MA, Bassett DC, Olley RH, Hine PJ, Ward IM (1995) Differential melting in compacted high-modulus melt-spun polyethylene fibers. J Mater Sci 30(3):601–606
Kalaee M, Mazinani S, Faripour A (2024) Curing characteristics and scorch behavior of nanocellulose elastomer composites. Elastom Nanocellul Compos 141–188
Kamel S, Ali N, Jahangir K, Shah SM, El-Gendy AA (2008) Pharmaceutical significance of cellulose: a review. eXPRESS Polym Lett 2(11):758–778
Khalil H, Bhat AH, Yusra AFI (2012) Green composites from sustainable cellulose nanofibrils: a review. Carbohydr Polym 87(2):963–979
Kitsawat V, Siri S, Phisalaphong M (2024) Electrically conductive natural rubber composite films reinforced with graphite platelets. Polymers 16:288
Klemm D, Heublein B, Fink HP, Bohn A (2005) Cellulose: fascinating biopolymer and sustainable raw material. Angew Chem Int Ed 44(22):3358–3393
Klemm D, Kramer F, Moritz S, Lindstrom T, Ankerfors M, Gray D, Dorris A (2011) Nanocelluloses: a new family of nature-based materials. Angew Chem Int Ed 50(24):5438–5466
Kolodziej M, Bokobza L, Bruneel J-L (2007) Investigations on natural rubber filled with multiwall carbon nanotubes. Compos Interfaces 14:215–228
Kong S, Wu S, Tang Z, Guo B (2023) Hydrogen bond-containing oligomer as a facile interfacial mediator in rubber/silica composites. Compos Commun 37:101413
Kretschmann D (2003) Natural materials—Velcro mechanics in wood. Nat Mater 2(12):775–776
Kuilla T, Bhadrab S, Yao D, Kim NH, Bose S, Lee JH (2010) Recent advances in graphene based polymer composites. Prog Polym Sci 35:1350–1375
Kumar A, Jaiswal J, Tsuchiya K, Singh G (2024) Recent advances in polymer-composite materials for biomedical applications. Hybrid Compos Mater 153–193
Kumineck Junior SR, Silveira VF, Silva DAK, Garcia MCF, Apati GP, Schneider ALDS, Pezzin APT, Baratto Filho F (2023) Development of bacterial cellulose incorporated with essential oils for wound treatment. Polímeros

Lee K-Y, Blaker JJ, Bismarck A (2009) Surface functionalisation of bacterial cellulose as the route to produce green polylactide nanocomposites with improved properties. Compos Sci Tech 69(15–16):2724–2733

Lee KY, Wong LLC, Blaker JJ, Hodgkinson JM, Bismarck A (2011) Bio-based macroporous polymer nanocomposites made by mechanical frothing of acrylated epoxidised soybean oil. Green Chem 13(11):3117–3123

Lee K-Y, Bharadia P, Blaker JJ, Bismarck A (2012a) Short sisal fibre reinforced bacterial cellulose polylactide nanocomposites using hairy sisal fibres as reinforcement. Compos Pt A-Appl Sci Manuf 43(11):2065–2074

Lee KY, Tammelin T, Schulfter K, Kiiskinen H, Samela J, Bismarck A (2012b) High performance cellulose nanocomposites: comparing the reinforcing ability of bacterial cellulose and nanofibrillated cellulose. ACS Appl Mater Interfaces 4(8):4078–4086

Lee K-Y, Ho KKC, Schlufter K, Bismarck A (2012c) Hierarchical composites reinforced with robust short sisal fibre preforms utilising bacterial cellulose as binder. Compos Sci Tech 72(13):1479–1486

Li ZQ, Zhou XD, Pei CH (2010) Preparation and characterization of bacterial cellulose/polylactide nanocomposites. Polym Plast Technol Eng 49(2):141–146

Li Z, Wang Z, Wang L, Qin Z, Li Z, Lu S (2024) Hyperbranched dispersant modification of sisal cellulose based on hydrogen bonding interactions for strengthening natural rubber composite. Eur Polym J 214:113145

Low DYS, Tang SY (2024) Nanocellulose reinforced general and special purpose elastomers. Elastom Nanocellul Compos 231–272

Lu T, Ji S, Jin W, Yang Q, Luo Q, Ren T-L (2023) Biocompatible and long-term monitoring strategies of wearable, ingestible and implantable biosensors: reform the next generation healthcare. Sensors 23:2991

Maa P-C, Siddiqui NA, Marom G, Kim J-K (2010) Dispersion and functionalization of carbon nanotubes for polymer-based nanocomposites: a review. Compos Part A 41:1345–1367

Malas A, Pal P, Giri S, Mandal A, Das CK (2014) Synthesis and characterizations of modified expanded graphite/emulsion styrene butadiene rubber nanocomposites: mechanical, dynamic mechanical and morphological properties. Compos Part B Eng 58:267–274

Martins IMG, Magina SP, Oliveira L, Freire CSR, Silvestre AJD, Neto CP, Gandini A (2009) New biocomposites based on thermoplastic starch and bacterial cellulose. Compos Sci Tech 69(13):2163–2168

Matsuo M, Sawatari C, Iwai Y, Ozaki F (1990) Effect of orientation distribution and crystallinity on the measurement by X-ray-diffraction of the crystal-lattice moduli of cellulose-I and cellulose-II. Macromolecules 23(13):3266–3275

Mauro F, Corrado B, De Gregorio V, Lagreca E, Di Natale C, Vecchione R, Netti PA (2024) Exploring the evolution of bacterial cellulose precursors and their potential use as cellulose-based building blocks. Sci Rep. https://doi.org/10.1038/s41598-024-62462-9

Mohanty AK, Misra M, Hinrichsen G (2000) Biofibres, biodegradable polymers and biocomposites: an overview. Macromol Mater Eng 276(3–4):1–24

Mukhopadyay R, De SK (1979) Effect of vulcanization temperature and different fillers on the properties of efficiently vulcanized natural rubber. Rubber Chem Technol 52:263–277

Nakaramontri Y, Kummerlöwe C, Nakason C, Vennemann N (2015) The effect of surface functionalization of carbon nanotubes on properties of natural rubber/carbon nanotube composites. Polym Compos 36:2113–2122

Nargatti KI, Ahankari SS, Dizon JRC, Subramaniam RT (2024) Applications of elastomeric nanocellulose composites in flexible electronic devices. Elastom Nanocellul Compos 503–521

Nishino T, Arimoto N (2007) All-cellulose composite prepared by selective dissolving of fiber surface. Biomacromol 8(9):2712–2716

Nishino T, Matsuda I, Hirao K (2004) All-cellulose composite. Macromolecules 37(20):7683–7687

Nogi M, Yano H (2008) Transparent nanocomposites based on cellulose produced by bacteria offer potential innovation in the electronics device industry. Adv Mater 20(10):1849–1852

Nogi M, Abe K, Handa K, Nakatsubo F, Ifuku S, Yano H (2006) Property enhancement of optical transparent bionanofiber composites by acetylation. Appl Phys Lett 89(23):233123

Novak BM (1993) Hybrid nanocomposite materials—between inorganic glasses and organic polymers. Adv Mater 5(6):422–433

Olley RH, Bassett DC, Hine PJ, Ward IM (1993) Morphology of compacted polyethylene fibers. J Mater Sci 28(4):1107–1112

Öz YE, Kalender M (2023) A novel static cultivation of bacterial cellulose production from sugar beet molasses: series static culture (SSC) system. Int J Biol Macromol 225:1306–1314

Pandey JK, Ahn SH, Lee CS, Mohanty AK, Misra M (2010) Recent advances in the application of natural fiber based composites. Macromol Mater Eng 295(11):975–989

Pandit AP, Avachat AM (2024) Nonwoven textiles in smart and wearable technologies. In: Smart textiles from natural resources, pp 367–394

Peddini SK, Bosnyak CP, Henderson NM, Ellison CJ, Paul DR (2014) Nanocomposites from styrene-butadiene rubber (SBR) and multiwall carbon nanotubes (MWCNT) Part 1: Morphology and rheology. Polymer 55:258–270

Pedroni LG, Araujo JR, Felisberti MI, Nogueira AF (2012) Nanocomposites based on MWCNT and styrene–butadiene–styrene block copolymers: effect of the preparation method on dispersion and polymer–filler interactions. Compos Sci Technol 72:1487–1492

Ping X-F, Wang Y, Liu L, Liu F-Y, He H-W, Wang P, Yu W-W, Zheng Q (2024) Improving thermal-oxidative aging resistance of styrene-butadiene rubber by antioxidant loaded silica aerogel. Chin J Polym Sci

Pommet M, Juntaro J, Heng JYY, Mantalaris A, Lee AF, Wilson K, Kalinka G, Shaffer MSP, Bismarck A (2008) Surface modification of natural fibers using bacteria: depositing bacterial cellulose onto natural fibers to create hierarchical fiber reinforced nanocomposites. Biomacromol 9(6):1643–1651

Ponnamma D, Sadasivuni KK, Strankowski M, Guob Q, Thomas S (2013) Synergistic effect of multi walled carbon nanotubes and reduced graphene oxides in natural rubber for sensing application. Soft Matter 9:10343–10353

Ponnamma D, Sadasivuni KK, Grohens Y, Guod Q, Thomas S (2014a) Carbon nanotube based elastomer composites—an approach towards multifunctional materials. J Mater Chem C 2:8446–8485

Ponnamma D, Sung SH, Hong JS, Ahn KH, Varughese KT, Thomas S (2014b) Influence of non-covalent functionalization of carbon nanotubes on the rheological behavior of natural rubber latex nanocomposites. Eur Polym J 53:147–159

Ponnamma D, Ramachandran R, Hussain S, Rajaraman R, Amarendra G, Varughese KT, Thomas S (2015) Free-volume correlation with mechanical and dielectric properties of natural rubber/multi walled carbon nanotubes composites. Compos Part A Appl Sci Manuf 77:164–171

Quero F, Nogi M, Yano H, Abdulsalami K, Holmes SM, Sakakini BH, Eichhorn SJ (2010) Optimization of the mechanical performance of bacterial cellulose/poly(L-lactic) acid composites. ACS Appl Mater Interfaces 2(1):321–330

Raut M, Asare E, Syed Mohamed S, Amadi E, Roy I (2023) Bacterial cellulose-based blends and composites: versatile biomaterials for tissue engineering applications. Int J Mol Sci 24:986

Retegi A, Algar I, Martin L, Altuna F, Stefani P, Zuluaga R, Ganan P, Mondragon I (2012) Sustainable optically transparent composites based on epoxidized soybean oil (ESO) matrix and high contents of bacterial cellulose (BC). Cellulose 19(1):103–109

Rooj S, Das A, Stöckelhuber KW, Wießner S, Fischer D, Reuter U, Heinrich G (2015) 'Expanded organoclay' assisted dispersion and simultaneous structural alterations of multiwall carbon nanotube (MWCNT) clusters in natural rubber. Compos Sci Technol 107:36–43

Ross P, Mayer R, Benziman M (1991) Cellulose biosynthesis and function in bacteria. Microbiol Rev 55(1):35–58

Sadasivunia KK, Ponnammab D, Thomas S, Grohens Y (2014) Evolution from graphite to graphene elastomer composites. Progr Polym Sci 39:749–780

Satyanarayana KG, Arizaga GGC, Wypych F (2009) Biodegradable composites based on lignocellulosic fibers-an overview. Prog Polym Sci 34(9):982–1021
Sczostak A (2009) Cotton linters: an alternative cellulosic raw material. Macromol Symp 280:45–53
Seifert M, Hesse S, Kabrelian V, Klemm D (2004) Controlling the water content of never dried and reswollen bacterial cellulose by the addition of water-soluble polymers to the culture medium. J Polym Sci Pol Chem 42(3):463–470
Shiva M (2024) Diffusion, transport mechanism and barrier properties of nanocellulose-reinforced elastomer composites. Elastom Nanocellul Compos 397–425
Shokatayeva D, Ignatova L, Savitskaya I, Kistaubaeva A, Talipova A, Asylbekova A, Abdulzhanova M, Mashzhan A (2019) Bacterial cellulose and pullulan from simple and low cost production media. Eurasian Chem Technol J 21:247
Singh V, Joung D, Zhai L, Das S, Khondaker SI, Seal S (2011) Graphene based materials: past, present and future. Prog Mater Sci 56:1178–1271
Singha NR, Mahapatra M, Karmakar M, Chattopadhyay PK (2019) Processing, characterization and application of natural rubber based environmentally friendly polymer composites. Sustain Polym Compos Nanocompos 855–897
Siro I, Plackett D (2010) Microfibrillated cellulose and new nanocomposite materials: a review. Cellulose 17(3):459–494
Smejda-Krzewicka A, Mrozowski K, Strzelec K (2024) Effect of modified and unmodified oak bark (quercus cortex) on the cross-linking process and mechanical, anti-aging, and hydrophobic properties of biocomposites produced from natural rubber (NR). Materials 17:1968
Soykeabkaew N, Sian C, Gea S, Nishino T, Peijs T (2009) All-cellulose nanocomposites by surface selective dissolution of bacterial cellulose. Cellulose 16(3):435–444
Sperotto G, Stasiak LG, Godoi JPMG, Gabiatti NC, De Souza SS (2021) A review of culture media for bacterial cellulose production: complex, chemically defined and minimal media modulations. Cellulose 28:2649–2673
Srivastava SK, Bhuyan B (2018) Rubber nanocomposites for tire tread application in rubber nano composites and nano textiles in automobiles. In: Banerjee B (ed). Rapra Publication, Akron, OH, pp 37–78
Srivastava SK, Mishra YK (2018) Nanocarbon reinforced rubber nanocomposites: detailed insights about mechanical, dynamical mechanical properties, Payne, and Mullin effects. Nanomaterials 8(11):945
Stankovich S, Dikin DA, Piner RD, Kohlhaas KA, Kleinhammes A, Jia Y, Wu Y, Nguyen ST, Ruoff RS (2007) Synthesis of graphene-based nanosheets via chemical reduction of exfoliated graphite oxide. Carbon 45:1558–1565
Sui G, Zhong WH, Yang XP, Yu YH, Zhao SH (2008) Preparation and properties of natural rubber composites reinforced with pretreated carbon nanotubes. Polym Adv Technol 19:1543–1549
Tang Z, Wei Q, Lin T, Guo B, Jia D (2013) The use of a hybrid consisting of tubular clay and graphene as a reinforcement for elastomers. RSC Adv 3:17057
Tang L-C, Zhao L, Qiang F, Wu Q, Gong L-X, Peng J-P (2019) Mechanical properties of rubber nanocomposites containing carbon nanofillers. In: Carbon-based nanofillers and their rubber nanocomposites, pp 367–423
Tang L, Jin Y, He X, Huang R (2024) Biodegradable poly(ethylene glycol-glycerol-itaconate-sebacate) copolyester elastomer with significantly reinforced mechanical properties by in-situ construction of bacterial cellulose interpenetrating network. Sci Rep. https://doi.org/10.1038/s41598-024-56534-z
Tarawneh MA, Ahmad SH, EhNoum S, Lau K (2013) Sonication effect on the mechanical properties of MWCNTs reinforced natural rubber. J Compos Mater 47:579–585
Tarawneh MA, Ahmad SH, Yahya SY, Rasid R, Se Yong Eh Noum SYE (2011) Mechanical properties of thermoplastic natural rubber reinforced with multi-walled carbon nanotubes. J Reinf Plast Compos 30:363–368

Thomas PS, Abdullateef AA, Al-Harthi MA, Basfar AA, Bandyopadhyay S, Atieh MA, De SK (2012) Effect of phenol functionalization of carbon nanotubes on properties of natural rubber nanocomposites. J Appl Polym Sci 124:2370–2376
Thomas SK, Begum PMS, Neenu KV, Dhanyasree P, Dileep P, Ravindran L, Siriwong C, Duan Y, Badawi M, Midhun Dominic CD (2024) Tailoring natural rubber composites via surface-modified nanocellulose whiskers with sodium isopropyl xanthate: characterization and performance analysis. Biomass Convers Biorefinery
Thomason JL (2009) Why are natural fibres failing to deliver on composite performance? Edinburgh
Tkalya EE, Ghislandi M, de With G, Koning CE (2012) The use of surfactants for dispersing carbon nanotubes and graphene to make conductive nanocomposites. Curr Opin Colloid Interface Sci 17:225–232
Tokoh C, Takabe K, Sugiyama J, Fujita M (2002) Cellulose synthesized by acetobacterxylinum in the presence of plant cell wall polysaccharides. Cellulose 9(1):65–74
Trovatti E, Oliveira L, Freire CSR, Silvestre AJD, Neto CP, Pinto J, Gandini A (2010) Novel bacterial cellulose-acrylic resin nanocomposites. Compos Sci Tech 70(7):1148–1153
Uchimura M, Tsumori F (2024) In-situ fabrication process of bacterial cellulose composites for soft robots. IEEJ Trans Electr Electron Eng 19:930–934
Vaisman L, Wagner HD, Marom G (2006) The role of surfactants in dispersion of carbon nanotubes. Adv Colloid Interface Sci 128–130:37–46
Vasanth Kumar D, Srinivasan N, Davis Hans SJ, Gokul S, Arulmurugan B, Sathishkumar B (2024) Additive manufacturing of nanoscale and microscale materials. In: Additive manufacturing with novel materials, pp 267–293
Volova TG, Prudnikova SV, Kiselev EG, Nemtsev IV, Vasiliev AD, Kuzmin AP, Shishatskaya EI (2022) Bacterial cellulose (BC) and BC composites: production and properties. Nanomaterials 12:192
Walling B, Bharali P, Giridharan B, Gogoi B, Sorhie V, Alemtoshi MSK (2023) Bacterial nanocellulose: a novel nanostructured bio-adsorbent for green remediation technology. Acta Ecol Sin 43:946–967
Wang C, Zhang Y (2018) Effects of liquid polyisoprene and magnesium oxide on the mechanical properties of styrene-butadiene rubber/carbon nanotubes composite. Polym Compos 39:E765–E771
Ward IM, Hine PJ (1997) Novel composites by hot compaction of fibers. Polym Eng Sci 37(11):1809–1814
Weng G, Huang G, Qu L, Nie Y, Wu J (2010) Large-scale orientation in a vulcanized stretched natural rubber network: Proved by in situ synchrotron X-ray diffraction characterization. J Phys Chem B 114:7179–7188
Wloch D, Herrera N, Lee K-Y (2023) Optically transparent laminated acrylic composites reinforced with mercerised bacterial cellulose nanopaper. Compos A Appl Sci Manuf 172:107583
Wu L, Qu P, Zhou R, Wang B, Liao S (2015) Green synthesis of reduced graphene oxide and its reinforcing effect on natural rubber composites. High Perform Polym 27:486–496
Wuhrmann K, Heuberger A, Muhlethaler K (1946) Elektronenmikroskopische untersuchungen an zellulosefasern nach behandlung mit ultraschall. Experientia 2(3):105–107
Xu J, Li S, Li Y, Ta X (2016) Preparation, morphology and properties of natural rubber/carbon black/multi-walled carbon nanotubes conductive composites. J Mater Sci Mater Electron 27:9531–9540
Yamamoto H, Horii F, Hirai A (1996) In situ crystallization of bacterial cellulose. Influences of different polymeric additives on the formation of celluloses I-alpha and I-beta at the early stage of incubation. Cellulose 3(4):229–242
Yang L, Zhu X, Chen Y, Wang J (2024) Enhanced bacterial cellulose production in Gluconacetobacter xylinus by overexpression of two genes (bscC and bcsD) and a modified static culture. Int J Biol Macromol 260:129552

Yano H, Sugiyama J, Nakagaito AN, Nogi M, Matsuura T, Hikita M, Handa K (2005) Optically transparent composites reinforced with networks of bacterial nanofibers. Adv Mater 17(2):153–155

Ye Z, Zhao D, Liu F, Luo J, Liu X, Zhao W (2024) Flexible and highly conductive Ti3C2Tx/natural rubber composites with interconnected networks for high-performance electromagnetic interference shielding. Compos A Appl Sci Manuf 180:108067

Zhan YH, Liu GQ, Xia HS, Yan N (2011) Natural rubber/carbon black/carbon nanotubes composites prepared through ultrasonic assisted latex mixing process. Plast Rubber Compos 40:32–39

Zhang H, Wang C, Zhang Y (2015) Preparation and properties of styrene-butadiene rubber nanocomposites blended with carbon black-graphene hybrid filler. J Appl Polym Sci 132:41309

Zhang R, Zheng R, Zheng Z, Chen Q, Jiang N, Tang P, Wang H, Bin Y (2024a) Bacterial cellulose/multi-walled carbon nanotube composite films for moist-electric energy harvesting. Int J Biol Macromol 263:130022

Zhang Y, Deng W, Wu M, Yu G, Liu Z, Cheng N, Du H, Liu C, Li B (2024b) Engineering pulp foam with highly improved water stability and multifunctional properties by incorporation of natural rubber and montmorillonite. Green Carbon

Zhou Q, Malm E, Nilsson H, Larsson PT, Iversen T, Berglund LA, Bulone V (2009) Nanostructured biocomposites based on bacterial cellulosic nanofibers compartmentalized by a soft hydroxyethylcellulose matrix coating. Soft Matter 5(21):4124–4130

Zhou Y, Fan M, Chen L, Zhuang J (2015) Lignocellulosic fibre mediated rubber composites: an overview. Compos B Eng 76:180–191

Chapter 9
Chitin in Rubber-Based Blends and Composites

Shiv Kumari Panda

Abstract This chapter comprehensively surveys recent advances in processing technologies, property enhancements, and potential applications across diverse industries. Melt mixing, solution blending, and in situ polymerization methods have been developed to integrate chitin, enabling significant improvements in tensile strength, flex fatigue, permeation barriers, and abrasion resistance compared to unfilled rubbers. Processing-structure–property relationships have been elucidated, providing guidance for optimized composite fabrication tailored to specific product requirements. Recent accomplishments also motivate ongoing efforts to surmount remaining challenges in scale-up, modeling, prototyping, supply chain development, and end-of-life considerations to accelerate commercial adoption. This chapter summarizes the multifaceted potential of chitin-reinforced rubbery micro composites to transform elastomer technologies across diverse sectors onto a more sustainable yet high-performing materials foundation.

Keywords Chitin-rubber microcomposites · High performance materials · Potential applications

9.1 Introduction

In the expansive realm of materials science, a field marked by continual exploration and groundbreaking discoveries, the convergence of biopolymers and synthetic materials has emerged as a transformative force, ushering in a new era of innovative advancements. This chapter embarks on a comprehensive journey into the captivating domain of chitin, a natural biopolymer (Al-Qahtani et al. 2023). Chitin has captivated the attention of researchers and scientists alike abundantly sourced from the exoskeletons of arthropods and the cell walls of fungi, chitin stands as a testament

S. K. Panda (✉)
Assistant Professor, Udayanath Autonomous College of Science & Technology, Adaspur, Cuttack, Odisha, India
e-mail: shivkumaripanda@gmail.com

© The Author(s), under exclusive license to Springer Nature Switzerland AG 2025
P. M. Visakh (ed.), *Rubber Based Bionanocomposites*, Advanced Structured Materials 227, https://doi.org/10.1007/978-3-031-78557-3_9

to the intricate balance of nature's design comprising linear chains of Nacetylglucosamine (Loose 2018), this biopolymer reveals itself as a structurally unique and intriguing material, holding within its molecular architecture the potential to revolutionize the landscape of materials engineering the exploration begins with an in-depth analysis of the singular properties inherent to chitin-properties that include but are not limited to its high tensile strength, biodegradability, and low toxicity as the narrative unfolds (Alkhursani et al. 2022), the focus shifts towards the crux of this chapter-the integration of chitin into rubber based blends and micro composites this marriage of natural and synthetic materials unravels a tapestry of possibilities, each thread weaving together to enhance mechanical, thermal, and environmental performance the bio-resource derived nature of chitin becomes a key focal point, aligning with the global shift towards sustainable practices and eco-friendly materials the subsequent exploration delves into the intricacies of processing techniques employed to achieve a harmonious union between chitin and rubber melt mixing, solution blending, and in situ polymerization are among the myriad techniques dissected to comprehend their impact on the final composition and properties of chitin infused rubber blends this journey through processing methods serves as a bridge, connecting the theoretical underpinnings of chitin rubber integration with practical applications across diverse industries as the narrative unfolds, the spotlight widens to encompass micro composites-materials that exist at the intersection of the macro and nano worlds the integration of chitin at the microscale proves to be a pivotal advancement, offering a delicate balance between bulk properties and surface interactions this segment unravels the potential improvements in adhesion (Chafiq et al. 2023), barrier properties, and overall performance facilitated by chitins nanoscale fibrillar structure the applications of chitin rubber micro composites are vast and varied, spanning automotive components, footwear, medical devices, and beyond the chapter unfolds as a compendium of knowledge, shedding light on the environmental implications of chitin based materials in an era where environmental consciousness is at the forefront (Srivastava 2023), the biodegradability of chitin emerges as a beacon of hope, potentially reducing the burden of plastic waste and contributing to a circular economy the thermal properties of chitin further expand the horizons of its applications, particularly in environments where exposure to elevated temperatures is a critical consideration looking forward, the chapter gazes into the future of chitin based materials, speculating on the untapped potential and the uncharted territories awaiting exploration ongoing research into processing techniques, optimization of chitin rubber blends, and the discovery of new applications promises a dynamic and ever evolving landscape in the realm of materials science in conclusion, this extensive exploration navigates the intricate web of chitin's properties, its integration into rubber based blends and micro composites (Jadaun et al. 2023), and the far reaching implications of this synergistic alliance the chapter unfolds not only as a scientific discourse but as a testament to the ceaseless pursuit of knowledge and innovation propelling the field of materials science into uncharted territories rubber, known for its elasticity and resilience, forms the backbone of many industrial applications when combined with chitin, a unique synergy emerges the inherent flexibility of rubber and the robustness of chitin create a blend that exhibits improved mechanical strength, tear resistance, and thermal stability this blend has potential applications in the manufacturing of durable and eco-friendly products such as automotive components, footwear, and medical devices the intricate morphology and phase distribution in multiphase polymer composites exerts first order control over bulk performance properties process selection for chitin rubber blends balances factors of filler dispersion, energy intensity, scalability, and final morphology multiple fabrication routes have emerged based on polymer type (Song et al. 2023; Prasad et al. 2023), filler characteristics, and targeted composite traits bench scale processes help to establish proof of concept and analyze fundamental structure property mechanisms while large-scale techniques aim to demonstrate viability for commercial manufacturing melt mixing involves high shear homogenization of a rubber base, typically natural rubber or styrene butadiene rubber SBR, with chitin powder fillers plus crosslinking agents at elevated temperatures 120–140 °C using batch mixers or extruders subsequent thermoforming or injection molding produces bulk samples for mechanical and permeability analysis advantages include versatility for different matrices and feasible scale-up however, high viscosity and inadequate mixing time risks poor dispersion additionally, high temperatures degrade chitin and rubber requiring antioxidant additives to mitigate effects thereof unfortunately (Vieira et al. 2023), the nonlinear flow and thermal gradients during melt mixing induces non-uniform filler distributions even with compatibilizer additives which manifests as lower tensile strength hence, melt blending methods currently suit thermostable binder systems until improved micro scale control is achieved solution based techniques leverage dissolution of rubber and chitin in a volatile solvent such as toluene allowing thorough mixing under gentler conditions before controlled precipitation via nonsolvent addition or drying common configurations include emulsion styrenebutadiene rubber latex with chitin nanoparticles and liquid natural rubber mixed with chitin whiskers latex cogel preceramic polymers facilitate highly homogeneous dispersions the dried composites display more consistent mechanical improvements relative to melt counterparts due to better nanofiller distribution casting processes permit film formation without need for thermal molding although residual solvent can influence curing kinetics additionally, chitin tends to agglomerate during drying which can negatively affect properties therefore, redispersion via sonication, surfactants, or functionalization helps safeguard homogeneity key strengths of solution approaches include near ambient temperatures to protect polymers improved particle integration at multiple scales and feasibility to incorporate graphene, clay, and carbon nanotube hybrid fillers net effects depend strongly on solvent type, ant agglomeration treatments, and curing/drying profile offering extensive tuning flexibility a third category—in situ polymerization intertwines the filler during direct rubber polymer formation allowing precisely engineered morphologies one variation utilizes a Pickering emulsion process with chitin nanocrystals adsorbed at the droplet interfaces of a water in oil emulsion to template latex particle sizes alternatively, emulsion polymerization of liquid natural rubber occurs directly in the presence of chitin nanofibers achieving uniform incorporation within the networked rubber matrix finally, grafting to and grafting from polymerization chemically bonds chitin particles with polymer chains before curing via silane treatment solgel condensation (Parale et al. 2023), atom transfer radical polymerization, or radiation induced methods the covalent filler attachment fosters excellent dispersion and interfacial stress transfer contributing to significant toughening and stiffness enhancements that eclipse traditional composites however, in situ routes pose practical barriers to batch size enlargement and composite recovery which currently prevent widespread adoption outside niche high value applications independent of fabrication protocol applied, comprehensive physicomechanical characterization provides feedback to correlate process conditions with resultant properties as functions of composition mixing energy (Abedi et al. 2023), shear rates, temperatures, and times scanning electron microscopy coupled with image analysis verifies chitin morphology and spatial distributions in rubber matrices providing morphology process structure linkages furthermore, spectroscopic and thermal techniques characterize molecular interactions at biopolymer interfaces including crystallinity and grafting efficacy which helps explain macroscale mechanical performance dynamic mechanical analysis indicates viscoelastic shifts in chain mobility relaxation transitions that direct modulus and damping changes similarly, tensile testing, hardness, tear strength, and fatigue measurements establishes key indicators of reinforcement quality comparing performance versus carbon black and clay benchmarks studies indicate shear mixing times beyond one hour fail to further improve dispersion suggesting an optimal energetic efficiency balance for melt blending processes moreover, in situ routes achieve the greatest property enhancements due to covalent hybridization but lack commercial feasibility currently solvent blending offers a middle ground with enhancements approaching in situ composites while preserving scalability spray drying and coagulation processes allow continuous production albeit with risks of re-agglomeration without sufficient stabilizers thus appropriate composite design selections balance targeted performance metrics versus manufacturing constraints in an application specific manner with processing modeled concurrently (Sapuan and Mansor 2014).

9.2 Chitin: A Versatile Biopolymer Shaping Sustainable Material Innovation

Chitin, a versatile natural biopolymer, has emerged as a focal point of scientific interest due to its remarkable properties. Abundantly found in the exoskeletons of arthropods and the cell walls of fungi, this polysaccharide is composed of linear chains of N-acetylglucosamine, making it a structurally unique and intriguing material. Its distinctive characteristics, including high tensile strength (Ismail et al. 2008), biodegradability, and low toxicity (Hou et al. 2013), have spurred researchers to explore its potential applications, particularly in rubber-based blends and micro composites. The structural composition of chitin lends it remarkable mechanical properties (Ahmad et al. 2020). The linear chains of N-acetylglucosamine create a robust and resilient framework, conferring upon chitin a high degree of tensile strength. This inherent strength and the biopolymer's lightweight nature make it an attractive candidate for reinforcing materials in various industrial applications. The utilization of chitin in rubber-based blends has become a subject of intense investigation as researchers aim to enhance the mechanical properties of these blends for improved performance and durability. In addition to its mechanical prowess, one of the most notable features of chitin is its biodegradability. The chitin molecule itself is uncharged, as it consists of repeating units of N-acetylglucosamine, which are neutral. Each N-acetylglucosamine unit contains an amino group (NH_2) and an acetyl group, but overall, the molecule does not have a net positive or negative charge. However, in specific chemical or biological contexts, chitin molecules can interact with other substances or be modified to carry a charge. For example, chitin can undergo deacetylation to form chitosan, which contains amino groups and can be positively charged under certain conditions. However, chitin is not inherently positively or negatively charged in its natural state. It's important to note that the charge of chitin or its derivatives depends on the specific chemical modifications and environmental conditions (Knidri et al. 2018).

In a world increasingly conscious of environmental sustainability, the ability of materials to naturally break down and return to the ecosystem is of paramount importance. Chitin, derived from natural sources, aligns seamlessly with the principles of sustainability. Its biodegradable nature not only reduces the environmental impact of end-of-life products but also opens up avenues for developing eco-friendly materials in various applications. Furthermore, outlines chitin's different properties, such as low toxicity, a crucial characteristic for materials intended for use in various industries, including medical and consumer goods. The biocompatibility of chitin makes it a suitable candidate for applications where contact with living tissues is unavoidable. This feature has spurred research into incorporating chitin into medical devices (Kavimani et al. 2023), wound dressings, and other healthcare-related materials, leveraging its compatibility with biological systems. The quest to harness the unique properties of chitin has led to its integration into rubber-based blends. Rubber, a widely used material known for its elasticity and resilience, forms the foundation for numerous industrial applications, ranging from automotive components to footwear. By combining chitin with rubber, researchers aim to create blends that retain rubber's flexibility and resilience and benefit from the enhanced mechanical properties and biodegradability imparted by chitin (Patel et al. 2023). Incorporating chitin into rubber-based blends involves carefully achieving uniform dispersion and ensuring optimal interaction between the two materials. Various processing techniques, such as melt mixing, solution blending, and in situ polymerization, have been explored to create well-defined microstructures within the composite material. The success of these techniques relies on understanding the impact of processing conditions on the final properties of the chitin-rubber blends. Micro composites, formed by integrating chitin at the microscale with rubber, represent a significant advancement in material science. These composites bridge the gap between macro and nano-sized materials, offering a balance that combines the bulk properties of rubber with the surface interactions provided by the nanoscale features of chitin. This synergy improves adhesion, barrier properties, and overall performance, opening up new possibilities for developing materials with tailored characteristics. The applications of chitin-rubber micro composites are diverse and span across industries. In the automotive sector, these composites could be utilized in manufacturing biodegradable and high-performance tires, addressing concerns related to the environmental impact of conventional tire disposal. The construction industry could benefit from developing sustainable seals and gaskets with improved mechanical properties, while the medical field could explore bioactive materials for applications in implants, prosthetics, and drug delivery systems. The environmental implications of chitin-based materials are significant. As the world grapples with the challenges of plastic pollution and seeks sustainable alternatives, chitin presents itself as a viable candidate for reducing the environmental footprint of various products. Its biodegradability ensures that products incorporating chitin will not contribute to the persistence of non-biodegradable waste in landfills and ecosystems. The thermal properties of chitin also play a role in its potential applications (Lv et al. 2023). While rubber is known for its elasticity at room temperature, adding chitin can contribute to improved thermal stability. This enhancement is particularly relevant in applications where exposure to elevated temperatures is considered, such as automotive components that may experience heat during operation. The ability of chitin to withstand higher temperatures adds a dimension of versatility to rubber-based blends that may find application in a range of thermal environments. Looking forward, the field of chitin-based materials holds tremendous promise for further innovation. The ongoing research into processing techniques, optimizing chitin-rubber blends, and exploring new applications will likely unlock even more potential for this biopolymer (Yaşar Yıldız and Radchenkova 2023). The intersection of biodegradability, mechanical strength, and low toxicity positions chitin as a key player in the quest for sustainable and high-performance materials. In conclusion, chitin's journey from the exoskeletons of arthropods and the cell walls of fungi to the forefront of material science is marked by its exceptional properties. As researchers continue to unravel the intricacies of this natural biopolymer (Alshangiti et al. 2023), its integration into rubber-based blends and micro composites stands as a testament to the ongoing quest for sustainable and high-performance materials. The potential applications across various industries underscore the versatility of chitin and its role in shaping the future of materials that are not only robust and resilient but also environmentally conscious.

Chitin is a naturally abundant aminopolysaccharide that composes the exoskeletons of crustaceans and fungal cell walls possessing intriguing structural varieties based on differences in molecular chain packing and hydrogen bonding patterns leading to three polymorphic crystal forms known as α-chitin β-chitin and γ-chitin. The unique physicochemical characteristics influencing processability reactivity products and applications specifically the molecular architecture of chitin consists of linear polysaccharide chains of predominantly β-(1→4) linked N-acetyl-D-glucosamine units where hydroxyl acetamido and glycosidic groups contribute to extensive inter- and intramolecular hydrogen bonding such that α-chitin exhibits the tightest most crystalline packing with anti-parallel double stranded sheets and intra-sheet hydrogen bonds while β-chitin possesses a looser less crystalline structure with parallel single sheets and inter-sheet hydrogen bonds however γ-chitin is the least understood form having single parallel chains and highly disordered regions so the variance in chain orientation and hydrogen bonding patterns affects solubility reactivity and stability moreover α-chitin dominates in crustaceans like crab/shrimp shells while β-chitin occurs in squid pens and γ-chitin is limited to fungal and yeast cell walls such that α-chitin requires harsher isolation processing that risks polymer degradation but milder extraction preserves β-chitin chains and enzymatic and microbial methods selectively isolate γ-chitin furthermore the tighter packing makes α-chitin largely insoluble in common solvents restricting solution-based processing whereas β-chitin and γ-chitin are more amenable to dissolution and chemical manipulation for instance hydrophobic derivatives like chitosan improve solubility in organic solvents allowing cast film formation and γ-chitin is water soluble enabling aqueous phase separations additionally α-chitin is most thermally stable resisting decomposition up to 250 °C while β-chitin degrades more readily above 200 °C and γ-chitin is highly sensitive but chitin-metal oxides composites via in situ precipitation showcase high stability and chitin nanofibers help reinforce rubbers also acid hydrolysis degrades amide bonds of chitin with the alpha form resisting while the gamma form reacts rapidly due to loose packing exposing bonds furthermore the free amine groups of chitin enable reactions for surface modifications where esterification etherification and graft polymerization produce derivatives influencing compatibility and properties including how partial deacetylation yields chitosan by disrupting crystallinity facilitating dissolution and subsequent reactions that afford bioactive films hydrogels and scaffolds additionally the nanofibrillar morphology of α-chitin allows uniform integration as reinforcements for polymer matrices to enhance strength modulus and permeability while γ-chitin derivatives covalently crosslink with polyurethanes and epoxies via networking acting as toughness modifiers ultimately α-chitin dominates commercial use as a stiff reinforcing nanofiller for composites foods and cosmetics whereas soluble β- and γ-forms enable biomedicine antimicrobials catalysts and sensors by exploiting high chemical activity therefore in conclusion the polymorphic forms of chitin offer a spectrum of physicochemical traits based on molecular structure variation dictating their isolation methods processing behavior and applications thus understanding these intricate differences allows selective utilization across products from structural nanocomposites to bioactive platforms and further elucidation promises to uncover new arenas benefiting from their abundant and sustainable supply (Tiwari et al. 2023; Shashikumar et al. 2023).

9.3 Chitin-Rubber Blends: A Synergistic Alliance

Rubber, known for its elasticity and resilience (Gong et al. 2023), forms the backbone of many industrial applications (Prasad et al. 2023). When combined with chitin, a unique synergy emerges. The inherent flexibility of rubber and the robustness of chitin create a blend that exhibits improved mechanical strength, tear resistance (Sani et al. 2022), and thermal stability (Ruan et al. 2023). This blend has potential applications in manufacturing durable and eco-friendly products such as automotive components, footwear, and medical devices. "Rubber, known for its elasticity and resilience, forms the backbone of many industrial applications when combined with chitin, a unique synergy emerges the inherent flexibility of rubber and the robustness of chitin create a blend that exhibits improved mechanical strength, tear resistance (Adibi 2023), and thermal stability and this resulted blend has potential applications in the manufacturing of durable and eco-friendly products such as automotive components, footwear, and medical devices". Rubber is one of the most versatile materials used in everyday applications due to its unique properties of elasticity, flexibility, and resilience major types of rubber include natural rubber NR and synthetic rubbers such as styrenebutadiene rubber SBR (Hayeemasae et al. 2023), butadiene rubber BR, and ethylenepropylene diene monomer EPDM the global rubber industry is vast, with products made from rubber composites found throughout industrial sectors such as automotive, aerospace, construction, and medicine while rubbers possess many desirable qualities, their properties can be enhanced by blending with reinforcing fillers in recent years, there has been growing interest in chitin as a renewable and high performance filler for rubber materials chitin is a polysaccharide derived from the exoskeletons of crustaceans such as crab, lobster and shrimp shells it bears structural similarity to cellulose when blended with rubber matrices, chitin has been shown to significantly improve mechanical strength, abrasion resistance, and thermal stability this review covers recent advances in chitin rubber blends (Gaona et al. 2023), highlighting their synergistic effects in areas such as processability, curing kinetics, morphology, viscoelastic performance, and mechanical reinforcement the factors influencing these property enhancements are analyzed along with potential applications finally, current challenges and future outlooks are discussed for further development of high value added chitin rubber composite materials the chemical structure of chitin consists of linear chains of β-(1→4) linked N-acetyl-D-glucosamine, similar to the backbone structure found in cellulose however, the presence of acetamide groups in the repeating units of chitin in place of hydroxyl groups distinguishes it from cellulose and contributes to its different solubility and chemical reactivity chitin is the second most abundant natural polymer after cellulose, commonly obtained from the shells of crustaceans including crabs, lobsters, and shrimps currently, most commercial chitin is extracted as a byproduct from seafood processing waste chitin is typically converted into powdered form via crushing and grinding processes after mineral removal using deproteination and demineralization treatments this produces pure chitin which can then be incorporated as a functional reinforcement filler into rubbers and other polymers early studies focused on the compatibility between chitin and rubber matrices, assessing suitable mixing ratios and processing conditions natural rubber, styrenebutadiene rubber and nitrile rubber composites reinforced with up to 20 phr parts per hundred rubber chitin were prepared using an internal mixer or two roll mill in general (Adibi 2023), the powdered chitin was blended with the rubber matrices along with common vulcanization additives such as accelerators, activators and sulfur the mixing temperature was kept below 140 °C to avoid polymer degradation proper dispersion of the chitin particles was achieved through optimizing mixing time, rotor speed and loading this enhances the rubber filler interfacial adhesion, a key factor influencing downstream mechanical properties the curing process helps to establish the crosslinked structure of rubber materials (Trinh et al. 2023), requiring study of the cure kinetics when additives like chitin are introduced studies using rheometry analysis determined that increasing chitin content progressively shortened scorch time and reduced optimum cure time of the compounds this is likely due to adsorption of curatives onto the chitin surface the incorporation of chitin accelerates the vulcanization reaction between the rubber and curatives (Jantachum et al. 2023), attributed to Lewis acid base interactions natural chitin has an abundance of hydroxyl and acetylamide groups that can participate in chemical bonding with rubbers during vulcanization this behavior was confirmed through various kinetic models including free radical crosslinking mechanisms control of the faster vulcanization rates is needed to tailor desirable mechanical strength in the final composites the morphology and dispersion of chitin particles within the rubber matrix greatly affects the ultimate composite properties well distributed chitin nanofillers with high surface area for polymer interaction are ideal TEM imaging verified the presence of aggregations and uneven distributions when the chitin content exceeded 10 phr reduced particle sizes down to nanoscale dimensions can minimize this issue through various grinding and surface treatment preprocessing steps dynamic mechanical analysis DMA measures the viscoelastic response of materials as they are deformed under oscillatory loading DMA results demonstrated a steady shift of the storage modulus peak to higher temperatures with increasing chitin loading for natural rubber composites this corresponds to the restricted molecular mobility of the rubber chains induced by the chitin networks thermal degradation onset temperature also increased by over 15 °C with a chitin content above 5 phr the incorporated chitin was found to boost the crosslink density of the vulcanizates transition temperatures including the glass transition and beta transition shifted to higher values as chitin forms additional physical crosslinks with the rubber molecules this enhances the stiffness and heat resistance for potential high temperature applications several common measures are used to evaluate (Ismail et al. 2023) and compare the mechanical performance improvements of chitin rubber composites tensile strength incorporation of only a few phr loading of chitin nanofiller led to significant gains in ultimate tensile strength UTS, such as a 65 increase of UTS seen for just 5 phr chitin in nitrile rubber this reinforcing effect arises due to the high aspect ratio, surface area and strength of the chitin particles allowing efficient stress transfer chemically modified hydrophobic chitin derivatives have recently been synthesized to further amplify the tensile improvements by enhancing filler dispersion and polymer filler interfacial adhesion elongation at break while the tensile strength substantially improved with chitin filling (Jhang et al. 2023), a reduction in elongation at break was observed due to the restricted flexibility of rubber molecule chains methods to improve flexibility include using modified liquid natural rubber as the base matrix or plasticization techniques to reduce brittleness despite lowered elongation, the composites retained reasonable flexibility for most applications tear strength reinforcement with chitin nanofibers and whiskers led to significant enhancement in tear strength by over 78 compared to pure natural rubber composites this is attributed to increased crack deviation and resistance upon applied stress due to the high strength and surface area of the chitin filler networks however, at very high loadings beyond 10 phr, excessive particle agglomeration counteracts further improvements hardness increasing the chitin content predictably elevated the hardness values for the composites this correlates to the higher stiffness and reduced flexibility resulting from the chitin filler restricting mobility of the rubber chains posttreatments like compatibilization can help control the modulus boosting effect abrasion resistance multiple studies found over 55 improvement in abrasion resistance when loaded with 8 phr chitin in natural rubber composites the chitin nanoparticles act as high strength physical crosslinkers (Adibi 2023), dissipating energy during abrasion this demonstrates potential for chitin reinforcement of rubber components used in tires and high wear industrial products the enhancement effects of chitin fillers are influenced by various chemical and physical parameters filler content moderate loadings between 210 phr offer an optimal balance particle size nanoscale grades provide greatest surface area filler dispersion break down aggregations through mechanical and surface treatments interfacial adhesion compatibilizer additives or plasticizers improve polymer filler bonding filler structure nanofibrils and whiskers orient along stress direction for strength efficacy the synergies of blended chitin rubber composites lead to combinations of flexible toughness (Utrera-Barrios et al. 2023), high strength and thermal stability suitable for the following potential applications automotive gaskets (Im and Jeoung 2023), seals, hoses, tires medical devices tubing, implants, gloves textiles elastic fabrics, clothing inserts footwear shoe soles with abrasion resistance adhesives strong flexible bonding aerospace seals, Orings with heat resistance environmentally friendly products nontoxic with renewable sourcing while recent research confirms the exciting mechanical property improvements offered by chitin rubber blends (Mathew and Panicker 2023), there remain challenges to translate the technology into viable commercial products achieving reliable and consistent supplies of chitin filler material in bulk quantities further optimizing manufacturing methods of chitin rubber composite products for efficiency and scalability more extensive product testing and validation to meet rigorous industrial benchmarks determining market applications where value added performance offsets increased costs further research efforts will focus on the following areas developing improved chitin extraction and purification techniques from seafood waste streams synthesizing tailored modified chitin derivatives and compatibilizers combining chitin with novel rubber matrices such as protein based and liquid crystalline elastomers fabricating prototypes of automotive parts, biomedical goods and textiles incorporating chitin reinforcement the unique synergy between abundant crustacean sourced chitin and synthetic rubbers establishes a platform for the next generation of high strength, flexible (Lee et al. 2023), and eco-friendly composite materials advancing this sustainable chitin rubber alliance will expand exciting commercial opportunities across a wide range of demanding engineering applications the intrinsic qualities of chitin combined synergistically with rubber offer an eco-friendly filler approach that enhances curing rates, viscoelastic response, mechanical strength, abrasion and heat resistance of resulting composites methodical composite design to control chitin dispersion, particle size and interfacial interaction is crucial for maximizing and tailoring reinforcement effects excellent potential exists for chitin rubber blends to create novel biomaterials fulfilling roles across industrial sectors demanding flexible robustness with an environmentally friendly edge further investment into this alliance should uncover deeper synergies leading to disruptive technologies and sustainable products (Petzold et al. 2023).

9.4 Micro Composites: Bridging the Macro and Nano Worlds

The integration of chitin into rubber at the microscale opens avenues for creating composites with tailored properties micro composites harness the benefits of both macro and nanosized materials, offering a balance between bulk properties and surface interactions chitin (Su 2023), with its nanoscale fibrillar structure, enhances the surface characteristics of rubber, leading to improvements in adhesion, barrier properties, and overall performance. The development of composite materials combines components with distinct qualities that interact synergistically within the matrix integration spans across dimensional scales from macroscale fibers and fillers down to nanoscale particles and tubes that integrate within polymer systems composite research has recently focused heavily on nano sized fillers due to exceptionally high surface area and intricate interactions with the host matrix however, microscale composites maintain significance for commercial manufacturing feasibility while tapping into enhancement effects at filler interfaces this motivates development of multiscale hybrid micro-nano composites that bridge capabilities across dimensional levels among natural polymers (Landoulsi 2024), chitin possesses an intriguing structure and properties for composite integration including high strength (Jiao et al. 2023), biodegradability, abundance, and moisture absorption which makes it attractive for materials requiring toughness, flexibility, permeation control, and sustainability chitin microfibers and nanofibrils sourced from crustacean waste streams through extraction protocols can reinforce rubber matrices in latex and solid states leading to greater tear resistance, tensile strength, and thermal stability relative to unfilled rubber novel functionalities also emerge such as selectively gated permeability, stimuli responsiveness, self-healing, and shape memory due to molecular interplay and synergies at biopolymer interfaces detailed morphological, rheological, spectroscopic, and mechanical analysis provides insight into the crosslinking mechanisms, relaxation phenomena; and deformations pathways within the composites this enables rational tuning of composition and processing strategies toward targeted property enhancements for niche applications surface characteristics play a major role in composite performance making nano structured chitin an intriguing reinforcement filler the high specific surface area and abundant hydroxyl groups on crystalline chitin nanofibrils promotes adhesion and interfacial stress transfer within non-polar rubber matrices which lack inherent surface functionality moreover (Yihun 2022), established chemical modification procedures allow tailoring surface activity of chitin with grafting of polymer chains and reactive moieties to further bolster filler matrix compatibility and uniform dispersion this significantly boosts composite strength toughness, fatigue resistance, and durability beyond traditional fillers through synergistic melding across multiple length scales in the micro-nano reinforcement hierarchy controlled integration of chitin nanostructured networks intertwined within rubber matrices effectively restricts polymer chain mobility which manifests as increased tensile modulus and hardness values that can meet application demands in high load bearing components like transmission belts and compressor pads without compromising flexibility and resilience furthermore, oriented stiff chitin nanofibers foster crack bridging and deflection contributing to tear resistance and abrasion resistance surpassing that of carbon black and silica benchmarks by over 25% as expected (Arunkumar et al. 2023), tradeoffs exist between strength enhancements and ultimate elongation due to limited extensibility of the integrated filler phase however, retention of 50% strain at break satisfies requirements for high toughness components and vibration damping elements relative to common unfilled/carbon black rubbers the nanoscale effects extend to thermal transport where chitin fillers reduce conductivity by 20% through phonon scattering while also delaying heat-induced decomposition by over 35 °C through hindered chain motion and radical termination this significantly expands working temperature range meeting demands in high heat environments such as automotive engine components chitin nanostructures likewise deliver selective barriers to gases and liquids dependent on penetrant size this gives rubber products enhanced protection against environmental pervaporation while maintaining water vapor permeability for breathability the tortuous diffusion pathways around anisotropic whiskers provide tunable permeability reductions up to 60% for small penetrant molecules while only limiting diffusion of larger species by 25% the hierarchical and multifunctional characteristics thereby bridge across nano and micro morphologies toward enhancing bulk composite performance furthermore, the sustainable sourcing of crustacean waste-derived chitin coupled with low mixing energy of latex-based processing enables competitive commercial-scale production with 75% lower embodied energy relative to oil-based fillers given societal pressures toward environmentally responsible manufacturing the micro-nanoscale synergies of chitin-rubber composites provide a timely platform for creating next generation components boasting novel properties to meet critical application demands across sectors spanning transportation (Bragg-Sitton et al. 2020), infrastructure, medicine, and defense while balancing advanced functionality, scalable production, and materials sustainability through integration across multiple length dimensions these bio-inspired "structural hybrids" effectively mimic nature's balancing of intricacy and efficiency opening doors for continued innovation toward stretching limits of electrical transports, resilient shelters, and protective safety equipment that contribute lasting societal benefit through harmonious merger of material dimensionality scales (Wu 2017).

The dimensional scale and distribution of fillers blended into polymer composites exerts critical impact on the resultant interface area interactive phenomena, achievable dispersion parameters (Idumah and Obele 2021), ultimate property enhancements, feasible manufacturing methods and costs such that blends in micro composites from 1 to 100 μm sit intermediate between nano grades under 0.1 μm and macro grades above 100 μm determining delicate balances in performance and process ability specifically nano composites maximize the specific interfacial area enabling extensive polymer filler interaction molecular rearrangement and matrix reinforcement effectively bolstering mechanical properties through stress transfer mechanisms achieving the highest recorded property improvements yet nanoparticle integration relies on specialized solution processing or in situ polymerization that currently pose barriers for bulk production scalability moreover dense nano filler differentiation risks embrittlement reducing ductility while the exceedingly high surface area also makes nano composites prone to photooxidation and environmental degradation issues as compared to micro composites that deliver an optimal combination of integration control tailored enhancement and manufacturing flexibility through powder dispersions and melt compounding towards laminates (Yuan et al. 2019), adhesives and foams albeit micro composites do still experience limits to extensibility and thermo-oxidative stability relative to conventional macro composites although macro composites formulated with conventional sized fibers and large particle fillers are easiest to blend into standard thermoplastics and thermosets using commercially scalable processes but this comes at the expense of well-controlled dispersion and reinforcement limited to surface roughness or crack pinning effects hence the bulk matrix inherently dominates the composite response thus on balance micro composites can specifically target material attributes through understanding links from dispersion states to hierarchical structure–property translations harnessing interfacial volume effects between particles and chains while avoiding nano deriviatization challenges or macro blend deficits consequently micro composite development compels multiobjective perspectives connecting designed manufacturing selections with application-specific thermomechanical requisites environmental resilience and responsible end-of-life pathways accordingly the tunability according to precise length scales and phases offers both strategic advantages and disadvantages across the spectrum of property enhancement risk factors that collectively inform R&D trajectories based on balancing performance cost sustainability and scale translating laboratory innovations into viable real-world solutions (Schot and Steinmueller 2018).

9.5 Processing Techniques for Chitin-Rubber Micro Composites

Several processing techniques have been explored to achieve uniform dispersion and optimal interaction between chitin and rubber methods such as melt mixing (Mincea et al. 2012), solution blending, and in situ polymerization have been employed to create well-defined microstructures understanding the impact of processing conditions on the final composite's properties is crucial for tailoring the material to specific applications. The intricate morphology and phase distribution in multiphase polymer composites exerts first order control over bulk performance properties process selection for chitin rubber blends balances factors of filler dispersion, energy intensity, scalability, and final morphology multiple fabrication routes have emerged based on polymer type, filler characteristics, and targeted composite traits bench scale processes help to establish proof of concept and analyze fundamental structure property mechanisms while large-scale techniques aim to demonstrate viability for commercial manufacturing melt mixing involves high shear homogenization of a rubber base, typically natural rubber or styrene butadiene rubber SBR (Faverjon et al. 2023), with chitin powder fillers plus crosslinking agents at elevated temperatures 120–140 °C using batch mixers or extruders subsequent thermoforming or injection molding produces bulk samples for mechanical and permeability analysis advantages include versatility for different matrices and feasible scale-up however, high viscosity and inadequate mixing time risks poor dispersion additionally, high temperatures degrade chitin and rubber requiring antioxidant additives to mitigate effects thereof unfortunately, the nonlinear flow and thermal gradients during melt mixing induces no uniform filler distributions even with compatibilizer additives which manifests as lower tensile strength hence, melt blending methods currently suit thermos table binder systems until improved microscale control is achieved solution based techniques leverage dissolution of rubber and chitin in a volatile solvent such as toluene allowing thorough mixing under gentler conditions before controlled precipitation via no solvent addition or drying common configurations include emulsion styrene butadiene rubber latex with chitin nanoparticles and liquid natural rubber mixed with chitin whiskers latex cogel preceramic polymers facilitate highly homogeneous dispersions the dried composites display more consistent mechanical improvements relative to melt counterparts due to better nanofiller distribution casting processes permit film formation without need for thermal molding although residual solvent can influence curing kinetics additionally (Lim et al. 2008), chitin tends to agglomerate during drying which can negatively affect properties therefore, redispersion via sonication, surfactants, or functionalization helps safeguard homogeneity key strengths of solution approaches include near ambient temperatures to protect polymers improved particle integration at multiple scales and feasibility to incorporate graphene, clay, and carbon nanotube hybrid fillers net effects depend strongly on solvent type (Zhang et al. 2012), antiagglomeration treatments, and curing/drying profile offering extensive tuning flexibility a third category—in situ polymerization intertwines the filler during direct rubber polymer formation allowing precisely engineered morphologies one variation utilizes a Pickering emulsion process with chitin nanocrystals adsorbed at the droplet interfaces of a water-in-oil emulsion to template latex particle sizes alternatively, emulsion polymerization of liquid natural rubber occurs directly in the presence of chitin nanofibers achieving uniform incorporation within the networked rubber matrix finally, grafting to and grafting from polymerization chemically bonds chitin particles with polymer chains before curing via silane treatment solgel condensation, atom transfer radical polymerization, or radiation induced methods the covalent filler attachment fosters excellent dispersion and interfacial stress transfer contributing to significant toughening and stiffness enhancements that eclipse traditional composites however, in situ routes pose practical barriers to batch size enlargement and composite recovery which currently prevent widespread adoption outside niche highvalue applications independent of fabrication protocol applied, comprehensive physico mechanical characterization provides feedback to correlate process conditions with resultant properties as functions of composition mixing energy, shear rates, temperatures, and times scanning electron microscopy coupled with image analysis verifies chitin morphology and spatial distributions in rubber matrices providing morphology process structure linkages furthermore, spectroscopic and thermal techniques characterize molecular interactions at biopolymer interfaces including crystallinity and grafting efficacy which helps explain macroscale mechanical performance dynamic mechanical analysis indicates viscoelastic shifts in chain mobility relaxation transitions that direct modulus and damping changes similarly (Wood et al. 2016), tensile testing, hardness, tear strength, and fatigue measurements establishes key indicators of reinforcement quality comparing performance versus carbon black and clay benchmarks studies indicate shear mixing times beyond one hour fail to further improve dispersion suggesting an optimal energetic efficiency balance for melt blending processes moreover, in situ routes achieve the greatest property enhancements due to covalent hybridization but lack commercial feasibility currently solvent blending offers a middle ground with enhancements approaching in situ composites while preserving scalability spray drying and coagulation processes allow continuous production albeit with risks of re-agglomeration without sufficient stabilizers thus appropriate composite design selections balance targeted performance metrics versus manufacturing constraints in an application specific manner with processing modeled concurrently given the vast phase space combinations yet to explore significant opportunities exist to push property barriers through multiobjective optimization of fabrication protocols toward extreme mechanical responses, permeation selectivity (Wang et al. 2020), energy savings, and environmental stability realizing the full potential of chitin reinforced rubbery micro composites compels a systems perspective interrelating composition processing structure order and ultimately functionality via high throughput experimentation and simulation tools development of next generation chitin filled elastomeric composites demands a convergence of materials expertise spanning multiple length scales sustainable chemistry and manufacturing principles elevated levels of property performance benchmarking and multiscale modeling to systematically link process structure property relationships with validated predictive capabilities thereby accelerating knowledge growth and rational design of tailored composites possessing both advanced mechanical characteristics and feasible large-scale production through crosstalk between microscopic morphologies and macroscopic processing signatures composite researchers play a critical role in this progression through elucidation of fundamental crash transfer mechanisms at bio-inspired filler interfaces detailed mapping of vulnerabilities to chemical and environmental degradation advancing creativity in designing processing configurations that offer specific advantages over status quo methods commitment to experiential data sharing within the scientific community engagement with industrial collaborators to receive feedback on performance requirements and transition promising labscale developments toward pilot studies and ultimately commercial manufacturing scaled implementation lastly (Liu et al. 2019), life cycle assessment and end of life considerations will grow in importance when evaluating process technologies guiding responsible decision-making to maximize sustainability longterm the interdisciplinary knowledge generated across these efforts collectively drives innovation in developing the next generation of advanced chitin based biocomposites while accelerating adoption into society through scalable manufacturing practices that balance performance enhancement goals environmental stewardship and societal value addition (Zhang and Cao 2020). The observations indicate that as chitin content increases, agglomeration occurs, impacting the foam's cell morphology. These visual representations are crucial for understanding the structural changes induced by varying chitin concentrations in the foam composite. The cells exhibit a regular and circular shape, characteristic of the foam structure (Dananjaya et al. 2023). The walls between the cells appear smooth and evenly distributed. With the addition of 1 phr (parts per hundred parts of rubber) chitin, the foam's microstructure. The foam cells maintain their circular shape, and the chitin particles seem evenly dispersed, having minimal impact on the overall cell morphology. At 2 phr chitin loading, cell shape is still similar to the pure NRLF; however, a closer inspection reveals the beginnings of chitin agglomeration along the cell walls. With a chitin loading of 3 phr, noticeable changes in cell shape become apparent. Chitin agglomeration on the walls is more pronounced, and the foam cells begin to deviate from their circular form. The presence of chitin seems to influence the foam structure. At 4 phr chitin loading, the agglomeration of chitin on the walls intensifies. The cells exhibit a more irregular and distorted shape, signifying a greater influence of chitin on the foam's microstructure. With the highest chitin loading of 5 phr, the foam cells show a significant departure from their original circular form. Chitin agglomeration is more pronounced, and the foam structure appears to be substantially altered by the increased chitin content.

Several processing techniques have been explored for fabricating chitin-rubber micro composites (Qiu and Wang 2017), each presenting distinctive advantages and challenges. Melt mixing involves blending chitin powder with rubber and crosslinking agents at elevated temperatures, allowing thermal phase mixing but risking uneven dispersion and thermal degradation. Solution blending co-dissolves chitin and rubber in solvents like toluene, enabling lower temperatures and improved particle distribution but introducing challenges such as residual solvent and re-agglomeration. Latex coagulation combines chitin nanoparticles with SBR latex through controlled coagulation, ensuring nano-dispersion but posing challenges in coagulant removal. In situ polymerization incorporates chitin during rubber polymer formation, providing a precisely engineered structure with covalent particle attachment, though scale-up remains a current challenge (Paliwal et al. 2014). Electrospinning utilizes electric fields to create non-woven mats with microscale fiber diameters, enhancing mechanical reinforcement through fiber orientation. Each method contributes to the versatile landscape of chitin-rubber micro composites, offering tailored solutions for specific applications (Gaona et al. 2023).

Here are some examples of processing techniques that have been explored for fabricating chitin-rubber micro composites.

9.5.1 Melt Mixing

Chitin powder is blended with rubber (natural or synthetic like SBR) and crosslinking agents using high-shear or extruders at elevated temperatures (120–140 °C). This allows the thermal phase mixing of components (Wiśniewska et al. 2023). Downsides can be uneven dispersion and thermal degradation. In the dynamic realm of material science, the fusion of chitin, a natural biopolymer, with rubber in melt mixing has emerged as a pivotal avenue for creating innovative and versatile chitin-rubber micro composites. This method involves combining chitin powder with rubber, be it natural or synthetic, like Styrene-Butadiene Rubber (SBR). The process unfolds under the influence of crosslinking agents, utilizing high-shear mixers or extruders operating at elevated temperatures, typically from 120 to 140 °C. This thermal journey through melt mixing facilitates the blending of components, creating a robust and unified material with distinctive properties. The intricate dance between chitin and rubber is at the heart of melt mixing. Chitin, sourced from the exoskeletons of arthropods and the cell walls of fungi, brings its unique structural composition into play. Comprising linear chains of N-acetylglucosamine, chitin forms a resilient framework characterized by high tensile strength. On the other hand, rubber, whether derived from natural sources or synthesized into SBR, introduces elasticity and flexibility into the composite. The process begins with carefully blending chitin powder with the chosen rubber base. This amalgamation is not a mere physical combination but a strategic integration aimed at harnessing the synergies between the distinct characteristics of chitin and rubber. The chitin chains intertwine with the rubber matrix, creating a material that inherits the tensile strength of chitin while retaining the elasticity and resilience intrinsic to rubber. Elevated temperatures, ranging between 120 and 140 °C, catalyze the thermal phase mixing of chitin and rubber (Liu et al. 2023). This temperature range allows the components to undergo a state of flux, blending intimately and forming a cohesive material. The high-shear mixers and extruders employed in this process play a pivotal role in ensuring the uniform distribution of chitin within the rubber matrix. The thermal energy acts as a facilitator, breaking down the physical barriers between chitin and rubber, encouraging molecular interactions, and promoting the creation of a homogeneous blend (Sun et al. 2023). This thermal synergy transforms the individual components into a unified entity, where the distinctive properties of chitin and rubber coalesce into a material that transcends the sum of its parts. However, challenges arise in achieving perfect harmony as with any intricate dance. One such challenge in the realm of melt mixing for chitin-rubber composites is the potential for uneven dispersion. Despite the efforts to achieve uniform distribution, the shear forces and elevated temperatures may not always guarantee homogeneity. Uneven dispersion can lead to variations in the mechanical and thermal properties of the final composite. Areas with higher concentrations of chitin may exhibit different characteristics compared to regions where chitin is sparsely distributed. This non-uniformity challenges ensuring consistent and predictable performance across the entire material. Thermal degradation stands as another hurdle in the journey of melt mixing. While essential for achieving the desired blending, elevated temperatures can also subject the components to thermal stress. Chitin and rubber may be susceptible to degradation or changes in their properties under prolonged exposure to high temperatures. To mitigate these challenges, researchers delve into the intricacies of the melt mixing process. They explore ways to optimize conditions, fine-tune parameters, and introduce additives that enhance dispersion and mitigate thermal degradation. The goal is to refine the dance, ensuring that the choreography of chitin and rubber in melt mixing results in a seamless and well-balanced performance. In the pursuit of overcoming challenges, researchers employ a multidisciplinary approach. Advanced techniques such as rheological studies, thermal analysis, and microscopy provide insights into the behavior of chitin and rubber during the melt mixing process. These studies enable a deeper understanding of the factors influencing dispersion and thermal stability. Additives are crucial in addressing uneven dispersion and thermal degradation (Davoodi et al. 2023). Antioxidants may be introduced to mitigate the impact of high temperatures on the components, preserving their integrity throughout the melt mixing process. Compatibility or coupling agents are explored to enhance the interaction between chitin and rubber, promoting a more even dispersion of chitin within the matrix. Innovative modifications to processing equipment and techniques also contribute to the refinement of melt mixing (Huang et al. 2023). For example, the design of mixing elements in extruders is optimized to achieve better dispersion without compromising the integrity of chitin or rubber. Process parameters such as mixing time, shear rate, and temperature profiles are systematically adjusted to strike the delicate balance required for a harmonious blend (Frey et al. 2023). As the journey through melt mixing unfolds, the future holds promise and potential. Researchers continue to explore novel solutions to challenges, seeking to unlock the full potential of chitin-rubber blends. The quest for improved dispersion, enhanced thermal stability, and consistent performance drives innovation in processing techniques and material formulations. The dance of chitin and rubber in melt mixing is not just a scientific endeavor but a choreography of sustainability and performance. With its biodegradability and low toxicity, Chitin aligns seamlessly with the growing demand for environmentally conscious materials. Incorporating chitin into rubber matrices through melt mixing opens avenues for developing sustainable and high-performance composites. The challenges encountered in melt mixing for chitin-rubber composites are not roadblocks but stepping stones toward refinement. Each hurdle invites exploration, experimentation, and the application of ingenuity. The researchers and scientists engaged in this dance are not merely observers; they are active participants, refining the steps, and elevating the performance of chitin-rubber blends. In conclusion, melt mixing stands as a transformative process in the creation of chitin-rubber micro composites. It is a dance where the partners, chitin and rubber, merge their unique attributes, creating a material that holds the promise of enhanced mechanical strength, thermal stability, and environmental consciousness (Mawardi et al. 2023). As researchers continue to refine the choreography, melt mixing emerges as a key enabler in the journey towards sustainable and high-performance materials that transcend conventional boundaries.

9.5.2 Solution Blending

In the ever-evolving symphony of material science, the solution blending technique emerges as a nuanced and intricate movement, orchestrating the fusion of chitin and rubber into finely tuned composites (Thapliyal et al. 2023). This method involves the co-dissolution of chitin and rubber in solvents such as toluene, followed by controlled drying to aggregate the composites. Unlike its counterpart, melt mixing, solution blending dances to a different tune, enabling the creation of chitin-rubber micro composites at lower temperatures with the promise of better particle distribution. However, within this harmonious composition, challenges arise in the form of residual solvent and the re-agglomeration specter, adding complexity layers to the melody of solution blending (Khan et al. 2023). Solution blending embarks on a delicate ballet where chitin and rubber find unity in the embrace of solvents. Chitin, sourced from the robust exoskeletons of arthropods and the resilient cell walls of fungi, brings its unique structural composition into the solvation sphere. The linear chains of N-acetylglucosamine dissolve gracefully alongside rubber (Li 2023), whether it be the elasticity of natural rubber or the synthetic resilience of Styrene-Butadiene Rubber (SBR). Toluene, a common solvent in this symphony, is the fluid medium where chitin and rubber engage in a nuanced dance of dissolution. Unlike the elevated temperatures in melt mixing, solution blending operates at lower thermal thresholds. This lower energy environment fosters a gentler interaction between the components, allowing them to harmonize without the stress of high temperatures. The choreography of solution blending reaches its crescendo with the act of controlled drying. The solvent, once the facilitator of dissolution, now takes a backseat as the composites aggregate through the controlled removal of this fluid medium. The drying process carefully orchestrates the formation of chitin-rubber micro composites, preserving the spatial arrangement achieved during dissolution. The controlled drying phase is pivotal in shaping the final structure of the composite. It allows for the aggregation of chitin particles within the rubber matrix, creating a well-defined microstructure. This control over the drying process contributes to the potential for better particle distribution, a characteristic that is significant in determining the resulting composite's mechanical and thermal properties. One distinctive feature of solution blending is its ability to operate at lower temperatures than melt mixing. This cooler environment is particularly advantageous for heat-sensitive materials, preventing thermal degradation and preserving the inherent properties of both chitin and rubber. The gentleness of this process aligns with the growing emphasis on sustainable and energy-efficient manufacturing practices. Lower temperatures also extend a welcoming hand to the environmental aspect of the composite's lifecycle. Solution blending, with its reduced energy requirements, echoes the broader narrative of eco-friendly material production. The dance between chitin and rubber unfolds in a cooler ambiance, showcasing the adaptability of solution blending to diverse material sensitivities. While the solution blending ballet unfolds gracefully, it leaves a lingering echo in the form of residual solvent challenges. Toluene, or the solvent of choice, may not gracefully exit the stage after its role in dissolution and composite formation (Yi et al. 2023). Residual solvent introduces complexity and necessitates meticulous attention to detail in the subsequent processing phases. The presence of residual solvent can influence the mechanical and thermal properties of the final composite. Addressing this issue is imperative to ensure that the material meets the desired specifications. Techniques such as post-treatment processes, prolonged drying, or the introduction of alternative solvents may be explored to mitigate the impact of residual solvents (Wang et al. 2023). As the curtain falls on the solution blending performance, there is an unwanted encore waiting in the wings—the potential for re-agglomeration. The gentle dissolution and controlled drying that crafted the composite structure may, under certain circumstances, lead to the re-clustering of chitin particles. This phenomenon can result in non-uniform distribution and compromise the homogeneity achieved during the blending process. Researchers engaged in solution blending grapple with strategies to prevent or minimize re-agglomeration. Redispersion techniques, such as sonication or the introduction of surfactants, may be employed to break down agglomerates and restore uniformity. These efforts contribute to refining the composition, ensuring the material retains the desirable characteristics envisioned in the initial choreography. Despite its challenges, solution blending remains a masterful technique in creating chitin-rubber micro composites. Researchers, armed with knowledge and innovation, continue to refine the symphony, introducing new movements to master the challenges that accompany this method. In residual solvent management, advancements in post-treatment processes and drying techniques strive to minimize the impact on the final composite. Sustainable solvent choices and efficient recovery methods form part of the conductor's repertoire, orchestrating an environmentally conscious performance. The specter of re-agglomeration is met with countermeasures that seek to maintain the harmonious distribution achieved during solution blending. Redispersion techniques, whether through mechanical means or the introduction of stabilizing agents, contribute to preserving a uniform structure within the composite. As solution blending continues to be a prominent technique in creating chitin-rubber micro composites, the overture to future innovations echoes through the halls of material science. Researchers delve into the intricacies of solvent dynamics, seeking alternative solvents with minimal environmental impact. The exploration of novel solvents contributes to the sustainability of the solution-blending process, aligning with the broader narrative of eco-friendly manufacturing. In parallel, advancements in analytical techniques play a crucial role in unraveling the mysteries of solution blending (Ahmad et al. 2024). High-resolution microscopy, spectroscopy, and computational modeling provide insights into molecular interactions, dispersion patterns, and composite structures. These tools empower researchers to refine their choreography (Smith et al. 2023), achieving greater precision in synthesizing chitin-rubber micro composites. In conclusion, solution blending in chitin-rubber micro composites encapsulates a harmony in progress. It is a dance where the partners—chitin, rubber, and solvent—move synchronously to create materials that bridge sustainability and performance. The challenges posed by residual solvent and re-agglomeration are not stumbling blocks but invitations for innovation and refinement. As the symphony of solution blending evolves, it sets the stage for a future where chitin-rubber micro composites find applications across diverse industries. The ongoing research endeavors, guided by the principles of sustainability and material excellence, promise to unravel new chapters in the fascinating narrative of solution blending. The dance continues, and with each movement, researchers bring us closer to a harmonious blend of chitin and rubber that resonates with the needs of a dynamic and conscious world.

9.5.3 Latex Coagulation

Natural rubber latex has limited applications due to its low tensile strength, but sulfur vulcanization used to reinforce it causes problems like toxicity and non-recyclability. Chitin nanoparticles combine with SBR latex via controlled coagulation chemistry to solidify composites from the emulsion state while securing nano-dispersion. Trade-offs exist with coagulant removal (Yang et al. 2023). In the grand tapestry of material science, the technique of latex coagulation emerges as a captivating chapter, weaving the intricate threads of chitin nanoparticles with Styrene-Butadiene Rubber (SBR) latex. This method orchestrates controlled coagulation chemistry to solidify composites from the emulsion state while ensuring nano-dispersion. A delicate dance ensues between chitin and latex, where the nuances of coagulation become the choreographer, directing the formation of finely tuned chitin-rubber micro composites (Gaona et al. 2023). However, trade-offs arise in pursuing this exquisite harmony, most notably in the realm of coagulant removal, adding complexity to the composition.

One of the hallmarks of latex coagulation is its ability to secure nano-dispersion. The controlled coagulation process, when executed masterfully, ensures that chitin nanoparticles are dispersed at the nanoscale within the rubber matrix. This nanoscale dispersion is a crucial contributor to the enhanced properties of the composite. Nano-dispersion creates a large interfacial area between chitin and rubber, promoting effective stress transfer and reinforcement. The intimate intermingling of chitin at the nanoscale with the rubber matrix results in a composite material that capitalizes on the unique properties of both constituents. This elegant distribution is a testament to the precision achievable through latex coagulation. By latex coagulation, the graphene oxide/chitin nanocrystal hybrids can effectively reinforce the natural rubber latex without sulfur vulcanization, offering a more sustainable alternative. Liu et al. (2021) discussed natural rubber latex and graphene oxide/chitin nanocrystal hybrids as reinforcements for high-performance elastomers without sulfur vulcanization. The authors introduce a small amount of graphene oxide/zwitter ionic chitin nanocrystal hybrids into natural rubber latex to reinforce it without sulfur vulcanization. The chitin nanocrystals act as molecular bridges between the graphene oxide sheets and natural rubber particles due to their attraction to both. They also improve the dispersion of graphene oxide in the natural rubber. The graphene oxide/chitin nanocrystal hybrids form uniform networks that absorb onto the natural rubber macromolecules, significantly enhancing the mechanical properties. The resulting natural rubber/hybrid composites exhibit a high tensile strength of 19.2 MPa while maintaining a sizeable breaking elongation of 825.9%, rivaling sulfur-vulcanized rubber. Natural latex gloves made from natural rubber/hybrid composites perform better than sulfur-vulcanized ones, including recyclability, higher water vapor permeability, and good biocompatibility. The preparation of natural rubber (NR) nanocomposites involved an eco-friendly aqueous solution dipping method (Bambo et al. 2023). Initially, graphene oxide (GO) and nanochitin (NC) were mixed, and the pH of the mixture was adjusted to 10 using a 0.5 M NaOH solution. Subsequently, the mixture underwent a 2-h sonication process to achieve a homogeneous suspension, resulting in graphene-chitin (GC).

The process of obtaining GC is a crucial step in the preparation. Following this, the homogeneous suspensions of GO, NC, and GC were individually added to pristine NR latex. The mixtures were stirred continuously at room temperature for 1 h to ensure proper integration. Glass molds, coated with coagulators, were then dipped into the latex mixtures to form composite films, namely NR/GO, NR/NC, and NR/GC. The final composite films were dried at 40 °C for 12 h in an oven and carefully removed from the glass molds using silica powders (Sharma et al. 2023). Notably, the content of each nanofiller (GO, NC, GC) introduced into the NR latex was maintained at 1 wt%, and the ratio of GO to NC in the GC hybrid was set at 2:1 unless specified otherwise. This method aligns with eco-friendly practices and provides a systematic approach to achieving well-dispersed nanofillers in NR matrices, paving the way for enhanced properties in the resulting nanocomposites. The schematic highlights a well-thought-out interface design that capitalizes on the synergistic effects of NC, GO, and NR, aiming to achieve improved mechanical, thermal, and functional properties in the resulting nanocomposites. The depiction underscores the innovative approach to creating a molecularly interconnected structure within the nanocomposite, paving the way for advanced materials with tailored characteristics (Ruan et al. 2023).

Latex coagulation sets the stage with the entrancing concept of emulsion—a colloidal suspension where chitin and SBR latex coexist in a dynamic equilibrium. With its structural elegance derived from the exoskeletons of arthropods, Chitin introduces its nanoparticles into the emulsion (Lu et al. 2023). SBR latex, a synthetic rubber known for its versatility, becomes the fluid medium that facilitates the intricate dance of dispersion. Controlled coagulation chemistry, the maestro of this process, guides the transition from emulsion to a solidified state. Coagulation refers to the aggregation or clumping of particles, and in the context of latex coagulation, it marks the transformation of the emulsion into a coherent composite. This controlled coagulation is essential for achieving uniform dispersion of chitin nanoparticles within the rubber matrix. This coagulation process is fundamental in the production of rubber, where latex is processed to obtain rubber sheets or other rubber products. Adding acid to destabilize the colloidal suspension is a key step in initiating coagulation. Latex is a colloidal suspension of rubber particles in water. Each rubber particle is enveloped by a protein membrane, creating a negatively charged barrier around the rubber molecules. The negatively charged protein membrane on the rubber particles creates a repulsive force between them, preventing them from coming together. As a result, latex remains in a liquid, dispersed state. Acid, which provides H+ ions, is added to the latex. The H+ ions neutralize the negatively charged protein membrane around the rubber particles. As the H+ ions neutralize the negative charge on the protein membrane, the repulsive forces between rubber particles are reduced. With the reduction in repulsive forces, the rubber particles are no longer strongly repelled from each other. They start to collide and come into proximity. The collisions and reduced repulsion lead to the breakdown of the protein membrane surrounding the rubber particles. The once-stabilized colloidal suspension begins to destabilize. The rubber molecules are no longer individually stabilized as the protein membrane breaks down. Instead, they start to clump together. The clumping and aggregation of rubber molecules result in the coagulation of latex (Suksaeree et al. 2023). The once stable liquid latex transforms into a coagulated or solid state. The heart of latex coagulation lies in the solidification from emulsion. Chitin may interact with the protein membrane surrounding the rubber particles. It can potentially adsorb onto the surface of rubber particles or interact with the protein components of the membrane. This interaction can influence the stability of the protein membrane and its response to coagulation agents. This transformation defines the spatial arrangement of chitin within the rubber matrix, contributing to the microstructure that governs the material's properties. The controlled coagulation process ensures that the chitin nanoparticles are evenly distributed within the rubber matrix (Liao et al. 2023), a critical aspect for achieving the desired mechanical and thermal characteristics. The choreography of coagulation dictates the spatial relationships between chitin and rubber, influencing the synergistic interactions that define the performance of the resulting chitin-rubber micro composite. While the dance of latex coagulation unfolds with grace, it introduces a complex choreography when it comes to coagulant removal. The coagulation process often involves the use of coagulants or precipitating agents that aid in the aggregation of particles. However, the challenge lies in ensuring that these coagulants are effectively removed from the final composite. Residual coagulants can have detrimental effects on the properties of the chitin-rubber micro composite. They may act as impurities, influencing the material's mechanical, thermal, and even biodegradable characteristics. Therefore, removing coagulants becomes critical in refining the composition and ensuring that the composite aligns with the desired specifications. Efficient coagulant removal methods are a focal point in the ongoing research on latex coagulation. Filtration, washing, and purification steps are employed to eliminate residual coagulants, but each method introduces challenges. Achieving thorough coagulant removal without compromising the integrity of the composite remains an intricate ballet in the synthesis process. The choice of SBR latex as the partner in this dance of latex coagulation is strategic and contributes to the versatility of the resulting composite. SBR latex, a synthetic rubber copolymer of styrene and butadiene, brings a spectrum of properties to the composition. Its flexibility, resilience, and compatibility with aqueous systems make it an ideal counterpart for chitin in the emulsion. The versatility of SBR latex extends beyond its role in the coagulation process. It forms a robust matrix that accommodates the dispersion of chitin nanoparticles while contributing to the composite's overall mechanical strength and flexibility (Dutta and Sit 2024). The interplay between chitin and SBR latex sets the stage for a composite material that leverages the best of both worlds.

9.5.4 In Situ Polymerization

Chitin fills rubber directly during polymer formation via Pickering emulsions or radiation grafting. Provides precisely engineered structure with covalent particle attachment but current barriers in scale-up. In the intricate realm of chitin-rubber micro composites, the in situ polymerization technique emerges as a captivating and innovative process (Monfared et al. 2024). This method, akin to a master weaver, integrates chitin seamlessly into the rubber fabric during polymer formation. The weaving is orchestrated through Pickering emulsions or radiation grafting methods, creating a precisely engineered structure with covalent particle attachment. Yet, amid the elegance of this synthesis, the composition encounters scale-up barriers, adding complexity to the potential proliferation of this remarkable approach. In situ polymerization opens with the promise of a transformative overture, where chitin becomes an integral part of the rubber matrix from its inception. Unlike other blending techniques, in situ polymerization takes shape during the creation of the rubber itself, offering a unique opportunity for a more intimate and tailored integration of chitin at the molecular level. The essence of in situ polymerization lies in creating what can be termed a "biohybrid" material. This biohybrid combines the inherent properties of rubber with the structural benefits conferred by chitin (Abdul Khalil et al. 2024), resulting in a composite that capitalizes on the strengths of both constituents. The amalgamation is achieved by carefully orchestrating the polymerization process, where chitin particles are entwined within the polymer chains.

9.5.5 a. Pickering Emulsions: A Symphony of Stability and Precision

One of the key pathways through which in situ polymerization manifests is the utilization of Pickering emulsions. Pickering emulsions, named after the British scientist Percival Spencer Pickering, involve stabilizing droplets of one phase in another using particles as emulsifiers (Al-Sakkaf and Onaizi 2024). In the context of chitin-rubber micro composites, chitin particles act as stabilizers for the emulsion. The process unfolds as chitin particles adsorb onto the interfaces of the rubber droplets within the emulsion. This adsorption stabilizes the droplets, preventing their coalescence and ensuring a finely dispersed system. As the polymerization reaction commences, chitin becomes entwined within the growing polymer chains, creating a composite with a high degree of control over the spatial distribution of chitin. Shuang et al. (2024) discussed the use of carboxylated chitin nanowhiskers (C-ChNWs) as a stabilizer for Pickering emulsions and high internal phase Pickering emulsions (HIPPEs). C-ChNWs were prepared from chitin powder using ammonium persulfate oxidation, introducing carboxyl groups onto the chitin surface. The C-ChNWs had an acicular or rod-like shape with an aspect ratio of 8.8 and a zeta potential of −48.9 mV, making them suitable as Pickering emulsion stabilizers (Hei et al. 2024; Aratani et al.

2024). The C-ChNWs could adsorb at the oil–water interface and form a network structure stabilizing the Pickering emulsions. Pickering emulsions with internal phase volume fractions from 0.5 to 0.8 were successfully prepared using the C-ChNWs. The HIPPEs stabilized by the C-ChNWs showed good texture characteristics, indicating potential application in 3D printing.

Perrin et al. (2014) discussed using chitin nanocrystals (ChiNC) as stabilizing particles for oil-in-water emulsions. ChiNC are rod-like nanoparticles derived from chitin, an abundant natural polymer. ChiNC has dimensions of around 160 nm by 16 nm and a positive surface charge density of 0.48 charge/nm^2 due to amine groups. ChiNC can stabilize oil droplets to form Pickering emulsions. The emulsions are stable over time despite natural creaming. The average droplet size is not significantly affected by ionic strength or pH (Khalesi et al. 2024). To form stable emulsions, an ionic strength of at least 2 mM is needed to screen the positive charge on the ChiNC and allow them to pack densely at the oil–water interface. Higher ChiNC concentrations result in smaller average droplet sizes up to a plateau where the interface is densely covered. ChiNC can also stabilize high internal phase emulsions (HIPE) containing 96% oil. The HIPE forms a gel whose texture can be modulated from soft to solid by adjusting ChiNC concentration, pH, and ionic strength. The ability of ChiNC to stabilize both uncomplicated and high internal phase emulsions makes them a versatile and stable emulsifying agent. ChiNC (Chitin Nanocrystals) demonstrated its ability to conform at the oil/water interface, as evidenced by the SEM (Scanning Electron Microscopy) (Li et al. 2024a). The coverage of the beads, representing the ratio of the particle area available to stabilize the interface, was calculated. The total area of the oil droplets, considering the measured droplet diameter for each emulsion, was considered in this calculation. The result indicated an approximate 100% coverage, highlighting the exceptional ability of these solid particles to cover and stabilize the interface.

9.5.6 b. Radiation Grafting: Bonding Chitin and Rubber at a Molecular Level

Another enchanting movement within the symphony of in situ polymerization involves radiation grafting. In this method, the polymerization occurs in the presence of radiation, leading to the grafting of chitin onto the polymer chains (Mincheva et al. 2024). The use of radiation as a catalyst allows for the creation of covalent bonds between chitin and rubber, establishing a molecular-level connection. Radiation grafting offers precision and control that is distinctive to in situ polymerization. The covalent bonding ensures a strong and stable linkage between chitin and rubber, creating a biohybrid material with enhanced mechanical properties. The process allows for tailoring the composite's structure at the molecular level, providing avenues for fine-tuning its properties to meet specific application requirements. A defining characteristic of in situ polymerization is the covalent attachment of chitin particles to the polymer chains. Unlike other blending methods where particles are dispersed within the matrix without a strong chemical bond, in situ polymerization forges a molecular symphony where chitin becomes an integral part of the polymer structure. Covalent bonds are among the most robust chemical connections, and their formation between chitin and rubber in the in situ polymerization process imparts remarkable strength and durability to the resulting composite. This covalent attachment ensures that chitin does not simply reside within the rubber matrix but becomes active in the overall structure, influencing the material's mechanical, thermal, and even biodegradable properties (Ding et al. 2024).

9.5.7 Electrospinning

Chitin-rubber blended solutions pass through electric fields to fabricate non-woven mats with fiber diameters in the microscale range. Fiber orientation helps mechanical reinforcement. In the grand tapestry of chitin-rubber micro composites, the electrospinning technique is a meticulous craftsman, weaving intricate non-woven mats with fibers in the microscale range. This method transforms blended solutions of chitin and rubber under the influence of electric fields, creating a symphony of fibers that not only showcase microscale elegance but also promise enhanced mechanical reinforcement. Let us embark on a journey through electrospinning, exploring its nuances, applications (Bhatia et al. 2024), and the harmonic potential it brings to the world of chitin-rubber composites. Electrospinning, often likened to cotton candy spinning, involves the controlled ejection of a polymer solution or melt through a fine nozzle in the presence of an electric field. The electric field induces a stretching and thinning of the solution, forming ultrafine fibers that are then collected on a grounded surface, creating a non-woven mat. This seemingly delicate yet intricately controlled process has found applications across various fields, including textiles, medicine, and, more recently, in the realm of composite materials (Siwach et al. 2024). Electrospinning in creating chitin-rubber micro composites represents a sophisticated approach to fabricating materials with tailored structures at the microscale. The resulting fibers, with diameters in the microscale range, offer unique properties that stem from the interplay between chitin and rubber, providing mechanical reinforcement and opening avenues for diverse applications. The electrospinning process commences with preparing a solution blending chitin and rubber. This solution, a harmonious blend of two distinct materials, sets the stage for the intricate dance of chitin and rubber fibers. The solution is carefully loaded into a syringe, and a high voltage is applied between the syringe and a grounded collector. As the electric field intensifies, the solution at the tip of the syringe undergoes a transformation. Charged ions within the solution respond to the electric field, initiating what can be described as a dance of delicate threads. The solution is drawn towards the collector in a fluidic ballet, with the electric field dictating the movements of the chitin and rubber components. The stretching and thinning of the solution result in the formation of ultrafine fibers delicately deposited on the collector, creating a non-woven mat. The beauty of electrospinning lies in the microscale elegance of the resulting fibers. The diameter of these fibers typically ranges from a few nanometers to several micrometers, presenting a level of intricacy and precision characteristic of the electrospinning process. In the context of chitin-rubber micro composites, the microscale fibers hold significance for several reasons. Firstly, the microscale diameter allows for a high surface area-to-volume ratio. This increased surface area can facilitate enhanced interactions between the chitin and rubber components, influencing properties such as adhesion, mechanical strength, and even the permeability of the resulting material. The intimate contact at the microscale creates a platform for synergistic effects, where the unique characteristics of chitin and rubber harmonize to produce a composite material with properties distinct from those of individual components (Mincheva et al. 2024). Secondly, the microscale fibers contribute to the overall structural integrity of the non-woven mat. The entangled network of fibers forms a cohesive structure akin to a musical composition where each note plays a role in the harmony of the entire piece. This structural integrity translates into mechanical reinforcement, imparting strength and resilience to the electrospun chitin-rubber composite. One of the key promises of electrospinning in the context of chitin-rubber composites is the mechanical reinforcement achieved by forming microscale fibers. Rubber, known for its elasticity, can benefit significantly from adding chitin as electrospun fibers (Angel et al. 2024). The entangled network of fibers provides a reinforcing scaffold within the rubber matrix, akin to the support beams in a grand concert hall. The mechanical properties of the electrospun chitin-rubber composite can be tailored by adjusting parameters such as the composition of the solution, the processing conditions, and the collector design. The resulting composite can exhibit improved tensile strength, flexibility, and tear resistance, which are crucial in various industrial applications. The orientation of the fibers within the non-woven mat further contributes to the mechanical properties of the composite. The controlled deposition of fibers during electrospinning allows for alignment and orientation, resembling the arrangement of instruments in an orchestra. This fiber orientation influences the anisotropic mechanical behavior of the composite, offering opportunities to design materials with specific directional properties. The applications of electrospun chitin-rubber composites extend across a spectrum of industries, each note in this symphony resonating with the potential for innovation and performance enhancement. The unique properties conferred by the microscale fibers find relevance in diverse fields, shaping the narrative of how these composites can contribute to advancements in materials science. In textiles, electrospun chitin-rubber fibers could be incorporated into fabrics to impart mechanical strength, tear resistance, and antimicrobial properties. The microscale structure of the fibers could create textiles with enhanced breathability and comfort, opening avenues for the development of advanced clothing materials. In biomedical applications, the mechanical reinforcement provided by electrospun chitin rubber composites could be useful in fabricating tissue scaffolds and medical implants. The controlled fiber orientation and the biocompatibility of chitin make these composites suitable for applications where interaction with biological systems is essential. While the electrospinning symphony brings harmonious melodies, it is not without its challenges and refrains. The solution's precise composition of chitin and rubber is crucial for obtaining the desired properties in the electrospun fibers (Fahanwi et al. 2024). The interplay between these components, influenced by solution viscosity, concentration, and electrospinning parameters, requires fine-tuning to orchestrate a well-balanced composition. One challenge lies in ensuring uniform dispersion of chitin within the solution. Agglomeration of chitin particles can lead to uneven fiber formation, akin to a discordant note in an otherwise melodious composition. Addressing this challenge involves exploring methods such as surface modification of chitin particles or the incorporation of dispersants to enhance their compatibility with the rubber matrix (Fazeli et al. 2024). The choice of solvents for the solution also plays a crucial role in the electrospinning process. The selection of solvents affects the solubility of both chitin and rubber, influencing the spinnability of the solution and the properties of the resulting fibers. Balancing the solubility of the components while maintaining the desired processing conditions requires a delicate calibration, much like tuning instruments in an orchestra to achieve perfect harmony. As researchers and engineers continue to refine the orchestration of electrospinning techniques, the applications of chitin-rubber composites are set to expand, reaching into realms yet unexplored. The symphony of electrospinning, with its ability to craft microscale fibers with precision, adds a distinctive tune to the chorus of materials science, ushering in a new era of tailored composites with enhanced properties and diverse applications. The final movement of this symphony is still being written, and with each note, the potential for innovation and discovery continues to resonate.

Various processing techniques for chitin-rubber micro composites highlight their advanced aspects and limitations. Melt Mixing involves blending chitin powder and rubber at elevated temperatures, offering versatility and feasible scale-up but facing challenges like uneven dispersion and thermal degradation. Solution Blending codissolves chitin and rubber, showing improved particle integration yet presenting challenges such as residual solvent influence and agglomeration. Latex Coagulation achieves nano-dispersion in latex, providing control over particle size, but faces difficulties in coagulant removal. In situ Polymerization offers covalent filler attachment for excellent dispersion and enhancements in toughness and stiffness (Xue et al. 2024), yet has practical barriers to batch size enlargement. Electrospinning creates microscale fibers with high surface area and enhanced mechanical properties, but challenges exist in achieving precise composition and scalability, as well as potential agglomeration of chitin during the process.

9.6 Applications and Future Prospects

The unique synergies arising from chitin-rubber composites lead to combinations of flexible toughness, high strength, and thermal stability suitable for diverse applications across industrial sectors (Ali et al. 2024). Specifically, the enhanced mechanical, rheological, and permeation properties benefit manufacturing durable, high-performance products to meet critical transportation, infrastructure, medicine, and defense needs. Furthermore, the sustainability of crustacean-derived chitin coupled with optimized material usage through lightweight leads to significant reductions in environmental impact relative to traditional composites and petroleum-based rubbers. Natural rubber has long served a vital role in tires due to its unmatched abrasion resistance and handling performance (Jašarević 2024). However, incorporating chitin as a secondary reinforcing filler has been shown to further augment wear resistance and cut growth resistance. Chitin-silica hybrid systems as an alternative to carbon black fillers display 25 and 65% improvements in tread wear and rolling resistance over conventional unfilled formulations. The hierarchical reinforcement effects on dynamic mechanical properties also help to lower vibration and noise. Nie et al. (2019) created a composite by mixing epoxidized natural rubber and chitin nanocrystals without conventional crosslinking agents. Hydrogen bonds formed between the chitin nanocrystals and epoxidized natural rubber, creating a supramolecular network. Some of the hydroxyl groups on the chitin nanocrystals reacted with the epoxy groups on the rubber, which helped form the supramolecular network and improve the composite's mechanical properties and self-healing ability. The tensile strength of the composite with 20% chitin nanocrystals was almost double that of the neat epoxidized natural rubber. The composite with 20% chitin nanocrystals showed a self-healing efficiency of 83% of its original strength and 95% of its original extensibility without any external stimulus at room temperature. Surface functionalization of the chitin particles creates additional chemical bonding with the rubber matrix to stabilize the interface. The resulting composites cope with repeated cyclic deformations and extreme stresses during high-speed subsonic flight. When blended into glycol-plasticized biobased polyurethane rubber, the chitin nanofibrils raise tensile strength by over 80% compared to the neat matrix, meeting safety margins for interior aircraft sealing applications. Due to its gas barrier properties, chitin foil membrane integration allows tires to maintain inflation pressure longer through suppressed diffusion and permeation. This effect enhances fuel economy ratings since underinflated tires increase rolling resistance and reduce tread life (Alkhalifa et al. 2024). The biocompatibility, non-toxicity, and intrinsic antimicrobial properties of chitin make chitin-rubber blends well-suited for various biomedical devices. Specifically, the materials avoid unfavorable pH changes, which help prevent bacterial infection and biofilm formation. Tubing, seals, and medical gloves fabricated with 10% chitin loading in silicone rubber increase tear resistance by over 40% without compromising flexibility and dexterity. Furthermore, the nanostructured surface topography showed reduced tissue attachment during biological evaluations (Zarei et al. 2024). Compared to traditional rubbers, the composites also better withstand sterilization treatments including autoclaving, chemical disinfection, and gamma irradiation. Radical scavenging additives such as vitamin E work synergistically with the chitin nanofillers to hinder polymer chain scission. For cartilage tissue replacements that require load-bearing resilience, chitin whale reinforcement of polydimethylsiloxane networks creates compressive modulus comparable to natural tissue while avoiding brittle fracture modes. Controlled swelling then allows cell adherence for regenerative medicine scaffolds supporting stem cell growth. In wound healing applications, chitin nanoparticles are loaded with anti-inflammatory and antimicrobial drugs embedded in silicone matrices with tuned adhesion strength, facilitating release profiles that accelerate healing with reduced scarring. The high drug payload capacity expands treatment options, particularly for burns, diabetic ulcers, and trauma injuries. A single composite pad application can achieve sustained release over 5–7 days. Chitin reinforcement of peroxide-cured chloroprene rubber produces lightweight, flexible materials with hardness and tensile strength exceeding most epoxy-based composites. 25% chitin loading withstands over 650,000 fatigue cycles before failure at 20% strain compared to under 100,000 cycles for carbon black fillers. Such remarkable durability suits sealing elements, insulation panels, and vibration dampers for infrastructure assets, including high-speed rail, bridges, wind turbines, dams, and marine platforms. Under large compressive loads, the energy absorption capability of chitin hydrogel/SBR hybrid foams surpasses even syntactic PU foams by over 30%, owing to efficient stress distribution and prevention of catastrophic collapse. The materials protect structures against earthquakes and blast/shock waves (Dong et al. 2024). Their high elongation return during off-loading cycles preserves dimensional stability for repeated loading events. The eco-friendly nature of crustacean waste-derived chitin combined with the absence of volatile organic compounds during processing enables manufacturing with over 80% lower carbon footprint and embodied energy than traditional glass/carbon composites and synthetic rubbers. This aligns with strict environmental regulations. Optimized material usage through potential lightweight using chitin micro reinforcements yields further sustainability benefits over status quo materials for construction and infrastructure.

9.7 Future Outlook and Challenges

The encouraging enhancements of chitin rubber composites across mechanical, barrier, and biological metrics validate the viability of the applications above. However, work remains to mature the technology through continued efforts in the following areas:

- Overcoming inconsistencies in sourcing, purification, and reliable bulk supply of suitable chitin grades. Investment in efficient biorefinery methods is vital.
- Advancing predictive multiscale modeling frameworks to accelerate knowledge transfer between processing, structure, and properties. Holistic experiment-simulation dialogue illuminates synergistic mechanisms and optimizes composite formulations/configurations toward targeted responses.
- More extensive product prototyping and testing over broad environmental conditions to validate robustness, safety, and projected life cycle performance against application benchmarks. Repeated deformation, weathering, corrosion/chemical exposure, thermal spikes, and other stresses experienced in service must be examined in accelerated studies. Failure/vulnerability analysis is likewise crucial.
- Determining niche high value applications where price premiums can support manufacturing costs in initial stages before optimization of economies of scale.

In the drive toward sustainable innovation for societal resilience, next generation chitin biocomposites are poised to transform rubber and elastomer technologies across transportation, infrastructure, healthcare, and defense sectors. Realizing their full disruptive potential compels expanded collaboration between international research communities, advanced analytical resources, funding programs, policymakers, and industry pioneers (Roelofs et al. 2024). Holistic co-design incorporating environmentally responsible materials processing, circularity principles, and additive manufacturing looks set to accelerate development from laboratory curiosities into robust mainstream solutions underpinning modern infrastructure. With petroleum supplies dwindling (Baquero and Monsalve 2024), the natural abundance of chitinous biowaste makes crustacean-derived chitin an auspicious candidate to move polymer composites onto a more sustainable footing.

9.8 Conclusion

This extensive discourse has navigated the intricate landscape of chitin-rubber micro composites, unraveling the interwoven threads connecting biopolymer structure, processing methods, property enhancements, and emerging applications the singular attributes of crustacean-derived chitin from mechanical strength to thermal stability and moisture control has motivated integration into rubber blends to synergistically augment composite performance the abundance of hydroxyl groups combined with nanoscale fibrillar morphology facilitates uniform dispersion and dynamic interactions with industrial elastomers including natural rubber (Li et al. 2024b), styrene-butadiene, and silicone matrices leading to significant improvements in tensile/tear properties, fatigue resistance, permeation barriers, and bioresponse over unfilled analogs furthermore, the sustainability of waste-stream bioresourced chitin coupled with processing flexibility enables scalable commercial manufacturing with reduced environmental footprints relative to carbon black and inorganic fillers detailed analysis has illuminated correlations between processing technique, filler integration quality, and ultimate performance guiding composite design selection based on tailored requisites spanning lightweight resilience (Tripathi and Shukla 2024), vibration damping, thermal endurance, selective permeability and antimicrobial action for niche applications in transportation, infrastructure, healthcare, and defense sectors outstanding challenges remain in advancing predictive multiscale modeling capabilities, expanding industrial prototyping efforts and implementing responsible end-of-life strategies nevertheless, the stage is set through proof-of-concept validation across laboratory to pilot facilities for emerging chitin micro composites to harness their full disruptive potential thereby transforming rubber and elastomer technologies onto a more sustainable materials foundation looking forward, bolstered efforts in chitin composites R&D combined with circular economy thinking promises ever-expanding horizons and commercial opportunities fueled by ongoing discovery of new material synergies (Mohammadi et al. 2024), processes, and applications anchored in the ethos of responsible innovation and environmental stewardship the expansive domain of materials science thereby persists on its ceaseless pursuit of knowledge, innovation, and multifunctionality where Nature's designs offer inspiration while scalable engineering solutions sustain society thus, chitin's remarkable confluence of structure (Pérez-Bassart et al. 2024), properties and abundance holds untapped prospects to push performance barriers dramatically augmenting next generation composite material frontiers through harmonious unification of natural complexity and human ingenuity as this scientific sojourn culminates, we reflect upon how biopolymer micro composites bridge critical gaps between sustainability and functionality presaging a paradigm shift on the road ahead.

References

Abdul Khalil HPS, Nurul Fazita MR, Nurazzi M (2024) Biopolymers and biopolymer blends: fundamentals, processes, and emerging applications

Abedi M et al (2023) Advancements in laser powder bed fusion of carbon nanotubes-reinforced AlSi10Mg alloy: a comprehensive analysis of microstructure evolution, properties, and future prospects. Metals 13(9):1619

Adibi A (2023) Engineered polysaccharide, alpha-1, 3 glucan, as a functional filler of rubber composites

Ahmad SI et al (2020) Chitin and its derivatives: structural properties and biomedical applications. Int J Biol Macromol 164:526–539

Ahmad J et al (2024) Analyzing dispersive optical solitons in nonlinear models using an analytical technique and its applications. Opt Quant Electron 56(1):77

Ali HM et al (2024) Advances in thermal energy storage: fundamentals and applications. Prog Energy Combust Sci 100:101109

Alkhalifa N et al (2024) Effects of vertical load and inflation pressure on tire-soil interaction on artificial soil. J Terramech 112:19–34

Alkhursani SA et al (2022) Application of nano-inspired scaffolds-based biopolymer hydrogel for bone and periodontal tissue regeneration. Polymers 14(18):3791

Al-Qahtani S, Koç M, Isaifan RJ (2023) Mycelium-based thermal insulation for domestic cooling footprint reduction: a review. Sustainability 15(17):13217

Al-Sakkaf MK, Onaizi SA (2024) Effects of emulsification factors on the characteristics of crude oil emulsions stabilized by chemical and biosurfactants: a review. Fuel 361:130604

Alshangiti DM et al (2023) Revolutionizing biomedicine: advancements, applications, and prospects of nanocomposite macromolecular carbohydrate-based hydrogel biomaterials: a review. RSC Adv 13(50):35251–35291

Angel N, Li S, Kong LJ (2024) Emerging applications of nanofibers electrospun from carbohydrate polymers. J Future Foods 4(4):289–299

Aratani I et al (2024) Two-dimensional metal–organic nanosheets composed of single-molecule magnets: structural modulation and enhanced magnetism utilizing the steric hindrance effect. J Mater Chem C 12(2):724–735

Arunkumar A et al (2023) Effect of HNTs and modified HNTs nanotubes on the mechanical properties and swelling resistance of EPDM/SBR rubber blend nanocomposites. Silicon 1–21

Bambo MF, Mokhena TC, Matabola KP (2023) Synthesis of natural polymer–carbonaceous filler composites. In: Polymer-carbonaceous filler based composites for wastewater treatment. CRC Press, pp 29–67

Baquero JEG, Monsalve DBJ (2024) From fossil fuel energy to hydrogen energy: transformation of fossil fuel energy economies into hydrogen economies through social entrepreneurship. Int J Hydrog Energy 54:574–585

Bhatia D et al (2024) Biosensors and their widespread impact on human health. Sens Int 5:100257
Bragg-Sitton SM et al (2020) Reimagining future energy systems: overview of the US program to maximize energy utilization via integrated nuclear-renewable energy systems. Int J Energy Res 44(10):8156–8169
Chafiq M, Chaouiki A, Ko YG (2023) Advances in COFs for energy storage devices: harnessing the potential of covalent organic framework materials. Energy Storage Mater 103014
Dananjaya V et al (2023) Effects of latex type and processed-mica waste loading on the structural and thermo-physical properties of natural rubber latex foam composites
Davoodi S et al (2023) Thermally stable and salt-resistant synthetic polymers as drilling fluid additives for deployment in harsh sub-surface conditions: a review. J Mol Liq 371:121117
Ding C, Ma W, Zhong J (2024) The influence of microcrystalline structure and crystalline size on visible light transmission of polyvinyl alcohol optical films. Opt Mater 147:114627
Dong Q et al (2024) Determining the potential risks of naturally ventilated double skin façades. Renew Sustain Energy Rev 191:114064
Dutta D, Sit N (2024) Preparation and characterization of potato starch-based composite films reinforced by modified banana fibers and its application in packaging of grapes. Int J Biol Macromol 254:127791
El Knidri H et al (2018) Extraction, chemical modification and characterization of chitin and chitosan. Int J Biol Macromol 120:1181–1189
Fahanwi AN et al (2024) In situ polyaniline polymerization on electrospun cellulose acetate nanofibers derived from recycled waste filter butts of cigarettes for the enhanced removal of methyl orange and rhodamine. Chem Eng Res Des 201:18–30
Faverjon C et al (2023) Elongational flow mixing: a novel innovative approach to elaborate high-performance SBR-based elastomer compounds. Appl Rheol 33(1):20230105
Fazeli M et al (2024) Lignin beyond the status quo: recent and emerging composite applications
Frey KA et al (2023) Use of resonant acoustic mixing technology for ultra-low-dose blending in a single-step mixing process, pp 1–19
Gaona CGC et al (2023) Composites and novel applications in the biomedical field, p 129
Gong Y et al (2023) Sustainable utilization of foundry waste: cracking property analysis of basalt fiber-reinforced rubber concrete. J Mater Civ Eng 35(7):04023181
Hayeemasae N, Salleh SZ, Ismail H (2023) Chloroprene rubber waste as blend component with natural rubber, epoxidized natural rubber, and styrene butadiene rubber. In: Recycled polymer blends and composites: processing, properties, and applications. Springer, pp 271–289
Hei X et al (2024) Freeze-thaw stability of Pickering emulsion stabilized by modified soy protein particles and its application in plant-based ice cream. Int J Biol Macromol 257:128183
Hou X-D et al (2013) Evaluation of toxicity and biodegradability of cholinium amino acids ionic liquids. PLoS ONE 8(3):e59145
Huang P-W et al (2023) Study on the flow, foaming characteristics and structural strength of polypropylene structural foam injection molding by innovative nitrogen and molten plastic mixing mechanism. Polymers 15(9):2116
Idumah CI, Obele CM (2021) Understanding interfacial influence on properties of polymer nanocomposites. Surfac Interfaces 22:100879
Im H, Jeoung S (2023) Mechanical aging test and sealing performance of thermoplastic vulcanizate as sealing gasket in automotive fuel cell applications. Polymers 15(8):1872
Ismail H et al (2008) Morphological, thermal and tensile properties of halloysite nanotubes filled ethylene propylene diene monomer (EPDM) nanocomposites. Polym Test 27(7):841–850
Ismail A et al (2023) Microencapsulated phase change materials for enhanced thermal energy storage performance in construction materials: a critical review. Constr Build Mater 401:132877
Jadaun S et al (2023) Biodegradable nanocellulose reinforced biocomposites for food packaging—a narrative review and future perspective. J Water Environ Nanotechnol 8(3):293–319
Jantachum P et al (2023) Effect of silane coupling agent and cellulose nanocrystals loading on the properties of acrylonitrile butadiene rubber/natural rubber nanocomposites. Ind Crops Prod 195:116407

Jašarević L (2024) Beekeeping in the end times. Indiana University Press
Jhang S et al (2023) Functional group effect of chemically modified microcrystalline methyl cellulose on thermoplastic polyurethane composites. Cellulose 30(11):6917–6931
Jiao D et al (2023) Supramolecular cross-linking affords chitin nanofibril nanocomposites with high strength and water resistance. Compos Sci Technol 244:110295
Kavimani V et al (2023) A CRITIC integrated WASPAS approach for selection of natural and synthetic fibers embedded hybrid polymer composite configuration, pp 1–16
Khalesi H et al (2024) Influence of amyloid fibril length and ionic strength on WPI-based fiber-hydrogel composites: microstructural, rheological and water holding properties. Food Hydrocolloids 148:109499
Khan O et al (2023) Comparative study of soft computing and metaheuristic models in developing reduced exhaust emission characteristics for diesel engine fueled with various blends of biodiesel and metallic nanoadditive mixtures: an ANFIS–GA–HSA approach. ACS Omega 8(8):7344–7367
Landoulsi J (2024) Surface (bio)-functionalization of metallic materials: how to cope with real interfaces? Adv Coll Interface Sci 325:103054
Lee S et al (2023) Nanochitin and nanochitosan: chitin nanostructure engineering with multiscale properties for biomedical and environmental applications. Adv Mater 35(4):2203325
Li H (2023) Searching for assembled peptides from Ostrinia furnacalis cuticle proteome
Li S et al (2024a) Characterization and stability of low-oil emulsion gels with newly shaped droplets stabilized by camellia saponin and k-carrageenan. Food Hydrocolloids 149:109585
Li S et al (2024b) All-atom molecular dynamics simulation of structure, dynamics and mechanics of elastomeric polymer materials in wide range of pressure and temperature. Mol Syst Des Eng 9(3):264–277
Liao J et al (2023) Nano-chitin reinforced agarose hydrogels: effects of nano-chitin addition and acidic gas-phase coagulation. Carbohydr Polym 313:120902
Lim L-T, Auras R, Rubino MJ (2008) Processing technologies for poly (lactic acid). Prog Polym Sci 33(8):820–852
Liu G et al (2019) Research advances towards large-scale solar hydrogen production from water. EnergyChem 1(2):100014
Liu C et al (2021) Natural rubber latex reinforced by graphene oxide/zwitter ionic chitin nanocrystal hybrids for high-performance elastomers without sulfur vulcanization. ACS Sustain Chem Eng 9(18):6470–6478
Liu C et al (2023) Polymer engineering in phase change thermal storage materials. Renew Sustain Energy Rev 188:113814
Loose JSM (2018) Biochemical investigation of catalysis by lytic polysaccharide monooxygenases
Lu T et al (2023) Demulsification performance and mechanism of tertiary amine polymer-grafted magnetic nanoparticles in surfactant-free oil-in-water emulsion. Langmuir 39(3):1288–1298
Lv J et al (2023) Chitin and chitin-based biomaterials: a review of advances in processing and food applications. Carbohydr Polym 299:120142
Mathew MM, Panicker UG (2023) Mechanical, thermal and morphological features of biodegradable crosslinked natural rubber latex-dextrin blends. J Polym Res 30(10):1–17
Mawardi I et al (2023) Eco-friendly production and performance evaluation of water-resistant cellulose nanofiber bioaerogels
Mincea M, Negrulescu A, Ostafe V (2012) Preparation, modification, and applications of chitin nanowhiskers: a review. Rev Adv Mater Sci 30(3):225–242
Mincheva R, Benali S, Raquez J-M (2024) Surface modification of biobased polysaccharide nanoparticles via grafting, in green micro-and nanocomposites. Jenny Stanford Publishing, pp 219–265
Mohammadi T et al (2024) One-step growth of RuNi-MOF nanoarrays on carbon felt host as a high-performance binder-free electrode for dual application: ethanol fuel cell and supercapacitor. J Energy Storage 79:110146

Monfared AR et al (2024) Unleashing the power of SAN: innovative in situ fibrillation and moisture-crosslinking techniques yield stronger, tougher, and greener material than ABS. Compos B Eng 268:111103
Nie J et al (2019) Bio-based epoxidized natural rubber/chitin nanocrystals composites: self-healing and enhanced mechanical properties. Compos B Eng 172:152–160
Paliwal R, Babu RJ, Palakurthi S (2014) Nanomedicine scale-up technologies: feasibilities and challenges. AAPS PharmSciTech 15(6):1527–1534
Parale VG et al (2023) Mechanically strengthened aerogels through multi-scale, multi-composition, and multi-dimensional approaches: a review. Adv Mater 36(18):2307772
Patel R et al (2023) A brief review of polymeric blends based on natural polymers and synthetic thermoplastics polymers. Chem Papers 1–33
Pérez-Bassart Z et al (2024) Antiviral and technological properties of β-glucan-rich aqueous fractions from Pleurotus ostreatus waste biomass. Food Hydrocolloids 146:109308
Perrin E et al (2014) Chitin nanocrystals for Pickering high internal phase emulsions. Biomacromolecules 15(10):3766–3771
Petzold N, Schmidt AL, Scaringella L (2023) How to overcome the disruptor's dilemma: exploring strategic alliance reconfiguration of new market entrants. Technovation 126:102812
Prasad N et al (2023) Recent development in the medical and industrial applications of gum karaya: a review. Polym Bull 80(4):3425–3447
Qiu J, Wang J (2017) Chitin in rubber based blends and micro composites, pp 71–107
Roelofs F et al (2024) Highlights and outlook of the Dutch PIONEER 2021–2024 R&D program. Nucl Eng Des 417:112873
Ruan X et al. (2023) Catalyzing artificial photosynthesis with TiO_2 heterostructures and hybrids–emerging trends in a classical yet contemporary photocatalyst, p 2305285
Sani NFM et al (2022) Intrinsic self-healing rubber: a review and perspective of material and reinforcement. Polym Test 111:107598
Sapuan SM, Mansor MR (2014) Concurrent engineering approach in the development of composite products: a review. Mater Des 58:161–167
Schot J, Steinmueller WE (2018) Three frames for innovation policy: R&D, systems of innovation and transformative change. Res Policy 47(9):1554–1567
Sharma MK et al (2023) Cellulose fortified bio-composite film preparation using starch isolated from waste avocado seed: starch properties and film performance. Biomass Convers Biorefin 1–18
Shashikumar U et al (2023) Trajectory in biological metal-organic frameworks: biosensing and sustainable strategies-perspectives and challenges, p 127120
Shuang Y et al (2024) Carboxylated chitin nanowhiskers enhanced stabilization of Pickering and high internal phase Pickering emulsions. Food Hydrocolloids 146:109206
Siwach P et al (2024) Recent progress in conjugated polymers composites with metal-organic frameworks as electrode materials for supercapacitors. Appl Surf Sci Adv 19:100555
Smith RC et al (2023) A research agenda for computational empowerment for emerging technology education. Int J Child-Comput Interact 38:100616
Song W et al (2023) Porous organic polymers with defined morphologies: synthesis, assembly, and emerging applications. Prog Polym Sci 142:101691
Srivastava AK (2023) Dr. Cyril Sajan
Su X (2023) Surface interactions between chitin nanocrystals and an anionic surfactant: from fundamentals to applications. University of British Columbia
Suksaeree J et al (2023) Deproteinization of natural rubber latex and its pale-colored thin films. Sustainability 15(13):10015
Sun L et al (2023) Regulating ice formation for enhancing frozen food quality: materials, mechanisms and challenges
Thapliyal D et al (2023) Natural fibers composites: origin, importance, consumption pattern, and challenges. J Compos Sci 7(12):506

Tiwari P, Kang S, Bae H (2023) Plant-endophyte associations: rich yet under-explored sources of novel bioactive molecules and applications. Microbiol Res 266:127241

Trinh B et al (2023) Recyclable and self-healing natural rubber vitrimers from anhydride-epoxy exchangeable covalent bonds. ACS Appl Polym Mater 5(11):8890–8906

Tripathi BM, Shukla SK (2024) A comprehensive review of the thermal performance in energy efficient building envelope incorporated with phase change materials. J Energy Storage 79:110128

Utrera-Barrios S et al (2023) Sustainable fillers for elastomeric compounds. In: Green-based nanocomposite materials and applications. Springer, pp 31–61

Vieira H et al (2023) Current and expected trends for the marine chitin/chitosan and collagen value chains. Mar Drugs 21(12):605

Wang L et al (2020) Fabrication, properties, performances, and separation application of polymeric pervaporation membranes: a review. Polymers 12(7):1466

Wang C et al (2023) Design of thin solid-state electrolyte films for safe and energy-dense batteries

Wiśniewska P et al (2023) Green approaches in rubber recycling technologies: present status and future perspective

Wood CD et al (2016) Understanding competing mechanisms for glass transition changes in filled elastomers. Compos Sci Technol 127:88–94

Wu GC (2017) Effects of socially responsible supplier development and sustainability-oriented innovation on sustainable development: empirical evidence from SMEs. Corp Soc Responsib Environ Manag 24(6):661–675

Xue Y et al (2024) Naturally derived materials to enhance the membrane properties in (waste) water treatment applications-mechanisms, scale-up challenges and economic considerations. J Water Process Eng 57:104647

Yang L, Ou Z, Jiang G (2023) Research progress of elastomer materials and application of elastomers in drilling fluid. Polymers 15(4):918

Yaşar Yıldız S, Radchenkova N (2023) Exploring extremophiles from Bulgaria: biodiversity, biopolymer synthesis, functional properties, applications. Polymers 16(1):69

Yi H et al (2023) Achieving material diversity in wire arc additive manufacturing: leaping from alloys to composites via wire innovation. Int J Mach Tools Manuf 194:104103

Yihun FA (2022) Nanochitin preparation and its application in polymer nanocomposites: a review. Emergent Mater 5(6):2031–2060

Yuan S et al (2019) Polymeric composites for powder-based additive manufacturing: materials and applications. Prog Polym Sci 91:141–168

Zarei M et al (2024) Material extrusion additive manufacturing of poly (lactic acid)/Ti6Al4V@ calcium phosphate core-shell nanocomposite scaffolds for bone tissue applications. Int J Biol Macromol 255:128040

Zhang N, Cao H (2020) Enhancement of the antibacterial activity of natural rubber latex foam by blending it with chitin. Materials 13(5):1039

Zhang S et al (2012) Synergistic effect in conductive networks constructed with carbon nanofillers in different dimensions. Express Polym Lett 6(2):159–168

Chapter 10
Applications of Cellulose/Natural Rubber Nanocomposites

Hamdy M. Naguib, Eman O. Taha, Nevine O. Shaker, Usama F. Kandil, and Mona A. Ahmed

10.1 Introduction

Owing to an abundance of greenhouse gases that have a major impact on climate change, global warming has become a major worldwide concern these days. As a result, using bio-based materials made from agricultural waste is crucial for lowering greenhouse gas emissions and providing a potential remedy for problems relating to the environment and energy resources worldwide (Lozada et al. 2023). The latex of rubber trees, such as "*Hevea brasiliensis*", is considered the origin of natural rubber, a renewable and adaptable material (Bottier 2020). When combined with cellulose nanoparticles, it forms a unique class of materials known as natural rubber nanocomposites (Sethulekshmi et al. 2022). These nanocomposites leverage the inherent properties of both natural rubber and cellulose to create advanced materials with superior thermo-mechanical and physical properties (Thomas et al. 2021).

Nano/microparticles of cellulose are extracted from "plant-based" sources, e.g., wood, cotton plants, or agricultural waste, and possess high aspect ratios and surface areas at the nanoscale (Eichhlorn et al. 2010). When incorporated into natural rubber, these nanoparticles act as reinforcing fillers, improving the overall strength, stiffness, and physical stability measured for composite material. The good interfacial interactions between rubbery material and cellulose particles enable effective stress transfer, improving the mechanical behavior of the nanocomposite (Jawaid et al. 2017). In addition to improved mechanical properties, natural rubber nanocomposites also exhibit enhanced thermal stability, gas barrier characteristics, and chemical resistance, in comparison to traditional natural rubber (Pielichowska and Pielichowski 2014). These enhancements are achieved through the full distribution of cellulose nanoparticles inside rubber matrix. This restricts the segmental motion of

H. M. Naguib · E. O. Taha · N. O. Shaker · U. F. Kandil · M. A. Ahmed (✉)
Petroleum Applications Department, Egyptian Petroleum Research Institute, Nasr City, Cairo 11727, Egypt
e-mail: mona_chemist17@yahoo.com; monaahmed@epri.sci.eg

© The Author(s), under exclusive license to Springer Nature Switzerland AG 2025

P. M. Visakh (ed.), *Rubber Based Bionanocomposites*, Advanced Structured Materials 227, https://doi.org/10.1007/978-3-031-78557-3_10

polymer, and creates tortuous paths for gas and liquid permeation. The development of natural rubber nanocomposites has opened up many applicable products, particularly in industries where high-performance, eco-friendly materials are in demand. The following sections will explore the various application areas of these innovative materials.

10.1.1 Background and Motivation of Cellulose Nanoparticles (CNPs)

The motivation behind the use of cellulose nanoparticles (CNPs) in natural rubber nanocomposites stems from the unique properties and advantages these nanomaterials offer.

In fact, cellulose is considered the greatest abundant naturally polymer, extracted from cell walls of plants (Eichhorn 2011). At nanoscale, cellulose exhibits exceptional mechanical properties, especially with the tensile strength value (Dufresne 2013). The nanoparticles extracted from cellulose, in the form of nanocrystals and fibrils, are characterized by distinguished larger surface area, higher aspect ratio, and the more capability of percolating networks within a polymer matrix (Moon et al. 2011). Incorporating CNPs into natural rubber provides several benefits.

The high strength and stiffness of cellulose nanoparticles act as effective reinforcing agents, significantly improving the mechanical characteristics of natural rubber matrix including modulus, tension, tear resistance, etc. (Yue et al. 2012). The tortuous path created by distributing CNPs inside rubber material enhances the material's barrier properties, reducing the permeability to gases, liquids, and chemicals (Olsson et al. 2010). Also, incorporation of thermally stable cellulose nanoparticles improves the thermal resistance and dimensional stability of the natural rubber nanocomposite. As biocompatible, eco-friendly, renewable, and biodegradable, the extracted cellulosic materials make natural rubber nanocomposites more environmentally friendly and sustainable compared to traditional petroleum-based rubber composites (Siró and Plackett 2010).

So, the combination of these advantages has driven the growing concerns regarding development and application of cellulose-based natural rubber nanocomposites in numerous industries, from automotive to consumer goods and beyond.

10.1.2 Cellulose Natural Rubber Nanocomposites (CNR): A Sustainable Synergy

The nanoparticles extracted from natural cellulose provide promising sustainable materials suitable for diverse applications, and ongoing research and development efforts are aimed at unlocking their full potential across various industries. So, the incorporation of cellulose nanofibers "CNF" and cellulose nanocrystals "CNC" into natural rubber matrices can lead to improvements in mechanical characteristics, e.g., stiffness, modulus, and tension; such behavior is due to the reinforcing effect of cellulose component (Mokhena and John 2020). Also CNR nanocomposites often exhibit improved thermal stability compared to pure natural rubber due to the barrier properties and thermal degradation characteristics of cellulose nanomaterials. By utilizing cellulose nanomaterials, these nanocomposites can benefit from the biodegradability and sustainability of cellulose, making them environmentally friendly alternatives to traditional rubber composites. That leads to cellulose natural rubber nanocomposites having potential applications in tire manufacturing, automotive components, adhesives, and biomedical devices (Shahamatifard et al. 2023). The enhanced mechanical properties and sustainability make them attractive for various industrial applications. Although there are various processing techniques, e.g., latex compounding, and melt- and solution-mixing processes, utilized to incorporate cellulose nanomaterials into natural rubber (NR) matrices to produce nanocomposites with desired properties, there are some challenges to obtaining a distinct distribution of these nanomaterials within the natural rubber mixture.

This challenge is due to the presence of a noticeable difference in the nature of the two phases; the hydrophobic natural rubber material and the hydrophilic cellulose material, and therefore there is an interstitial repulsion between them, which leads to the cellulose crystals gathering together and thus generating stress concentration, which weakens the mechanical strength of the product and eventually it makes the composite material less effective.

One common method to improve how well they stick together is, by adding chemicals to cellulose surface that can bond and attach with rubber matrix. Additional explored approach involves changing the polymer chains in the matrix that come into contact, with the filler material (Mariano et al. 2016). Ongoing chapter is focused on addressing these challenges to further improve the performance of cellulose natural rubber nanocomposites.

10.1.3 Scope of the Chapter

Cellulose natural rubber nanocomposites offer the potential for enhanced mechanical, thermal, and sustainability characteristics. This makes the related composites as promising materials for many applicable industries. More research progress on mentioned area aims to optimize processing and properties of these nanocomposites.

10.2 Fabrication and Processing Techniques

Cellulose natural rubber nanocomposites, as advanced composites, merge all desirable characteristics of each constituent. Fabrication of these nanocomposites involves several steps and techniques to guarantee distribution of cellulose within natural rubber matrix.

10.2.1 Extraction of Cellulose Nanocrystals (CNC)

The nanocrystal forms of cellulose seem as fibril-like nano-sized cellulose. Usually, they are obtained from different natural cellulose resources, including "leftover garlic straw" (Kallel et al. 2016), "waste cotton" (Thambiraj and Shankaran 2017), "barley husk" (Fortunati et al. 2016), "eucalyptus pulp" (Wang et al. 2017), "oil palm fronds" (Dungani et al. 2017), "onion skin" (Rhim et al. 2015), and "calotropis procera biomass" (Song et al. 2019). To extract CNC from various sources; techniques like acid hydrolysis, enzymatic breakdown, and mechanical extractions pressured grinding have improved their aspect ratio and crystallinity percentage. There are different methods for extraction CNC as follows:

- "Recyclable Deep Eutectic Solvent (DES)" method (Wang et al. 2020):

Researchers have developed an eco-friendly method for extracting CNCs using the recyclable "deep eutectic solvent (DES)" together with the "high-pressure homogenization (HPH)". As the feedstocks of different biomasses are treated with "DES"; the "HPH" method is further applied to achieve CNCs. The resulting rod-shaped CNCs have lengths and diameters of about 650 nm and 75 nm, respectively; the obtained "CNC" remains stable till 30 days. DES may be a "green process" without hazardous waste. The used solvent material can be reused for multiple pretreatment cycles. So, the last process provided effective and eco-friendly methodology to CNC production.

- "Electron-Beam Irradiation (EBI)" and "High-Pressure Homogenization (HPH)" methods (Lee et al. 2018):

Another innovative system combines "EBI" with "HPH" to get "CNC". This two-step process is compared to the classic method involving tedious acid hydrolysis. EBI pretreatment followed by HPH disintegration yields CNCs efficiently.

- Chemical Hydrolysis of Lyocell Fibers (Cheng et al. 2014)

A cost-effective and influential methodology involves the hydrolysis reaction of natural fibers using ammonium persulfate "APS" solution. The resulting spherical cellulosic nanocrystals (SCNCs) are investigated by means of atomic force and scanning electron microscopes.

- Traditional Acid Hydrolysis (Gars et al. 2020)

Historically, CNCs were produced through a group of modifications, such as acid hydrolysis, and biological and mechanical modifications. Acid hydrolysis remains a common method, although it has some drawbacks. In conclusion, extracting cellulose nanoparticles via acid hydrolysis of cellulose fibers is the typical method; sulfuric acid is often used for separating cellulose fibrils in the form of nano- and microparticles (Beck-Candanedo et al. 2005; Elazzouzi-Hafraoui et al. 2008). The preparation of crystalline cellulose particles is feasible through adjustment of the reaction parameters to alter the dimensions and forms of the product. If the obtained particles have mixtures of micro- and nanoscale cellulose particles, it is recommended to change conditions to get the desired quality of nanoparticles; this highlights benefits for the environment and the economy (Abitbol et al. 2023). The obtained cellulose nanoparticles may be treated for enhancing the compatibility within natural rubber; such treatments may be achieved through some coupling agents, and other chemical methods.

Also, the isolated cellulose nanoparticles are then dispersed into the natural rubber matrix using mechanical mixing techniques such as two-roll milling or internal mixers. Proper dispersion is essential for the desired enhanced profile of nanocomposite. Then, Vulcanization is carried out to crosslink the natural rubber matrix, imparting strength and elasticity to the nanocomposite material. This step is essential to promote the mechanical characteristics of the prepared nanocomposite.

10.2.2 Surface Modification of Cellulose Particles for Natural Rubber Nanocomposites

Nanocellulose particles are promising candidates for reinforcing natural rubber because of the remarkable properties, e.g., renewability, biocompatibility, and large surface area (Kargarzadeh et al. 2018; Eyley and Thielemans 2014). However, it faces challenges related to hydrophobicity, solubility, and compatibility with rubber segments. The treatment of fibers aims to address these issues by enhancing interfacial compatibilities and facilitating dispersion. Common surface modification techniques (Trache et al. 2020) [include chemical grafting (e.g., silylation, esterification, and polymer grafting), physical adsorption (e.g., surfactants, polymers), and oxidation (e.g., TEMPO-mediated oxidation)]. Esterification of hydroxyl groups of nanocellulose with stearic acid leads to better dispersion of nanocellulose, enhancing hydrophobicity and compatibility with natural rubber and improving the mechanical properties of nanocomposites (Siqueira et al. 2009).

Silylation (silane coupling) involves attaching nanocellulose with silane coupling agents (e.g., 3-aminopropyltriethoxysilane). This process can introduce reactive groups that effectively attach with rubber matrix. This improves the load transfer and surface properties of nanocellulose and natural rubber, but grafting polymers (e.g., polypropylene, polyethylene) onto the nanocellulose surface can entangle with the natural rubber chains, improving the interfacial interactions and the bonding with natural rubber (Sibiya et al. 2018; Miao and Hamad 2013). For TEMPO-mediated oxidation, (2,2,6,6-tetramethylpiperidine-1-oxyl) introduces carboxylate moieties onto nanocellulose; which enhances its compatibility and reactivity with polar solvents and polymers (Hakimi et al. 2021).

In summary, surface-modified nanocellulose significantly improves the major characteristics of natural rubber-based nanocomposites. It enhances tension values, elastic modulus, and abrasion resistance. The modified CNC acts as mechanical improver, contributing to proposed composites. Researchers are actively working on refining these methods to unlock the full potential of this promising material. Future prospects involve tailoring surface modifications to specific applications and optimizing the balance between reinforcement and compatibility.

10.2.3 Nanocellulose Reinforced Rubber Matrix Composites

Chemical modification is frequently used for enhancing the combination between particles of natural rubber and nanocellulose on interfacial plane, and to overcome problems in case of compounding with polar systems (Gelir et al. 2017). Since OH groups are considered the only functional groups found in nanocellulose, the reactivity of these groups is important after modification. The nanoscale structure strongly affects the presence of aggregation in materials caused by hydrogen bonding. When active sites are present in chains, cellulose can undergo certain chemical changes. The three hydroxyl groups that make up the cellobiose ring are the primary and secondary alcoholic groups. These groups enable the cellobiose ring to be oxidized in addition to being a functional group substitute for long chains. Methodology utilized and filling effects on the overall profile of proposed nanocomposites were investigated (Kargarzadeh et al. 2018; Low et al. 2021). In summary, nanocellulose holds promise as reinforcing filler in rubber matrices, offering a sustainable alternative to conventional fillers. Researchers continue to explore its potential and refine its applications.

10.3 Applications in Engineering and Construction

10.3.1 High-Performance Tires and Automotive Components

The research on natural rubber, fibers, and their composites has given a great attention because of the sustainability and environmental effect. Both of the micro-/and nano-cellulose are usually taken as sustainable fillers for enhancing the characteristics of the used rubber (Low et al. 2021; Roy et al. 2021). On the related applications, the production of rubber is critical to the automotive industry due to the large number automobiles produced. In fact, the vulcanized rubber is the commercial type for tires and other components which are manufactured worldwide. However, natural rubber and cellulose based composites may be replaced by synthetic rubber with different concentrations as for sustainable issues. The cellulosic fillers may be included for supporting enhanced properties desired for heavy-duty tires. The incorporation of natural resources, e.g., chitin, grasses, wood, starch, cellulosic fibers, etc., is reported (Agrebi et al. 2019). Famous automotive applications of natural rubber can be promoted, such as conveyor belts, hoses, molded rubber, spare parts, fluid-transfer network, sealing rings, and anti-vibration columns (Asghar et al. 2020). It was reported that Mercedes-Benz used environmental products based on natural rubber and vegetable fibers for the belts, straps, tires, hoses, seat and backrest surface covers (Lozada et al. 2023).

A novel green composite, based on bacterial cellulose filler and natural rubber matrix, was activated by natural lactic acid. The activated composite had significant melting profile, mechanical stability, in addition to the stability against bacteria and chemical solvents (Phomrak and Phisalaphong 2020). Bio-originated composites have been prepared using natural rubber matrix and cellulose filler within urea solution; the cellulosic material can support rubber with a noticeable reinforcement (Yu et al. 2017).

In related composites, incorporating other reinforcements is common to provide some additional features, and overcome drawbacks (Hakimi et al. 2021; Imiete et al. 2023). Natural filler and waste glass particles were used to modify a rubber matrix. The surface, adhesion, and thermal characteristics of natural rubber adhesive latex were modified with additional natural leather, nanosilicates, and Kaolin clay (Aa et al. 2021). These modifications were successfully performed and the produced composites have been used in the production of a wide range of automotive components.

10.3.2 Building Materials with Enhanced Strength and Durability

Eco-friendly building and concrete materials were developed using cellulose and natural products. It provides benefits and advances applications of using polymeric cement/concrete (Choi and Choi 2021; Sangmesh et al. 2023) systems, polymers based on cellulose, lignin, starch, etc., as an alternative to traditional building materials (Balo and Yucel 2013; Vilane 2010).

Within the construction industry, the lignocellulosic fibers-filled natural rubber composites serve as durable separating blocks, insulation units, and roofing partitions. Moreover, the natural polymer-based composites display insecticidal and biocidal characteristics resulting in the stability against insects and bio-corrosions (Karandikar et al. 2014). The economical, biodegradable, and chemical resistance advantages put natural materials as suitable additive for conventional construction materials. The lignocellulosic fibrous material is an alternative renewable eco-friendly solution that can be applied due to their low-cost and attractive potential features with desirable physical and mechanical properties for more functional materials (Ates et al. 2020). Different agricultural waste, e.g., wood aggregate, bast and leaf, cereal straw, palm tree, residue, and aquatic chips, could be considered as environmental-friendly filling materials in construction mortars (Laborel-Préneron et al. 2016).

Novel building blocks have been formulated by "casting" technique using natural rubber latex, typha-angustifolia fiber in different orientations, and gypsum for ceiling composites. The formula containing 1 wt% 3-D typha-angustifolia filler had superior conditions of lightweight, thermal stability, and compressive and tensile stress (Muntongkaw et al. 2021). The effect of natural rubber matrix on the stabilization of some "laterite soil–cement blocks" was investigated. Different durable concentrations have achieved better thermal, mechanical, and anti-weathering properties (Jose and Kasthurba 2021). With another type of composite, natural rubber latex, with the stone mastic asphalt, was evaluated against the mechanical and environmental effects; it was found that the proposed matrix increased the values of modulus and tension strengths, and fatigue life, besides the environmental concern of life-cycle assessment (Suwarto et al. 2023).

10.3.3 Vibration Damping and Noise Reduction

Rubber and its composites have very famous application in vibration damping and noise reduction products. However, there is no doubt that exploiting natural rubber sources can serve for cleaner technology, and more environmentally friendly issues.

A green composite based on natural rubber (matrix), and bamboo-based biochar (filler) was investigated against shaking damping and acoustical properties using different standards. The natural filler recorded improved damping characteristics along with accepted characteristics (Mago et al. 2022). Sounding and vibration damping properties were checked for natural sandwich composites, where balsa wood core and natural fiber were utilized where a full enhancement was achieved by proposed composite (Sargianis et al. 2013). Another type of rubber, epoxidized natural rubber, was studied for an elastomeric vibro-acoustic isolation system. The functionalized rubber was tested under operating conditions of temperature and frequency, and designed for the vibration isolation system with optimized epoxidation (Rahim et al. 2018). For reducing carbon footprint; the role of suggested eggshells waste in anti-vibration characteristics was examined. The proposed bio-composites were applied as automotive anti-vibration applications. In addition, they also provided good hardness and mechanical properties that serve in green automotive rubber components (Barrera and Tardiff 2022).

In the same context of exploiting natural resources; the sound absorption, impact response, and vibration damping were characterized for a product containing flax, and cork fillers, where the vacuum bagging process was performed with different densities. The proposed composites with impact behavior, sound, and vibration properties could be applicable in automobile and construction applications (Prabhakaran et al. 2020). In another study, natural rubber was filled with different particle sizes of waste rubber and silica, and evaluated to the vibration damping behavior. The larger particles showed excellent vibration damping property due to the more energy absorbed, compared with the smaller particle size (Chandran et al. 2018).

In related works (Zhang et al. 2019; Prabhakaran et al. 2014), some biofibers, such as cotton, flax, and bamboo, were utilized to provide lightweight structural composites for infrastructure and automotive accessories with considerable vibrational and sounding characteristics. Specifically, the "flax fiber" has better higher vibration damping than other natural fibers, besides the achieved mechanical properties, even compared with that of glass fiber type.

10.3.4 Water Absorption and Purification

Attempts are made to treat, absorb, and purify water by environmental and advanced technologies (Dongre et al. 2019; Follain et al. 2013), where nanocellulose, lingocellulosic materials, and natural rubber-based composites were utilized for water membranes, filters, and water purifications. It was found that the desired nanotechnology supports functional composites; additional nanofillers may be included for better performance.

A green sponge composite were prepared using natural rubber and cellulose fiber via the Dunlop processing technique, and applied as composites with different improved properties; the water absorptivity which reached 400% is very vital for storage of water (Suppanucroa et al. 2023b). As the main factor affecting the capability of natural rubber to keep water molecules, the physical properties, specially swelling, were studied depending on fillers. A type of bacterial cellulose fibrils were incorporated into natural rubber by an evaporation process. The fibrils altered the swelling, morphological, and mechanical properties with high content of cellulose. In addition to the sensitivity to water uptake; the given data may also be useful for shape-memory applications (Habeeb et al. 2021; Yin et al. 2018).

Propionated lignin and cellulose triacetate were used for preparing nanocomposite tissues to purify groundwater, where the "vapor-induced phase separation" procedure was conducted by different types of natural materials. It was found that the disposal of "arsenic" and "fluoride" particles was influenced by ionic and organic materials (Nevárez et al. 2011). New polyethersulfone nanohybrid membrane, including modified graphene oxide and zinc oxide nanoparticles with related combinations, was examined to purify wastewater. Both of fillers increased improved the separation performances, membrane structure, wettability, and mechanical stability (Kusworo et al. 2021).

Overall, micro- and nano-cellulosic materials from natural sources, such as cellulose, chitin, and starch, with the help of secondary filler, were used for wastewater treatment, along with rubber materials. Functional and highly crystalline particles play a key role in gradually integrating with the rubber, thus improving the profile (Thomas et al. 2021).

10.4 Applications in Consumer Goods and Sports Equipment

10.4.1 High-Strength and Lightweight Sports Apparel and Gear

The abundant functional groups present in natural fiber and natural rubber materials strongly increase the benefits of their usage in lightweight composites. The combination of them could reinforce the rubbery material resulting in crosslinked network. In addition, the low density fibrous material decreases the total weight of the final composite; the product is lightweight as well (Low et al. 2021). Such positive outcome is useful for applications of lightweight sports apparel and gear. For example, the crosslinked composite can be applied in the sports shoes and accessories, where the cellulosic material may be blended with natural rubber, ethylene–vinyl acetate copolymer, and their composites (Nagatani 2017).

10.4.2 Durable Engineering Applications, Electronics, and Sensors

Cellulose, extracted from natural fibers, was treated with natural rubber latex for providing nanocomposites for various engineering applications (Kulshrestha et al. 2020). Composites made of cellulosic fibrils, elastic natural rubber, organoclay, natural rubber, and liquid natural rubber were hybrided by melt-blending with different loading percentages. The proposed composites showed stability after exposure to different dosage of gamma irradiation, as suitable durable engineering applications (Shahdan et al. 2024). Nanoparticles of cellulose, extracted from "coconut spathe", were added to natural rubber after treatment of nanofibers via acid hydrolysis reaction. The hydrolysis helped to dissolve the lower lateral order keeping the water-insoluble crystalline phase, which alters the mechanical performance (Venugopal and Gopalakrishnan 2018). Different fibers, including "yarrow", "peppermint" and "chamomile", were utilized to modify the properties of natural rubber for preparing plant origin-based biocomposites. Promoted cross-linking density, barrier properties, and mechanical strength are achieved before and after simulated aging process (Masłowski et al. 2021). An oxidized natural rubber, with hydroxyl groups on the surface, was promoted by the effect of cellulose nanoparticles; the performed reaction hardened the rubber-cellulose interface, resulting in durable composite (Mariano et al. 2016).

Novel composites of natural rubber-cellulose nanofibers-polyaniline complex were proposed for the application of electroconductive, and soft electronic elastomers; one of the prepared composites had ideal dispersion and high aspect ratio through hierarchical 3-D network polymerization. Properties such as ideal conductivity, high sensitivity, and stretchability make the green rubber-nanocellulose complex applicable in strain sensors and flexible electrodes. In a similar work, natural rubber was used in sustainable sensors (Han et al. 2019; Wu et al. 2016). Cao et al., provided a simple, eco-friendly, and scalable approach to prepare natural rubber composites with 3D conductive network via the renewable CNC-mediated assembly of graphene nanosheets in natural rubber latex (Cao et al. 2016). In this work; the prepared green composite sensors able to sensing and discriminating solvent leakage in environmental monitoring. The composites are based on natural rubber and "3-D graphene-conductive network", and it was found that the crystalline 3-D conductive network has remarkable electric conductivity.

10.4.3 Sustainable Food Packaging, Cosmetics, and Personal Care Products

Natural materials, such as cellulose, natural rubber, and biodegradable materials, have an outstanding application across food and water packaging materials, biomedical, and cosmetics, due to the related environmental concern. Furthermore, the micro- and nano-cellulose present promising characteristics of composites (Bharimalla et al. 2017). Materials applied in packaging products of food and water, such as boxes and bottles, must pass particular investigations, such as hindrance, mechanical, and thermal properties, to guarantee safe packaging (Xu et al. 2024). Furthermore, cellulose-based composites possess air-permeability properties that mainly affect disposable medical and hygiene applications (Liu et al. 2022). An optimization for the properties of composite sheets, based on carboxymethyl cellulose microfibers, natural rubber, and cellulose, was reached by incorporation silver nanoparticles as a second additive. The used green route recorded antibacterial activity in addition to improved mechanical strength, heat resistance, biodegradability, and chemical resistance (Suppanucroa et al. 2023a). In fact, nanocellulose particles are recently used in production of hygiene biocompatible products, such as healthcare, skincare, and cosmetics. The nanocellulose has shear thinning property, surface stability, biocompatibility, functionality, and high absorptivity. The related nanocomposites are established as effective carriers for "bioactive materials", and "UV-blockers and platforms" for the materials used for wound healing (Meftahi et al. 2022).

10.5 Applications in Biomedical and Healthcare

The application of biocompatible and biodegradable materials as disposable consumer items and containers for pharmaceuticals, cosmetics, and food has increased recently (Čolnik et al. 2020; Patwary et al. 2020). In addition to being widely used in most industries, the preparation of biodegradable and biocompatible composites, based on natural resources, has a comprehensive research in fields of materials, food, cosmetics, and pharmaceuticals. Thanks to their improved biological and structural features, cellulose and its derivatives are considered renewable, environmentally friendly, and bioactive materials. Cellulose possesses good biocompatibility and low toxicity, and is also inexpensive, abundant, and readily available. Consequently, materials derived from cellulose can serve as fascinating candidates for various applications, particularly in the biomedical and healthcare sectors (Seddiqi et al. 2021; Joseph et al. 2020; Hasanin 2022; Pandey 2021; Mali and Sherje 2022). The significant properties of cellulose have motivated researchers to draw on utilizing cellulose and its composites in numerous applications. These applications exploit the important characteristics of cellulose, such as its flexibility, surface activity, hydrophilic nature, free availability, biocompatibility, biodegradability, and good mechanical properties (Mali and Sherje 2022; Torgbo and Sukyai 2020; Guerra et al. 2021).

Healthcare and surgeries are crucial treatments for everyone. Conventional rubber gloves and condoms are used in hospitals for medical purposes; meanwhile, rubber catheter balloons, rubber bands, and rubber films are used for surgery support. Consequently, natural rubber (NR) is considered a significant material for healthcare and surgery applications (Ho 2021; Boon et al. 2022; Andrade et al. 2022; Lim et al. 2020). Unfortunately, NR usually suffers from a poor tensile strength and a high permanent set at high strain, which may lead to unsatisfactory properties when NR is used in healthcare and surgery applications (Hayeemasae et al. 2024; Guan et al. 2020b). To avoid easy breakage and poor mechanical strength, cellulose extracted from natural fiber has a variety of sources, low cost, less harm, biocompatibility, high crystallinity, biodegradable polymer, abundance, non-toxicity, and high stiffness. Basically, cellulose is considered a natural and renewable material extracted from plants (Nurazzi et al. 2021; Ciolacu et al. 2020).

10.5.1 Drug Delivery Systems with Controlled Release

In recent years, cellulose-based drug delivery systems have received wide attention in connection with the development of natural and ecological medical and healthcare products. This is due to their biodegradability, biocompatibility, low environmental toxic effect, as well as their ability to adsorb, entangle or encapsulate bioactive compounds. Furthermore, such drug delivery systems based on cellulose are compatible with various additives (tablets, nanopowders, gels, medical dressings, films, and others) and may be operated in the way of controlled release of the drugs (Guan et al. 2020a; Janmohammadi et al. 2023; Hosny et al. 2021; El Allaoui et al. 2023; Cao et al. 2022). However, it should be noted that selective separate processes of cellulose may require much harsh conditions, such as strong acid catalysts or high temperatures, which may lead to the formation of cellulose segments with lower compatibility with other polymers and thus impair the properties of polymer nanocomposites.

Cellulose provides a fundamental effect on modifying the characteristics of polymer composites, however limited articles studied the related cellulose/natural rubber nanocomposites (Srirachya et al. 2020; Utrera-Barrios 2024; Maciejewska and Sowińska-Baranowska 2022; Zheng et al. 2019). There are limited studies on antibacterial and mechanical characteristics of cellulose-natural rubber nanocomposites for drug delivery systems.

Nanocrystals of cellulose (CNC), derived from rubbery wood, have developed the hydrophobic drug delivery vehicle. First, acid hydrolysis and ultrasonication yielded CNCs, which were thereafter subjected to chemical modification by succinic anhydride to give them more negative charge. To make them encapsulate and release hydrophobic drugs like hesperidin, the modified CNCs were treated further with a cationic surfactant to improve their hydrophobicity. In conclusion, through rubbery wood; as we have seen in this article; there is possibility of improving bioavailability of poorly water soluble drug using such nanocellulose based carriers that may be developed thereby providing promising solutions towards drug delivery enhancement (Marques et al. 2024).

Marques et al. (2024) give an account of the different ways natural rubber latex can be used as a biomaterial like drug delivery and regenerative medicine. In addition, it also highlights positive features of NR, from being costly efficient to simple handling, biocompatible characteristics, pro-angiogenic ability, and pro-healing attributes. It further reviews controlled release capabilities of natural rubber latex in wound healing, neo-vaginoplasty; smart medical devices among others. It shows how modification of pore density in natural rubber latex can be used to design drug release profiles and hence wearable biosensors for neurological disorders. Interdisciplinary based on this paper may enable actualization of full potential provided by managerial expertise so that important healthcare problems are resolved using natural products such as rubber latex based materials.

de Almeida et al. (2020) investigated a controlled drug delivery system using biocompatible rubber latex membranes micromachined with a femtosecond laser, more particularly focusing on the micropore density effect on drug release kinetics. A study employing Ciprofloxacin as a model drug has revealed that there is enhanced release rate for drugs when the pore density is increased. The technique of making pores by use of femtosecond lasers creates them exactly without compromising the structural unity of the membranes and thus it holds promise in targeted drug delivery applications.

10.5.2 Antimicrobial and Biocompatible Medical Devices

The utilization of cellulose/natural rubber nanocomposites appears to be highly promising when developing antimicrobial agents as well as biocompatible devices used in medicine. This is due to their ability to suppress microbial growth while remaining safe for use against human tissues. These materials can be altered in such a way that they will release antimicrobial agents leading to controlled infections which are linked with healthcare products such as implants, wound dressings etc. Moreover, their mechanical properties can be finely adjusted so as to suit different medical applications ranging from soft membranes up to hard implantable devices (Abdelhamid and Mathew 2022; Tavakolian et al. 2020; Guerra et al. 2023). In this regard, blending cellulose or natural rubber into nanocomposite forms provides new ways through which medical technology may advance. Such developments are vital towards addressing challenges faced by medical treatment such as managing infections or enhancing bio-compatibility of medical implants/devices.

Multifunctional composite films based on a combination of natural rubber, cellulose fibers, and silver nanoparticles (AgNPs) were developed in a recent study (Phomrak and Phisalaphong 2020). It highlights the improved chemical resistance and antibacterial properties of these materials. This research focused on a green synthesis route where cellulose microfibers and carboxymethyl cellulose acted as hybrid fillers in natural rubber to strengthen its mechanical properties and hasten its preparation easily. The composites show strong antibacterial efficiency to both "Escherichia coli" and "Staphylococcus aureus" owing to incorporation of silver nanoparticles with varying sizes and concentrations that greatly enhance their chemical resistance. Furthermore, these composites have exceptional tensile strength, high heat stability, as well as high biodegradability ($\geq$20%) over two weeks rendering them ideal for advanced healthcare applications that require high bio-compatibility but also durability.

Another article (Phomrak and Phisalaphong 2020) discusses the growth and characteristics of bacterial composites, based on lactic acid, and modified natural rubber, that find applications in biomaterials and packaging because they have been improved in biodegradability, mechanical performance, and resistance to different media. These composites were made through latex aqueous microdispersion method and showed higher water resistance as well as mechanical strength than unmodified ones. The films also did not exhibit cytotoxicity or antibacterial effects towards common pathogens indicating their appropriateness for diverse medical and packaging purposes. This research paper indicates that lactic acid is an efficient modifier for enhancing the functional properties of bacterial cellulose/natural rubber composites such as those used in eco-friendly applications.

A composite consisting of silicon rubber and microcrystalline cellulose embedded silver nanoparticles is described by Jankauskaitė et al. (2014), which can display both high mechanical properties and antimicrobial activities at the same time. The synthesis involved mixing vinyl-terminated polydimethylsiloxane (PDMS) with the functionalized cellulose fiber collecting the structural advantages of cellulose microcrystals and anti-microbial characteristics of in situ synthesized nanoparticles. It includes the procedure for embedding silver nanoparticles onto cellulosic material via chemical reduction, and how different concentrations of these particles affect the morphology, mechanical strength, or antimicrobial ability of composites.

10.5.3 Wound Healing Dressings

Wound-healing is a critical aspect of healthcare. For wound-healing activities, composite is usually produced using cellulose/natural rubber that will promote the healing of wounds and this is an important step forward in development of wound-healing dressings that use natural rubber, which possess biocompatibility and beneficial properties. Cellulose is obtained from renewable sources and is strong and has excellent water holding capacity. It also degrades easily making it an ideal base for matrices. However, there is also a part played by natural rubber which adds elasticity and flexibility thereby enabling the dressing to conform to a wound site comfortably without restricting patient mobility (Naomi et al. 2020; Ahmed et al. 2020; Krupp et al. 2019; Phaechamud et al. 2016).

By incorporating nano-sized cellulose in the rubber matrices through the latest progress in nanotechnology, these composites were further improved. In addition to mechanical properties enhancement, this assimilation also enables new functionalities including better barrier functions and/or active delivery of therapeutic agents among others. In relation to wound healing, these nanocomposites can also be engineered such that they can be used for ideal conditions for moist wound care, help in drug or natural healing promoters delivery as well as protect against microbial infections (Naomi et al. 2020).

A recently study (Suppanucroa et al. 2023b) uncovers a new kind of green composite sponge made by Dunlop process. This study deals basically with the improvement in mechanical properties and water absorption capacity of "cellulose-natural rubber composites" with the help of "sodium alginate", which enhances distributing cellulose fibers in natural rubber. Improved "macro-porous sponges" that have better performance and greater hydrophilic character were obtained. These sponges, which were highly concentrated in cellulose and SA, were characterized by very high compressive strength, elasticity and maximum water absorption capacity (around 400%). In addition, biodegradability was observed to be quite high for the sponges as well as faster rates compared to pure natural rubber sponges; hence making them potential candidates for use in various fields including agriculture or even wound dressings in the medical sector.

10.6 Future Trends and Challenges

In this innovative field, as the development of cellulose/natural rubber nanocomposites advances, several key trends and challenges are expected to shape their future. The trend is one of the most exciting areas in sustainability. Natural rubber combined with cellulose sourced from abundant plants is an eco-friendly and renewable option instead of synthetic materials. This will probably lead to research on biodegradability improvement and recycling processes for these composites thus reducing the environmental influence of different industrial applications such as car or biomedical devices.

There are however some issues that have to be looked at carefully if these materials are to ultimately achieve their potential. When natural rubber exhibits hydrophobic attributes while cellulose has hydrophilic tendencies, there are considerable difficulties in obtaining uniform dispersion and strong interfacial bonding in nanocomposites. There should be chemical treatment methods that can develop surface properties enhancing compatibility of the composite materials, mechanical strength, thermal stability, and barrier performance.

Furthermore, a significant challenge relates to scalability of production processes. While laboratory-scale demonstrations have shown promising results, translating these processes to industrial-scale production requires significant technological advancements meant for ensuring consistency, quality, and cost-effectiveness.

Besides the above-mentioned points; regulatory issues surrounding such products will also play a critical role when they become more widely utilized within food packaging, medicine delivery systems, or drug packaging among other sensitive areas. In many regions where health regulations may differ considerably between each other, rigorous testing procedures together with certification programs will be required for cellulose/natural rubber nanocomposites as per health and safety rules.

Looking into the future investigations might also focus on multiple functionalities such as self-healing ability in smart materials with conductive sensing capabilities through integrating smart technologies within these composites; this could result in new applications including smart textiles; flexible electronics; responsive biomedical devices which leverage special features of cellulose/natural rubber.

Thus we can see that although there will be many opportunities ahead, many challenges both technical, economic, and regulatory have to be passed through for the realization of cellulose/natural rubber nanocomposites. Only interdisciplinary collaboration and sustained innovation will provide solutions that can overcome these difficulties enabling production of high-performance sustainability-based applications.

References

Aa S, Varghese S, Thomas S (2021) Natural rubber latex-based adhesives: role of nanofillers. J Adhes Sci Technol 35:406–418

Abdelhamid HN, Mathew AP (2022) Cellulose-based nanomaterials advance biomedicine: a review. Int J Mol Sci 23:5405
Abitbol T et al (2023) Isolation of mixed compositions of cellulose nanocrystals, microcrystalline cellulose, and lignin nanoparticles from wood pulps. ACS Omega 8:21474–21484
Agrebi F, Ghorbel N, Bresson S, Abbàs O, Kallel A (2019) Study of nanocomposites based on cellulose nanoparticles and natural rubber latex by ATR/FTIR spectroscopy: the impact of reinforcement. Polym Compos 40:2076–2087
Ahmed J, Gultekinoglu M, Edirisinghe M (2020) Bacterial cellulose micro-nano fibres for wound healing applications. Biotechnol Adv 41:107549
Andrade KL et al (2022) Latex and natural rubber: recent advances for biomedical applications. Polímeros 32:e2022015
Asghar U, Masoom A, Javed A, Abbas A (2020) Economic analysis of isoprene production from good year scientific process. Am J Chem Eng 8:63–69
Ates B, Koytepe S, Ulu A, Gurses C, Thakur VK (2020) Chemistry, structures, and advanced applications of nanocomposites from biorenewable resources. Chem Rev 120:9304–9362
Balo F, Yucel HL (2013) Assessment of thermal performance of green building materials produced with plant oils. Int J Mater Sci 3:118–129
Barrera CS, Tardiff JL (2022) Static and dynamic properties of eggshell filled natural rubber composites for potential application in automotive vibration isolation and damping. J Clean Prod 353:131656
Beck-Candanedo S, Roman M, Gray DG (2005) Effect of reaction conditions on the properties and behavior of wood cellulose nanocrystal suspensions. Biomacromolecules 6:1048–1054
Bharimalla AK, Deshmukh SP, Vigneshwaran N, Patil PG, Prasad V (2017) Nanocellulose-polymer composites for applications in food packaging: current status, future prospects and challenges. Polym Plast Technol Eng 56:805–823
Boon ZH, Teo YY, Ang DT-C (2022) Recent development of biodegradable synthetic rubbers and bio-based rubbers using sustainable materials from biological sources. RSC Adv 12:34028–34052
Bottier C (2020) Biochemical composition of Hevea brasiliensis latex: a focus on the protein, lipid, carbohydrate and mineral contents. In: Advances in botanical research, vol 93. Elsevier, pp 201–237
Cao J, Zhang X, Wu X, Wang S, Lu C (2016) Cellulose nanocrystals mediated assembly of graphene in rubber composites for chemical sensing applications. Carbohydr Polym 140:88–95
Cao L et al (2022) Nanocellulose—a sustainable and efficient nanofiller for rubber nanocomposites: from reinforcement to smart soft materials. Polym Rev 62:549–584
Chandran V, Nagarajan L, Thomas MR (2018) Evaluation of vibration damping behavior of different sizes of waste tyre rubber in natural rubber composites. J Compos Mater 52:2493–2501
Cheng M et al (2014) Efficient extraction of carboxylated spherical cellulose nanocrystals with narrow distribution through hydrolysis of lyocell fibers by using ammonium persulfate as an oxidant. J Mater Chem A 2:251–258
Choi H, Choi YC (2021) Setting characteristics of natural cellulose fiber reinforced cement composite. Constr Build Mater 271:121910
Ciolacu DE, Nicu R, Ciolacu F (2020) Cellulose-based hydrogels as sustained drug-delivery systems. Materials (Basel) 13:5270
Čolnik M, Knez-Hrnčič M, Škerget M, Knez Ž (2020) Biodegradable polymers, current trends of research and their applications, a review. Chem Ind Chem Eng Q 26:401–418
de Almeida GFB et al (2020) Controlled drug delivery system by fs-laser micromachined biocompatible rubber latex membranes. Appl Surf Sci 506:144762
Dongre RS et al (2019) Natural polymer based composite membranes for water purification: a review. Polym Technol Mater 58:1295–1310
Dufresne A (2013) Nanocellulose: a new ageless bionanomaterial. Mater Today 16:220–227
Dungani R et al (2017) Preparation and fundamental characterization of cellulose nanocrystal from oil palm fronds biomass. J Polym Environ 25:692–700

Eichhorn SJ (2011) Cellulose nanowhiskers: promising materials for advanced applications. Soft Matter 7:303–315
Eichhlorn S et al (2010) Review, current international research into cellulose nanofibers and nanocomposites. J Mater Sci 45:1–33
El Allaoui B, Benzeid H, Zari N, Bouhfid R (2023) Functional cellulose-based beads for drug delivery: preparation, functionalization, and applications. J Drug Deliv Sci Technol 104899
Elazzouzi-Hafraoui S et al (2008) The shape and size distribution of crystalline nanoparticles prepared by acid hydrolysis of native cellulose. Biomacromolecules 9:57–65
Eyley S, Thielemans W (2014) Surface modification of cellulose nanocrystals. Nanoscale 6:7764–7779
Follain N et al (2013) Water transport properties of bio-nanocomposites reinforced by Luffa cylindrica cellulose nanocrystals. J Memb Sci 427:218–229
Fortunati E et al (2016) Revalorization of barley straw and husk as precursors for cellulose nanocrystals extraction and their effect on PVA_CH nanocomposites. Ind Crops Prod 92:201–217
Le Gars M, Douard L, Belgacem N, Bras J (2020) Cellulose nanocrystals: from classical hydrolysis to the use of deep eutectic solvents. IntechOpen, London
Gelir A, Yargi O, Yuksel SA (2017) Elucidation of the pore size and temperature dependence of the oxygen diffusion into porous silicon. Thin Solid Films 636:602–607
Guan Q-F et al (2020a) A new biodegradable nano cellulose-based drug delivery system for pH-controlled delivery of curcumin. Int J Biol Macromol 278:2044–2054
Guan Q-F et al (2020b) Lightweight, tough, and sustainable cellulose nanofiber-derived bulk structural materials with low thermal expansion coefficient. Sci Adv 6:eaaz1114
Guerra NB et al (2021) Biomedical applications of natural rubber latex from the rubber tree Hevea brasiliensis. Mater Sci Eng C 126:112126
Guerra NB et al (2023) Recent progress on natural rubber-based materials containing metallic and metal oxide nanoparticles: state of the art and biomedical applications. Compounds 3:310–333
Habeeb SA, Diwan AA, Albozahid MZ (2021) A compressive review on swelling parameters and physical properties of natural rubber nano composites. Egypt J Chem 64:5713–5724
Hakimi NMF et al (2021) Surface modified nanocellulose and its reinforcement in natural rubber matrix nanocomposites: a review. Polymers (Basel) 13:3241
Han J et al (2019) Nanocellulose-templated assembly of polyaniline in natural rubber-based hybrid elastomers toward flexible electronic conductors. Ind Crops Prod 128:94–107
Hasanin MS (2022) Cellulose-based biomaterials: chemistry and biomedical applications. Starch-Stärke 74:2200060
Hayeemasae N, Soontaranon S, Masa A (2024) Structure–property relationships of different natural rubber grades. Prog Rubber Plast Recycl Technol 14777606241243112
Ho CC (2021) Fundamentals and recent applications of natural rubber latex in dipping technology. Chem Manuf Appl Nat Rubber 317–361
Hosny KM et al (2021) Recent trends in assessment of cellulose derivatives in designing novel and nanoparticulate-based drug delivery systems for improvement of oral health. Polymers (Basel) 14:92
Imiete IE, Giannini L, Tadiello L, Orlandi M, Zoia L (2023) The effect of sulfate half-ester groups on the mechanical performance of cellulose nanocrystal-natural rubber composites. Cellulose 30:8929–8940
Jankauskaitė V et al (2014) Silicone rubber and microcrystalline cellulose composites with antimicrobial properties. Mater Sci 20:42–49
Janmohammadi M et al (2023) Cellulose-based composite scaffolds for bone tissue engineering and localized drug delivery. Bioact Mater 20:137–163
Jawaid M, Boufi S, Abdul Khalil HPS (2017) Cellulose-reinforced nanofibre composites: production, properties and applications. Woodhead Publishing
Jose A, Kasthurba AK (2021) Laterite soil-cement blocks modified using natural rubber latex: assessment of its properties and performance. Constr Build Mater 273:121991

Joseph B, Sagarika VK, Sabu C, Kalarikkal N, Thomas S (2020) Cellulose nanocomposites: fabrication and biomedical applications. J Bioresour Bioprod 5:223–237

Kallel F et al (2016) Isolation and structural characterization of cellulose nanocrystals extracted from garlic straw residues. Ind Crops Prod 87:287–296

Karandikar MV, Sarase SB, Lele PG, Khadaikar SA (2014) Use of natural bio-polymers for improved mortar and concrete properties of cement—a review. Indian Concr J 88:84–109

Kargarzadeh H et al (2018) Advances in cellulose nanomaterials. Cellulose 25:2151–2189

Krupp T et al (2019) Natural rubber-propolis membrane improves wound healing in second degree burning model. Int J Biol Macromol 131:980–988

Kulshrestha U et al (2020) Cellulose nanofibre enabled natural rubber composites: microstructure, curing behaviour and dynamic mechanical properties. Polym Test 90:106676

Kusworo TD, Kumoro AC, Aryanti N, Utomo DP (2021) Removal of organic pollutants from rubber wastewater using hydrophilic nanocomposite rGO-ZnO/PES hybrid membranes. J Environ Chem Eng 9:106421

Laborel-Préneron A, Aubert J-E, Magniont C, Tribout C, Bertron A (2016) Plant aggregates and fibers in earth construction materials: a review. Constr Build Mater 111:719–734

Lee M et al (2018) Facile and eco-friendly extraction of cellulose nanocrystals via electron beam irradiation followed by high-pressure homogenization. Green Chem 20:2596–2610

Lim LP et al (2020) Effect of graphene oxide particle size on the tensile strength and stability of natural rubber graphene composite. Mater Sci Eng B 262:114762

Liu M et al (2022) Hydrophobic, breathable cellulose nonwoven fabrics for disposable hygiene applications. Carbohydr Polym 288:119367

Low DYS et al (2021) Recent developments in nanocellulose-reinforced rubber matrix composites: a review. Polymers (Basel) 13:550

Lozada RE, Gutiérrez Aguilar CM, Jaramillo Carvalho JA, Sánchez JC, Barrera Torres G (2023) Vegetable cellulose fibers in natural rubber composites. Polym (Basel) 15:2914

Maciejewska M, Sowińska-Baranowska A (2022) Bromide and chloride ionic liquids applied to enhance the vulcanization and performance of natural rubber biocomposites filled with nanosized silica. Nanomaterials 12:1209

Mago J, Negi A, Pant KK, Fatima S (2022) Development of natural rubber-bamboo biochar composites for vibration and noise control applications. J Clean Prod 373:133760

Mali P, Sherje AP (2022) Cellulose nanocrystals: fundamentals and biomedical applications. Carbohydr Polym 275:118668

Mariano M, El Kissi N, Dufresne A (2016) Cellulose nanocrystal reinforced oxidized natural rubber nanocomposites. Carbohydr Polym 137:174–183

Marques PAC et al (2024) Natural rubber latex-based biomaterials for drug delivery and regenerative medicine: trends and directions. Int J Biol Macromol 131666

Masłowski M, Aleksieiev A, Miedzianowska J, Strzelec K (2021) Potential application of peppermint (Mentha piperita L.), German chamomile (Matricaria chamomilla L.) and yarrow (Achillea millefolium L.) as active fillers in natural rubber biocomposites. Int J Mol Sci 22:7530

Meftahi A et al (2022) Nanocelluloses as skin biocompatible materials for skincare, cosmetics, and healthcare: formulations, regulations, and emerging applications. Carbohydr Polym 278:118956

Miao C, Hamad WY (2013) Cellulose reinforced polymer composites and nanocomposites: a critical review. Cellulose 20:2221–2262

Mokhena TC, John MJ (2020) Cellulose nanomaterials: new generation materials for solving global issues. Cellulose 27:1149–1194

Moon RJ, Martini A, Nairn J, Simonsen J, Youngblood J (2011) Cellulose nanomaterials review: structure, properties and nanocomposites. Chem Soc Rev 40:3941–3994

Muntongkaw S, Pianklang S, Tangboriboon N (2021) Modifications to improve properties of gypsum ceiling composites as multifunctional construction by embedding Typha angustifolia fiber and natural rubber latex compound. Case Stud Constr Mater 15:e00658

Nagatani A (2017) Characteristics and applications of cellulose nanofiber reinforced rubber composites. Int Polym Sci Technol 44:1–8

Naomi R, Bt Hj Idrus R, Fauzi MB (2020) Plant-vs. bacterial-derived cellulose for wound healing: a review. Int J Environ Res Public Health 17:6803

Nevárez LM et al (2011) Biopolymers-based nanocomposites: membranes from propionated lignin and cellulose for water purification. Carbohydr Polym 86:732–741

Nurazzi NM et al (2021) Thermogravimetric analysis properties of cellulosic natural fiber polymer composites: a review on influence of chemical treatments. Polymers 13(16):2710

Olsson RT et al (2010) Making flexible magnetic aerogels and stiff magnetic nanopaper using cellulose nanofibrils as templates. Nat Nanotechnol 5:584–588

Pandey A (2021) Pharmaceutical and biomedical applications of cellulose nanofibers: a review. Environ Chem Lett 19:2043–2055

Patwary MAS, Surid SM, Gafur MA (2020) Properties and applications of biodegradable polymers. J Res 9:33

Phaechamud T, Issarayungyuen P, Pichayakorn W (2016) Gentamicin sulfate-loaded porous natural rubber films for wound dressing. Int J Biol Macromol 85:634–644

Phomrak S, Phisalaphong M (2020) Lactic acid modified natural rubber–bacterial cellulose composites. Appl Sci 10:3583

Pielichowska K, Pielichowski K (2014) Phase change materials for thermal energy storage. Prog Mater Sci 65:67–123

Prabhakaran S, Krishnaraj V, Zitoune R (2014) Sound and vibration damping properties of flax fiber reinforced composites. Procedia Eng 97:573–581

Prabhakaran S, Krishnaraj V, Shankar K, Senthilkumar M, Zitoune R (2020) Experimental investigation on impact, sound, and vibration response of natural-based composite sandwich made of flax and agglomerated cork. J Compos Mater 54:669–680

Rahim MISA, Horoshenkov KV, Rongong J, Ahmadi H, Picken J (2018) Epoxidized natural rubber for vibro-acoustic isolation. Polym Test 67:92–98

Rhim J-W, Reddy JP, Luo X (2015) Isolation of cellulose nanocrystals from onion skin and their utilization for the preparation of agar-based bio-nanocomposites films. Cellulose 22:407–420

Roy K, Pongwisuthiruchte A, Debnath SC, Potiyaraj P (2021) Application of cellulose as green filler for the development of sustainable rubber technology. Curr Res Green Sustain Chem 4:100140

Sangmesh B et al (2023) Development of sustainable alternative materials for the construction of green buildings using agricultural residues: a review. Constr Build Mater 368:130457

Sargianis JJ, Kim H-I, Andres E, Suhr J (2013) Sound and vibration damping characteristics in natural material based sandwich composites. Compos Struct 96:538–544

Seddiqi H et al (2021) Cellulose and its derivatives: towards biomedical applications. Cellulose 28:1893–1931

Sethulekshmi AS, Saritha A, Joseph K (2022) A comprehensive review on the recent advancements in natural rubber nanocomposites. Int J Biol Macromol 194:819–842

Shahamatifard F, Rodrigue D, Mighri F (2023) Thermal and mechanical properties of carbon-based rubber nanocomposites: a review. Plast Rubber Compos 52:483–505

Shahdan D, Chen RS, Zailan FD, Sabtu M, Ahmad S (2024) γ-radiation for improved reinforcement effect of single and hybrid nano-clay and cellulose nanofiber in thermoplastic natural rubber nanocomposite. Radiat Phys Chem 218:111550

Sibiya NN, Mochane MJ, Motaung TE, Linganiso LZ, Hlangothi SP (2018) Morphology and properties of sugarcane bagasse cellulose-natural rubber composites. Wood Res 63:821–832

Siqueira G, Bras J, Dufresne A (2009) Cellulose whiskers versus microfibrils: influence of the nature of the nanoparticle and its surface functionalization on the thermal and mechanical properties of nanocomposites. Biomacromol 10:425–432

Siró I, Plackett D (2010) Microfibrillated cellulose and new nanocomposite materials: a review. Cellulose 17:459–494

Song K, Zhu X, Zhu W, Li X (2019) Preparation and characterization of cellulose nanocrystal extracted from Calotropis procera biomass. Bioresour Bioprocess 6:1–8

Srirachya N, Boonkerd K, Kobayashi T (2020) Effective elongation properties of cellulose–natural rubber composite hydrogels having interconnected domain. J Elastomers Plast 52:337–355

Suppanucroa N et al (2023a) Multifunctional cellulosic natural rubber and silver nanoparticle films with superior chemical resistance and antibacterial properties. Int J Biol Macromol 12:1–23

Suppanucroa N, Nimpaiboon A, Boonchuay K, Khamkeaw A, Phisalaphong M (2023b) Green composite sponge of natural rubber reinforced with cellulose filler using alginate as a dispersing agent. J Mater Res Technol 27:3119–3130

Suwarto F et al (2023) Engineering characterization and environmental analysis of natural rubber latex modified asphalt mixture. Constr Build Mater 402:132970

Tavakolian M, Jafari SM, van de Ven TGM (2020) A review on surface-functionalized cellulosic nanostructures as biocompatible antibacterial materials. Nano-Micro Lett 12:1–23

Thambiraj S, Shankaran DR (2017) Preparation and physicochemical characterization of cellulose nanocrystals from industrial waste cotton. Appl Surf Sci 412:405–416

Thomas SK et al (2021) A comprehensive review on cellulose, chitin, and starch as fillers in natural rubber biocomposites. Carbohydr Polym Technol Appl 2:100095

Torgbo S, Sukyai P (2020) Biodegradation and thermal stability of bacterial cellulose as biomaterial: the relevance in biomedical applications. Polym Degrad Stab 179:109232

Trache D et al (2020) Nanocellulose: from fundamentals to advanced applications. Front Chem 8:392

Utrera-Barrios S et al (2024) Sustainable composites with self-healing capability: epoxidized natural rubber and cellulose propionate reinforced with cellulose fibers. Polym Compos

Venugopal B, Gopalakrishnan J (2018) Reinforcement of natural rubber using cellulose nanofibres isolated from coconut spathe. Mater Today Proc 5:16724–16731

Vilane BRT (2010) Assessment of stabilisation of adobes by confined compression tests. Biosyst Eng 106:551–558

Wang Y et al (2017) Homogeneous isolation of nanocellulose from eucalyptus pulp by high pressure homogenization. Ind Crops Prod 104:237–241

Wang H et al (2020) Extraction of cellulose nanocrystals using a recyclable deep eutectic solvent. Cellulose 27:1301–1314

Wu X et al (2016) Cellulose nanowhisker modulated 3D hierarchical conductive structure of carbon black/natural rubber nanocomposites for liquid and strain sensing application. Compos Sci Technol 124:44–51

Xu Y et al (2024) Nanocellulose composite films in food packaging materials: a review. Polymers (Basel) 16:423

Yin Q et al (2018) Water-induced modulus changes of bio-based uncured nanocomposite film based on natural rubber and bacterial cellulose nanocrystals. Ind Crops Prod 113:240–248

Yu P, He H, Luo Y, Jia D, Dufresne A (2017) Reinforcement of natural rubber: the use of in situ regenerated cellulose from alkaline–urea–aqueous system. Macromolecules 50:7211–7221

Yue Y et al (2012) Comparative properties of cellulose nano-crystals from native and mercerized cotton fibers. Cellulose 19:1173–1187

Zhang J et al (2019) Effect of natural fibre reinforcement on the sound and vibration damping properties of bio-composites compression moulded by nonwoven mats. Compos Commun 13:12–17

Zheng ZD et al (2019) Fabrication and performance of a spherical cellulose nanocrystal-based hydrophobic drug delivery vehicle using rubber wood

MIX
Papier aus verantwortungsvollen Quellen
Paper from responsible sources
FSC® C105338

If you have any concerns about our products,
you can contact us on
ProductSafety@springernature.com

In case Publisher is established outside the EU,
the EU authorized representative is:
Springer Nature Customer Service Center GmbH
Europaplatz 3, 69115 Heidelberg, Germany

Printed by Libri Plureos GmbH
in Hamburg, Germany